Insects as alternative sources of protein for food and feed

It is widely recognised that agriculture is a significant contributor to global warming and climate change. Agriculture needs to reduce its environmental impact and adapt to current climate change whilst still feeding a growing population, i.e. become more 'climate-smart'. Burleigh Dodds Science Publishing is playing its part in achieving this by bringing together key research on making the production of the world's most important crops and livestock products more sustainable.

Based on extensive research, our publications specifically target the challenge of climate-smart agriculture. In this way we are using 'smart publishing' to help achieve climate-smart agriculture.

Burleigh Dodds Science Publishing is an independent and innovative publisher delivering high quality customer-focused agricultural science content in both print and online formats for the academic and research communities. Our aim is to build a foundation of knowledge on which researchers can build to meet the challenge of climate-smart agriculture.

For more information about Burleigh Dodds Science Publishing simply call us on +44 (0) 1223 839365, email info@bdspublishing.com or alternatively please visit our website at www.bdspublishing.com.

Related titles:

Advances in cultured meat technology
Print (ISBN 978-1-80146-376-8); Online (ISBN 978-1-80146-377-5, 978-1-80146-378-2)

Advances in pig nutrition
Print (ISBN 978-1-80146-694-3); Online (ISBN 978-1-80146-695-0, 978-1-80146-696-7)

Advances in poultry nutrition
Print (ISBN 978-1-80146-731-5); Online (ISBN 978-1-80146-732-2, 978-1-80146-733-9)

Developing animal feed products
Print (ISBN 978-1-78676-463-8); Online (ISBN 978-1-78676-465-2, 978-1-78676-466-9)

Seaweed and microalgae as alternative sources of protein
Print (ISBN 978-1-78676-620-5); Online (ISBN 978-1-78676-622-9, 978-1-78676-623-6)

Chapters are available individually from our online bookshop:
https://shop.bdspublishing.com

climate•SMART•publishing
IN AGRICULTURAL SCIENCE

BURLEIGH DODDS SERIES IN AGRICULTURAL SCIENCE
NUMBER 154

Insects as alternative sources of protein for food and feed

Edited by Adriana Casillas, Tebrio, Spain

burleigh dodds
SCIENCE PUBLISHING

Published by Burleigh Dodds Science Publishing Limited
82 High Street, Sawston, Cambridge CB22 3HJ, UK
www.bdspublishing.com

Burleigh Dodds Science Publishing, 1518 Walnut Street, Suite 900, Philadelphia, PA 19102-3406, USA

First published 2025 by Burleigh Dodds Science Publishing Limited

Library of Congress Control Number: 2024940649

British Library Cataloguing in Publication Data
A catalogue record for this book is available from the British Library

ISBN 978-1-80146-584-7 (print)
ISBN 978-1-80146-585-4 (ePub)
ISBN 978-1-80146-586-1 (PDF)
ISSN 2059-6936 (print)
ISSN 2059-6944 (online)

DOI: 10.19103/AS.2024.0139

Typeset by Deanta Global Publishing Services, Dublin, Ireland

Contents

Part 3 General issues in use of insect protein

Series list

Title	Series number
Achieving sustainable cultivation of maize - Vol 1 From improved varieties to local applications Edited by: Dr Dave Watson, CGIAR Maize Research Program Manager, CIMMYT, Mexico	001
Achieving sustainable cultivation of maize - Vol 2 Cultivation techniques, pest and disease control Edited by: Dr Dave Watson, CGIAR Maize Research Program Manager, CIMMYT, Mexico	002
Achieving sustainable cultivation of rice - Vol 1 Breeding for higher yield and quality Edited by: Prof. Takuji Sasaki, Tokyo University of Agriculture, Japan	003
Achieving sustainable cultivation of rice - Vol 2 Cultivation, pest and disease management Edited by: Prof. Takuji Sasaki, Tokyo University of Agriculture, Japan	004
Achieving sustainable cultivation of wheat - Vol 1 Breeding, quality traits, pests and diseases Edited by: Prof. Peter Langridge, The University of Adelaide, Australia	005
Achieving sustainable cultivation of wheat - Vol 2 Cultivation techniques Edited by: Prof. Peter Langridge, The University of Adelaide, Australia	006
Achieving sustainable cultivation of tomatoes Edited by: Dr Autar Mattoo, USDA-ARS, USA and Prof. Avtar Handa, Purdue University, USA	007
Achieving sustainable production of milk - Vol 1 Milk composition, genetics and breeding Edited by: Dr Nico van Belzen, International Dairy Federation (IDF), Belgium	008
Achieving sustainable production of milk - Vol 2 Safety, quality and sustainability Edited by: Dr Nico van Belzen, International Dairy Federation (IDF), Belgium	009
Achieving sustainable production of milk - Vol 3 Dairy herd management and welfare Edited by: Prof. John Webster, University of Bristol, UK	010
Ensuring safety and quality in the production of beef - Vol 1 Safety Edited by: Prof. Gary Acuff, Texas A&M University, USA and Prof. James Dickson, Iowa State University, USA	011
Ensuring safety and quality in the production of beef - Vol 2 Quality Edited by: Prof. Michael Dikeman, Kansas State University, USA	012
Achieving sustainable production of poultry meat - Vol 1 Safety, quality and sustainability Edited by: Prof. Steven C. Ricke, University of Arkansas, USA	013
Achieving sustainable production of poultry meat - Vol 2 Breeding and nutrition Edited by: Prof. Todd Applegate, University of Georgia, USA	014

Introduction

With significant concerns about the environmental impact of conventional crop and livestock production, there is a growing interest in utilising novel alternative sources of protein in both human and animal diets. Insects have long been consumed in by an array of species, including birds, fish and mammals, including humans. This book provides a comprehensive overview of the recent research undertaken in this rapidly-expanding area. The book reviews the production and application of black soldier flies, yellow mealworms and other insects as alternative protein sources in poultry, pig and fish feed, as well as their nutritional benefits and challenges that can arise as a result of their use. The book also explores how the safety of insect-derived protein can be ensured across the supply chain, as well as how we can optimise its use as a sustainable feed and food source to contribute to a more resilient food system.

Part 1 Production and applications of black soldier flies

The first group of chapters addresses key issues in the production and use of black solider fly (BSF, *Hermetia illucens* (L.), Diptera: Stratiomyidae) as a protein source. Chapter 1 focuses on advances in substrate source composition for rearing BSF. As the chapter points out, over the last decade, interest in rearing BSF has grown exponentially. BSF larvae (BSFL) can be reared on any waste substrate, but the type of substrate markedly impacts the overall efficiency of the process as well as the composition of the larval biomass. It also influences the safety and overall environmental impact of the generated larval protein. This chapter looks at available waste streams for BSFL rearing in the EU and globally. It describes the different process parameters, in addition to substrate properties, that influence process efficiency, and how different substrates affect the process, the composition of the larval biomass as well as environmental impact. Finally, the chapter concludes by discussing challenges faced by the insect industry and how these challenges can be met.

Chapter 2 reviews pathogens and diseases affecting large-scale production of black soldier fly (BSF), including viruses, bacteria, fungi, protists and nematodes. The chapter looks at ways of managing rearing conditions to reduce disease risk as well as specific disease prevention and control measures. The chapter emphasises the importance of understanding the natural resistance mechanisms of BSF, including the production of antimicrobial peptides (AMF). It emphasises the need to balance conditions optimising mass production (including temperature and moisture) with the need to avoid overloading insect immune systems through excessive pathogen loads.

The next group of chapters review the use of black soldier fly larvae meal (BSFLM) in livestock feed, starting with Ch 3 which discusses poultry. As the chapter points out, successful inclusion of (BSFLM) in the diet of broilers and laying hens requires a precise determination of its nutritional value. These issues are discussed in the first part of this chapter together with the current understanding of amino acid content, amino acid digestibility and metabolizable energy of BSFLM. The chapter continues with an overview of the effects of dietary inclusion of different types of BSFLM on the performance and health of broilers and laying hens. It concludes with an assessment of the current use of insects as feed and the future prospects of these promising alternative protein sources in poultry feed.

Chapter 4 focuses on current research on including BSF in the diets of pigs as an alternative protein source to conventional proteins such as soya bean meal (SBM) or fishmeal (FM). BSF reared on organic substrates is rich in crude protein and fats and has a nutritional profile comparable to that of SBM in terms of amino acid profile, offering excellent soya-replacement potential. The chapter discusses the benefits and the challenges of using BSF in pig diets and reviews what data is still lacking for widespread use of this novel protein.

Chapter 5 reviews the potential use of BSF-derived proteins in aquaculture. In particular, it focuses on the effect of BSF larva meal on fish performance and health as well as on economic and environmental parameters. The chapter reviews current research on the use of BSF for a wide variety of finfish and shellfish species, including rainbow trout, Atlantic salmon, perch, tilapia, carp and shrimp. As the chapter points out, using BSF in aquaculture not only provides valuable protein to aquaculture species, but also has positive impacts on heath due to the presence of bioactive compounds (antimicrobial peptides, chitin, and lauric acid) able to exert antimicrobial or prebiotic effects or to Boost animal immune systems. BSF also has the potential to boost beneficial compounds, such as omega-3 fatty acids, carotenoids and selenium. The chapter also explores the potential benefits of using BSF in improving the environmental performance of aquaculture, depending on factors such as substrate source (e.g. use of by-products or waste) and energy consumption for larval growth and processing.

Part 2 Production and application of yellow mealworms and other insects

The next group of chapters moves away from BSF to discuss other insect species such as the yellow mealworm (*Tenebrio molitor*). Chapter 6 reviews advances in mass breeding of yellow mealworms as a protein source. As it shows, the status of the yellow mealworm has undergone a significant transformation in recent years, from a global pest to a prominent insect species for feed and food. However, since farming mealworms is still relatively, there are ample opportunities for innovation and optimization. After providing an overview of the mealworm life cycle, the chapter discusses ways of designing rearing facilities, setting up an appropriate rearing environment as well as optimising rearing practices. The chapter also discusses challenges in scaling up operations.

Building on Chapter 6, Chapter 7 reviews developments in substrates for rearing yellow mealworms. As highlighted earlier in the book, different growing substrates have different effects on the growth and high-quality protein content of yellow mealworm as a potential feed source. Substrate proteins, fats, carbohydrates and fibre appear to have a crucial role in determining growth performance and nutritional characteristics. The species also offers the opportunity to turn agro-industrial co-products into high-quality biomass, including the potential to degrade polymer bio-plastics, thus helping to address the global plastic waste problem.

Given the relatively recent development of the sector, information on pathogens of insects in mass production systems remains limited. Chapter 8 brings together what we currently know about diseases in yellow mealworm as well as assessing the risk for existing, emerging and as-yet unknown pathogens which may cause lethal and sub-lethal effects in yellow mealworm production facilities. The chapter summarizes some of the most promising ways in which host-microbe interactions in *T. molitor* can be better understood and the health and the immune system of the species can be improved and better prepared to fight off infections. Finally, the authors highlight the need for coordinated and shared research and resources in an effort to protect this growing industry from future pathogenic threats.

As Chapter 9 indicates, based on nutritional analysis, yellow mealworm larvae are comparable to conventional poultry protein sources. This makes them an appropriate substitute for soybean meal and fishmeal as they have a similar influence on growth performance without having a negative impact on meat or egg quality. Chapter 9 reviews the nutritional benefits of yellow mealworms, including protein content, essential amino acids (e.g. lysine, methionine and tryptophan), essential polyunsaturated fatty acids, and presence of minerals such as iron and magnesium. As the chapter points out, mealworm larvae are rich in fibre which helps to improve feed utilisation in the gut while the chitin present in insects has a prebiotic effect, helping to improving the health of birds by reducing the accumulation of harmful microbial populations in the gut. The chapter also reviews the influence of processing and storage methods on the nutritional values of yellow mealworm larvae as well as available studies on their effects on parameters such as poultry feed intake and body weight gain.

Yellow mealworm has been found to be a safe and sustainable alternative protein source for aquafeed, given advantages such as high protein content and better amino acid profile when compared to alternative plant-protein sources. Chapter 10 provides an overview of the current use of *T. molitor* as an alternative ingredient to fishmeal. It summarises recent reviews on the use of these invertebrates in aquafeeds and their impact on key performance indicators (KPIs) associated with growth performance and feed efficiency, product nutritional quality, as well as their immunomodulatory potential and effect on host gut microbiota.

The common housefly (*Musca domestica*), a ubiquitous presence in various environments, has been a subject of scientific investigation for over a century. Chapter 11 explores the potential use of housefly larvae as a sustainable animal feed ingredient. Their ease of rearing, rapid growth, high nutritional value and environmental benefits in nutrient recycling have revived interest in using them as an alternative protein source for food and feed. Despite these advantages, housefly larvae have not achieved significant commercial viability. The chapter examines key challenges and ways of achieving successful commercial production, including optimising production parameters such as temperature, stocking density, propagation, feed substrate and harvesting.

Although long practiced, cricket farming is now becoming a rapidly-growing, modern agribusiness, driven by the increasing demand for alternative food and feed proteins. Chapter 12 highlights the current status of cricket farming, including key areas such as common species utilized, processing issues, nutritional composition, safety and applications. The chapter reviews the nutrient composition of edible cricket species, safety and disease issues in cricket production and regulations relating to edible crickets.

Part 3 General issues in use of insect protein

The final Part of the book reviews general issues relating to insect protein production. Chapter 13 discusses chemical and pathogenic hazards related to industrial production of insect species such as the yellow mealworm (*Tenebrio molitor*) and house cricket (*Acheta domesticus*), both of which are authorized for human food and animal feed in the European Union (EU). The chapter also considers black soldier fly larvae (BSFL) (*Hermetia illucens*) which are only allowed as animal feed in Europe. The chapter reviews hazards associated with rearing, processing and storage of insects and derived products, focussing particularly on chemicals and pathogens known to be of risk for human consumption. The chapter also discusses techniques related to hazard detection and monitoring.

The final chapter in the book, Chapter 14, echoes themes in earlier chapters in pointing to the growing interest in using insects as novel protein sources to meet the increasing global demand for sustainable and nutritious food. While insects have long been consumed in certain cultures, in others, particularly those where insects are not traditionally part of the human diet, the idea of eating insects has been met with scepticism and strong resistance. As

Chapter 14 shows, the interaction between contextual, cognitive and affective factors helps to shape consumers' responses towards insects as a food and ingredients. Lack of familiarity and knowledge regarding entomophagy can lead to a negative affective response (e.g. disgust) which acts as a significant obstacle to consumer acceptance. While price plays an important role in determining consumer willingness to purchase "insect based" products, future research should prioritize sensory and food characteristics (e.g., substrates and different species used) when developing novel foods. This is because these features have a major influence on attitudes and emotional responses towards insect-based products.

Part 1

Production and applications of black soldier flies

Chapter 1

Advances in substrate source composition for rearing black soldier fly larvae as a protein source

Cecilia Lalander[1] *and Ivã Guidini Lopes*[2]*, Swedish University of Agricultural Sciences, Sweden*

1 Introduction

Over the past decade, the black soldier fly (BSF, *Hermetia illucens* (L.), Diptera: Stratiomyidae) has become one of the most commonly used insects for feed production worldwide (van Huis, 2020). This is primarily due to the wide range of substrates that the larvae can be reared on, along with the high waste-to-biomass conversion efficiency of the larvae (Gold et al., 2018). This versatility of substrate use by fly larvae has made them nature's own waste managers (Fowles and Nansen, 2020).

1 Department of Energy and Technology, Swedish University of Agricultural Sciences, Box 7032, 75007 Uppsala, Sweden.
2 Department of Biosystems and Technology, Swedish University of Agricultural Sciences, Alnarp, Sweden.

http://dx.doi.org/10.19103/AS.2024.0139.02

With the reported decline in terrestrial insect abundance, questions have emerged regarding the impact this will have on the different ecosystem services they provide (van Klink et al., 2020; Sánchez-Bayo and Wyckhuys, 2019). One of many ecosystem services that insects contribute to is the degradation and recycling of organic matter (Dangles and Casas, 2019) and nutrients (Woelber-Kastner et al., 2021; Yang and Gratton, 2014). The ability of detritivore insects, such as the BSF, to degrade organic matter and recycle nutrients is one of the main reasons they are of interest not only for feed production but also for waste management purposes (Čičková et al., 2015). During this process, two products are generated: a larval biomass that can be used in animal feed (Lu et al., 2022), and a treatment residue known as frass or entomocompost that can be used as an organic fertilizer or soil amendment (Lopes et al., 2022). As two valuable products are generated in the process, this waste management technology fits well within the concept of a circular economy (Ojha et al., 2020), in which the by-product/waste stream in one process becomes the resource in another (Ellen Macarthur Foundation, 2015).

2 Bio-waste streams available for rearing black soldier fly larvae

2.1 Europe

BSF larvae (BSFL) have an extraordinary ability to feed on almost any bio-waste stream, including food waste, vegetable waste, human faeces, animal manure (pig, poultry, cattle), fish and fish offal, to mention a few (Hopkins et al., 2021; Lalander et al., 2019). They can also be reared on agro- and food industry side-streams, including spent grains, bread surplus, beet molasses, banana peels and maize distillers (Hopkins et al., 2021; Chia et al., 2020; Isibika et al., 2019). However, both the process efficiency (Lalander et al., 2019) and larval composition (Barragan-Fonseca et al., 2017) are heavily influenced by the substrate the larvae are reared on. The environmental impact of insect rearing also largely depends on the substrate the insects are reared on, with residual streams such as municipal waste exerting a considerably smaller impact than generic protein feed, while beet pulp resulted in a significantly greater environmental impact (Smetana et al., 2016).

In the European Union (EU), this issue is problematic for insect breeders, as insects are considered farmed animals and thus are not allowed to be fed any animal by-products, which limits the insect sector to using primarily pre-consumer bio-waste streams, such as agro- and food industry wastes (European Commission, 2009). This regulatory shift occurred in the 1980–90s after the outbreak of bovine spongiform encephalopathy (BSE) in Europe, during which European food production went from being fairly circular to becoming almost exclusively linear (Vågsholm et al., 2020a). The main fear of European legislators is the possible occurrence of misfolded prions, which caused the BSE outbreak, in animal by-products (Lalander and Vinnerås, 2022). Unfortunately for the European insect sector, Bosch et al. (2019) demonstrated that currently legal substrates for insect breeding in the EU have a higher environmental impact than those that are not permitted. In fact, they found that on a per produced protein basis, BSFL reared on permitted substrates that could be used directly as animal feed for other farmed animals, such as brewery distiller's grains, had an even higher environmental impact than conventional protein feeds such as fishmeal and soybean meal. The use of more post-consumer residual streams, such as catering waste (bio-waste from households and restaurants), would be important not only from an environmental point of view but also in providing a sufficient feed source for the larger volumes of insects that the feed industry requires. It is in the interest of the EU to increase the local production of high-quality protein feed.

In 2015, the EU launched an action plan to facilitate the transition from a linear to a circular economy (European Commission, 2015). This action plan has been followed up by the European Green Deal, in which the emphasis lies in making Europe climate neutral (fossil free) by 2050 (European Commission, 2020a). A crucial part of this Green Deal is the Farm to Fork Strategy (European Commission, 2020b), which aims to make European food production systems fairer, healthier and more environmentally friendly. This will be accomplished by, among other things, reducing nutrient losses by 50% and fertilizer use by 30%, and ensuring stronger food security within the EU. One of the weaknesses of food security within the EU, identified by the EU Commission, is the high level of imports of feed ingredients, such as oil seed meal and soy (European Commission, 2021b). The EU currently imports 76% of oil seed meal (rapeseed and sunflower seed meal; used for both livestock feed and in food processing) and 20% of animal feed, primarily in the form of soy (European Parliamentary Research Service, 2023; European Commission, 2021b). This is not in line with the European Green Deal and jeopardizes EU food security.

Step by step, the EU has started to return to a higher level of circularity in the food industry sector, to encourage a more environmental and resilient food system. The most recent change occurred in 2021, when the EU Commission decided to allow the use of processed animal proteins (PAPs) from pigs and insects to feed poultry, and from poultry and insects to feed pigs, lifting the ban following the BSE crisis (European Commission, 2021a). However, it has not been determined if insects can be fed PAPs. In 2019, 931 million tons of food waste were generated in Europe (United Nations Environment Programme, 2021). In high-income countries in Europe, 118 kg of food waste is generated per person per year, of which 70% is mixed household, i.e. post-consumer waste (United Nations Environment Programme, 2021). However, insects can only be fed agro- and food industry waste, so-called pre-consumer waste. This means that approximately 30% of total food waste in Europe is available for insect production (this does not include animal manure).

2.2 Global outlook

In many parts of the world, unlike in the EU, there are no legal limitations on feeding insects with different bio-wastes. The United Nations Environment Programme (2021) found that 931 million tons of food waste were generated globally in 2019, of which the majority (~60%) was from households. In contrast to previous reports, it also found that the generation of food waste on a household level is roughly the same in all income groups: in high-income countries, the average food waste is close to 80 kg capita^{-1} year^{-1}, while in lower- to middle-income countries, it is around 90 kg capita^{-1} year^{-1} (no data for low-income countries available). While it is highly desirable to avoid food loss, there will always be a proportion of food waste that is unavoidable, such as fruit peels, inedible parts of vegetables and fruits, eggshells and bones. However, the proportion of edible to inedible fractions in food waste is very hard to define and can be the topic of debate.

Given the large amounts of available food waste, there should be an ample amount of substrate for rearing BSFL. However, when Gold et al. (2021) conducted a systematic assessment of available bio-wastes for BSFL rearing in Nairobi, Kenya, they found that most bio-waste streams in Nairobi are unavailable, as no waste segregation occurs at the source, most bio-waste streams are contaminated with inorganic material, or are already used for animal feed purposes. They recommend using a mixture of human faeces, animal manure, fruit/vegetable waste and unsegregated food waste. Purkayastha and Sarkar (2022) investigated the efficiency of rearing larvae on municipal solid waste (MSW) and investigated compositions of MSW with a 43–58% biodegradable fraction. They found that, in India, it was possible to generate US$15–43 per ton of MSW processed from the generated BSFL protein, demonstrating a potential method for rearing BSFL from MSW.

In addition to food waste, there are also large streams of bio-waste from food industries and agriculture. The total amount of agricultural bio-waste is hard to estimate but comprises harvest side-streams (inedible parts left on the field) as well as harvest and slaughter losses. The latter is a fraction that has received surprisingly little attention. WWF UK (2021) estimated that around 15% of all food produced is wasted at farm level, totalling 1174 million tons per year globally. The harvest losses include crops that are never harvested or are spoiled after harvesting due to poor storage conditions. Even though these harvest losses are an interesting potential substrate for insect rearing, from a sustainability perspective, focus should be placed on unavoidable harvest losses and food industry waste streams.

However, there are a number of challenges in using these waste streams. They may, e.g. contain a high proportion of lignocellulosic material such as wheat straw, sugarcane bagasse, maize distillers, spent barley and spent sorghum, among others, which are not easy for fly larvae to digest and efficiently convert into biomass (Beyers et al., 2023; Hopkins et al., 2021). There may also be other issues involved with these waste streams. Isibika et al. (2023) conducted a survey on available food industry waste streams in Tanzania and found, in accordance with the findings of Gold et al. (2021), that there was competition for brewery waste (spent grains) and sunflower press cake, as these were already being used as animal feed. The food industry waste streams available for BSFL rearing included fruit peels, mango seeds, coffee husks and fish waste (comprising in this case of the fins and all internal content of the fish, including gills, liver, kidney, intestines, heart, stomach content and swim bladder). None of these sources alone have been found to be a good substrate for BSFL rearing; however, the authors recommend blending the different substrates to meet the nutritional requirements of the fly larvae.

3 Parameters affecting larval rearing efficiency: larval density

Many factors are known to affect the rearing of insects, sometimes referred to as mini-livestock, including the quantity of animals placed in a designated area, which correlates with their growth and development over time (Barragan-Fonseca et al., 2018). The most common term used for how many BSFL are placed in the rearing units is 'larval density' (Dzepe et al., 2020; Barragan-Fonseca et al., 2018), but other terms such as 'areal density' (Guillaume et al., 2023) and 'stocking densities' (Delvendahl et al., 2022) have also been used. Regardless of the terminology, these terms refer to the number of individuals present per unit area, which is most commonly expressed as larvae per square centimetre (larvae cm^{-2}). We have opted to use the term larval density.

In general, it is assumed that the higher the larval density, the lower the final weight of each individual, as larvae will compete for the available resources with more individuals, restricting growth (Barragan-Fonseca et al., 2018). However, when it comes to insect larvae, which are commonly reared in containers into which their feed substrate is placed, density alone does not account for the impact on process efficiency. Other parameters that are interconnected and impacted by larval density include nutrient availability per larva, larval feeding does/rate, substrate depth and moisture (Lopes et al., 2023). In addition, a key issue to consider is the type of container used to rear the larvae, as going from small-scale set-ups often used in research, to more industrial set-ups, is typically not linear (Yang and Tomberlin, 2020). Many studies have been carried out in quite small-scale set-ups, and it is therefore necessary to bear in mind the need for validating previously reported results on a larger scale.

Parra Paz et al. (2015) investigated the impact of larval density (2, 4 or 6 larvae cm^{-2}) on growth on a very small scale (using 49 cm^2 rearing boxes) in a factorial design with three feeding rates (60, 130 and 200 mg of feed per larvae per day). It was demonstrated that density was the most influential variable in that study, with the authors concluding that high densities and feeding rates would be counterproductive as the process efficiency would probably be reduced. Barragan-Fonseca et al. (2018) confirmed that increasing larval density reduced individual larval growth. They evaluated four densities (0.31, 0.62, 1.23 and 2.47 larvae cm^{-2}) in 155 cm^2 containers and obtained a linear response for reduced larval growth when density increased. Nevertheless, the total yield per container increased linearly with density, highlighting that even though individual larvae grew less, the total output can be higher when increasing larval density. The same trend was reported by Dzepe et al. (2020), who assessed the impact of density and substrate moisture on larval growth and development. The authors evaluated densities of 1–10 larvae cm^{-2} and verified lower individual weight and body thickness (in mm, measured using a calliper) in larvae placed at higher densities (6–10 larvae cm^{-2}), conducting the experiments in very small cylindrical containers with an area of 15.05 cm^2. They reported significantly negative correlations between substrate reduction and survival rate with increasing larval density. A negative impact of increasing densities (from 1 larvae cm^{-2} to 10 larvae cm^{-2}) on survival was also reported by Opare et al. (2022) when rearing BSFL in round containers with a total area of 86 cm^2, even when feeding the larvae *ad libitum*. Survival rates decreased by approximately 10–12% when density was increased from 1 larvae cm^{-2} to 5 larvae cm^{-2} and from 5 larvae cm^{-2} to 10 larvae cm^{-2}.

Using slightly larger containers for larval rearing (502 cm^2), Jiang et al. (2022) tested densities based on the wet weight of the waste stream being treated, in this case, swine manure. The evaluated densities were represented by mass fraction based on pig manure and larval weight initially, with the amount of added larvae being 0.08%, 0.24% and 0.40% of the manure's mass in each treatment, with the exact number of larvae not being disclosed. The authors found similar trends in comparison to the studies mentioned earlier, with lower individual weight and higher yield at higher densities. In a more realistic set-up, Miranda et al. (2020) used BSFL for bio-digesting manure (swine, dairy and poultry) in experimental units of 2289 cm^2, at a density of 4.37 larvae cm^{-2}, resulting in individual larval weights above 150 mg, comparable to their control reared on the Gainesville diet. Extreme values of individual weight can be obtained when either low or high densities are used. For instance, Lalander et al. (2019) reared BSFL using multiple substrates at a density of 0.6 larvae cm^{-2} and registered individuals weighing more than 300 mg after 10–12 days when fed abattoir waste and human faeces, while Opare et al. (2022) obtained larvae that did not reach 80 mg when reared at a density of 10 larvae cm^{-2}.

Opare et al. (2022) suggested that there is a need to find a balance between the trade-offs (biological and economical) when rearing insects under varied densities and conditions. Lopes et al. (2023) reached a similar conclusion when investigating the individual and combined effects of substrate properties and rearing conditions on process efficiency in BSFL rearing across various substrates. They examined the influence of varying feed doses (0.1–0.5 g volatile solids per larvae) and densities (3.03–8.33 larvae per cm^2), resulting in distinct depths within the rearing boxes (1.0–6.5 cm) and assessed their impact both individually and in combination. While a treatment with a medium depth (around 3 cm) and a medium larval feed dose (approximately 0.2 g of volatile solids per larvae) was suggested as a good starting point when initiating BSFL rearing on a new substrate, these factors may need adjustment to account for the specific characteristics of the substrate. In addition, Yakti et al. (2022) demonstrated significant differences in larval composition when using rearing boxes of varying size (2060, pp. 964 cm^2, 466 cm^2 and 194 cm^2), despite maintaining consistent process parameters across the scales. These differences were attributed to the elevated temperatures reached in the largest boxes, which exceeded 39°C, while it did not surpass 35°C in the other sized boxes.

Density is often discussed as a matter of welfare in livestock production. According to Dossey et al. (2016), the adoption of high larval densities is not problematic when it comes to fly larvae, as insects generally 'live in large groups in small amounts of space' in nature. However, this argument is not sufficient to justify rearing the insects without considering their behaviour and performance. It is known that under certain conditions (e.g. poor ventilation, uncontrolled moisture content and blends of substrates), temperatures inside rearing units can easily surpass 55°C (personal communication), which can result in reduced process efficiency. In addition, Vogel et al. (2022) suggested that high densities might increase the risk and spread of infectious diseases, as observed for other insect species (e.g. *Drosophila melanogaster*), even though the authors did not present direct evidence for BSFL.

4 Parameters affecting larval rearing efficiency: substrate depth

Some of the studies cited earlier failed to provide crucial information on the experimental set-up such as substrate moisture level, properties of the substrate at the end of the bioconversion process (whether it could be sieved or not), physical traits such as bulk density and the presence of resistant fibres, and substrate pH at the beginning and during the trials. A key factor that is known to interfere with larvae performance over time is the depth to which the substrate is placed in the treatment boxes. Industrial rearing of BSFL typically involves 1–3 feedings provided throughout the bioconversion process, and the feed substrate depth changes over time with such feeding regimes (personal communication).

Dortmans et al. (2017) suggested that substrate depth should not exceed 5 cm, in order to guarantee the larvae will reach the bottom of the treatment unit and process all provided material. Brits (2017) evaluated substrate depths of 5–20 cm and demonstrated that BSFL did not access the substrates in the bottom of cylindrical containers of 11.2 cm diameter and 28.2 cm height. Larvae survival was highly affected when depth increased, decreasing from around 95% (5 cm depth) to about 50% (20 cm depth), and consequently, the waste reduction and bioconversion of the feed substrate were reduced, demonstrating considerable loss of efficiency.

Abduh et al. (2018) evaluated the conversion of Philippine tung seed by BSFL at varying depths of 4, 6, 8 and 10 cm, in small-scale experimental units (416 cm^2). The authors reported that up to 6 cm depth, larval productivity was good (increasing by 56% from 4 cm to 6 cm depth); however, once the depth increased to 8 cm, it was significantly reduced by around 16%. When the substrate depth increases, larvae tend to place their head part into the material to allow their posterior spiracles to stick out of the material to improve their breathing, limiting their movement to the substrate surface, resulting in the material at the bottom of the container being left undigested, as reported by Lalander et al. (2020). In addition, greater depths (>5 cm) can result in anaerobic conditions, especially at the bottom of the container, which not only causes undesirable odours due to possible anaerobic decomposition but also affects the bioconversion process as a whole (Barrett et al., 2023).

Achieving an adequate substrate depth is challenging, especially when considering that the composition of feed substrates may change over time and that multiple feeding occasions increase the cost of treatment. An interesting observation was made by Bekker et al. (2021) when treating chicken feed at varying moisture contents (45–85%) with BSFL. At 45% moisture, the chicken feed would form lumps at the surface of the boxes, resulting in the larvae crawling to the bottom half; on the other hand, at 55–65% moisture, the chicken feed formed a crust at the surface and larvae were found at all depths. Conversely, at 75% moisture, a thin water film was observed on the surface of the chicken feed and larvae were predominantly in the upper half of the box. The specific moisture content at which these observations become apparent may vary in different feed substrates, but the general trend is likely to be similar.

5 Parameters affecting larval rearing efficiency: substrate moisture content

The initial moisture content of the feed substrate greatly influences the efficiency of the process. It not only impacts upon efficiency but also the possibility of harvesting the larvae at the end of treatment. At a high initial substrate moisture content, the treatment residue at the end of the treatment may be too wet to use a dry separation technique (Lalander et al., 2020; Cheng et al., 2017; Dortmans et al., 2017). In addition, it must be noted that using high-moisture waste streams may result in the transportation of water (contained in the waste) to the treatment plant, as well as the need to evaporate significant amounts of water throughout the process, which places higher demands on the ventilation system (Lalander et al., 2020).

There is no 'ideal' initial moisture content of the feed substrate, but values ranging from 45% to 80% are considered adequate for good process efficiency. Generally, in a substrate with an initial moisture content over 80%, larvae development and survival might be hampered, and the processed (or not so well processed) substrate could end up being very humid, preventing dry separation (Cheng et al., 2017). However, once effective ventilation is adopted, it becomes possible to treat substrates with water content exceeding 90%, even though it may sometimes not be feasible to do so, thus necessitating dewatering of the substrate (Lalander et al., 2020), or alternatives. It has been demonstrated that challenging wastes, such as faecal sludge from pit latrines with 90% moisture, can be bio-converted (material reduction above 60%) by BSFL without any dewatering step (Tokwaro et al., 2023). However, in that specific case, the authors did not mention the bioconversion efficiency or the sievability of the substrate at the end of the trial.

Varying substrate moisture contents between 45% and 75%, Bekker et al. (2021) demonstrated that BSFL growth and development remained unaffected, even though other process parameters were impacted. For instance, low moisture content promoted microbial degradation of the substrate and resulted in shorter larval growth time and small prepupae compared to those in higher moisture conditions. Similarly, Dzepe et al. (2020) found that higher moisture content (40–80%) led to longer BSFL development time (12.2–18.8 days) and a final larval weight of 0.12–0.19 g larva^{-1}. In addition, Liu et al. (2023) demonstrated that larvae pupation depended on both substrate type and moisture content, with increased pupation being observed in substrates with a higher moisture content (>60%).

Substrate moisture content also correlates closely with the capacity for treating varying volumes of waste within the same space. This is particularly important in upscaling; e.g. treating 200 kg of waste per day within a given area differs significantly from treating 500 kg in the same space. As demonstrated by

Lalander et al. (2020), the higher the initial moisture content, the higher the water evaporated from the substrate, with up to 840 g of water having to be removed from every kilogram of the substrate during the bioconversion process of very wet substrates. This means that more powerful ventilation systems are required for upscaling processes, or effective dewatering processes are needed, which both will depend on evaluating the economic and technical feasibility of the process as a whole.

6 Nutritional composition of feed substrates: protein, carbohydrate and lipid content

BSFL have been reported to thrive on a wide range of feed substrates, from feed mixtures (e.g. chicken feed and grain mixtures) to waste streams with varied and heterogeneous composition. BSFL can digest almost any organic material due to their high plasticity in terms of nutrient utilization and developmental time. As a result, most organic waste streams can serve as feed substrate for BSFL, regardless of how challenging they are in terms of bioconversion efficiency.

Nutritional balance is required for all livestock production – including insects. As extensively discussed by Seyedalmoosavi et al. (2022), both the quantity and quality of a specific dietary component are crucial in larval development and bioconversion efficiency, as well as the availability of nutrients in the larval diet. Even though BSFL are highly plastic in converting feed substrates of varying composition (Barragan-Fonseca et al., 2021), there are a number of known thresholds affecting how much of each dietary compound (e.g. proteins, carbohydrates and lipids) should be provided to maintain good process performance.

Considerable research has been directed to understand the protein (Pr) and carbohydrate (CHO) requirements of BSFL. Barragan-Fonseca et al. (2019) investigated multiple diets containing 10–24% Pr and 35–55% CHO and various combinations. In another study, the same authors evaluated 25 artificial diets with varying Pr+CHO and Pr:CHO proportions (Barragan-Fonseca et al., 2021). Both studies emphasized that the adequate dietary balance of Pr and CHO enabled better larval performance compared to diets based on Pr:CHO ratios. This means that by separately respecting the limits for both Pr and CHO, an efficient BSFL-rearing process – in terms of bioconversion efficiency, yields, larval growth and survival – can be achieved. The authors also highlighted the great plasticity that BSFL have when consuming feed substrates, as they thrive on diets containing Pr levels ranging from 10% to 15% and CHO levels from 10% to 60%. Fuso et al. (2021) demonstrated that once the minimal critical amount of Pr in the larval diet is reached, there are no advantages to increasing the Pr in the larval diet, as the process performance will not increase while its related costs might.

Barragan-Fonseca et al. (2018, 2019) showed that the higher the Pr content in the larval diet, the higher the protein and fat accumulation in the larvae. Lopes et al. (2020b) also reported that even small inclusions of fish to bread waste resulted in higher Pr accumulation in the BSFL, while (Ewald et al., 2020) observed that larvae accumulated more fat when fed diets with higher Pr content. Gold et al. (2020) found that larvae reared on vegetable canteen waste containing $12\%_{DM}$ Pr had a Pr content of $25\%_{DM}$, while the Pr content of BSFL reared on mixed canteen waste with a Pr content of $32\%_{DM}$ was $36\%_{DM}$. However, the biomass conversion efficiency when rearing the larvae on vegetable canteen waste was considerably higher than when rearing them on mixed canteen waste: $23\%_{DM}$ compared with $15\%_{DM}$. This could be explained by the observation made by Lalander et al. (2019): when it comes to larval development time, both the daily feeding rate of Pr and CHO are important, while only the CHO daily feeding rate matters when it comes to the final larval weight. The larvae reared on a CHO-rich and Pr-poor substrate (vegetable waste) grew quickly but took longer to turn into prepupae, probably because they had to accumulate sufficient Pr. Maximizing biomass conversion may thus not always yield the highest value, if the BSFL produced has a low Pr content.

It is important to note that yields may vary depending on the quality of the feed substrate and other compounds in the substrate. It is possible to produce BSFL with a 'tailored composition', as demonstrated by Barroso et al. (2019). The authors replaced a control (chicken feed-based) diet with fish waste over varying periods of time, meaning that some larvae consumed fish for 1 day, some for 4 days and others for the entire process period of 12 days. It was observed that the treatments with most fish waste resulted in lower Pr accumulation in the larvae, while treatment with lower inclusions (i.e. providing fish waste to larvae for shorter times (1–4 days)) resulted in higher Pr accumulation. The authors did not report the bioconversion efficiency or the sieving performance for larval separation, which can sometimes be affected by an unbalanced addition of substrates such as fish waste (Lopes et al., 2020b).

While most studies on BSFL nutrition focus on Pr and CHO and the interaction between these two factors, a very limited number of studies have focused on lipids. It is known that larval development can be shortened by providing lipid-rich diets to BSFL (Nguyen et al., 2013) and that lower-lipid diets can reduce fly emergence rates (Bellezza Oddon et al., 2022a). A thorough evaluation of lipid requirements by BSFL was conducted by Bellezza Oddon et al. (2022b), in which inclusion levels of dietary lipids were varied from 1% to 4.5% (dry basis). The authors found that a 1% level of lipids on a dry matter basis had a negative impact on growth performance resulting in the smallest larval size and longer development times, while a 4.5% level yielded better growth performance. However, the authors recommend further investigation of lipid levels exceeding 4.5%.

While balancing nutrients can be a strategic approach for producing BSFL of tailored composition, treating wastes of varying composition over time can be quite challenging. Nonetheless, there are ways of tackling such challenges. Gold et al. (2020) evaluated several waste streams (mill by-products, canteen waste, human faeces, poultry slaughterhouse waste, cow manure and vegetable waste) as feed substrate, either alone or in combination, with combinations achieving a Pr:CHO ratio of approximately 1:1. They found positive correlations between waste reduction and bioconversion efficiency with dietary lipids and negative correlations between the same process parameters and fibre dietary content. Their general conclusion was that blending different waste streams achieves higher bioconversion than treating materials alone and that a simplistic approach, such as balancing nutrients and other dietary compounds, can significantly increase the predictability of the bioconversion process, which is the target of every enterprise working with insect production. Similar findings were made by Isibika et al. (2021), who found that the biomass conversion of a banana and orange peel mixture could be more than doubled by including 25% fish waste. No improvement was found when altering the Pr:CHO ratio from 0.4 (25% fish waste inclusion) to 0.9 (50% fish waste inclusion), supporting the findings of Fuso et al. (2021).

7 How to deal with 'poor' substrates

Many suboptimal waste substrates for rearing BSFL have been explored as feed substrates. The reason why the substrates are not ideal for BSFL rearing can be substrates characterized by high lignocellulosic content (e.g. wheat straw and cattle manure) and/or are low in protein (e.g. fruit peels and vegetable wastes). Bioconversion efficiencies are quite low for these sort of waste streams, $<5\%_{DM}$ for brewery waste, wheat straw, sugar cane bagasse and orange and banana peels (Theron, 2022; Isibika et al., 2021; Liu et al., 2018), compared with more promising substrates, such as food waste, with conversion efficiencies $>20\%_{VS}$ (Lindberg et al., 2022a). Lignocellulosic substrates have a high lignin, hemicellulose and crystalline cellulose content that are difficult to degrade, which means larvae are unable to assimilate nutrients and convert them into their own biomass (Peguero et al., 2022).

In order to improve the bioconversion of these waste streams, different physical, chemical and biological pre-treatments have been assessed, aiming at degrading lignin and hemicellulose and reducing the crystalline structure of cellulose. Physical pre-treatments include heat treatments, with and without pressure, and mechanical pre-treatments, such as grinding, which is used to reduce particle size and increase porosity. Grinding is a widely used pre-treatment that is generally incorporated in most BSFL treatments, even for more suitable substrates (Dortmans et al., 2017).

Heat treatments have not shown promising results for BSFL applications. Isibika et al. (2019) found that the biomass conversion efficiency of banana peels was reduced by almost 20% when heating banana peels to 120°C under 2 bar pressure for 1 h compared with the control and stipulated that it was due to the release of tannins. Theron et al. (2023) did not observe any reduction in conversion efficiency when steam pre-treating wheat straw and sugar cane bagasse at different temperatures, ranging from 140°C to 185°C, but also found no improvement. At 200°C and 215°C however, conversion efficiency was greatly reduced, probably due to increased levels of furfural acids in the pre-treated material. Lower temperature pre-treatments (around 50–60°C) may, however, be suitable prior to BSFL rearing.

When it comes to chemical pre-treatments, mostly alkaline pre-treatments have been assessed prior to BSFL rearing. The addition of $1\%_{WW}$ ammonia to banana peels under 1 week did not result in any significant improvement in conversion efficiency (Isibika et al., 2019). At an addition of $5\%_{DM}$ for 3 days, Peguero et al. (2023) demonstrated that the conversion efficiency of cow manure was halved, even though the fibre content was reduced. They concluded that ammonia pre-treatment was not suitable for BSFL rearing due to the toxic nature of ammonia at elevated concentrations.

Microbial pre-treatments have been shown to be more efficient prior to BSFL rearing. Isibika et al. (2019) demonstrated a great increase in the conversion efficiency of banana peels for both fungal (*Trichoderma reesei* and *Rhizopus oligosporus*) and microbial (isolated BSF gut bacteria) pre-treatments, but only for pre-treatments that lasted 14 days, with no improvement for the substrates pre-treated for only 7 days. In accordance with this, Mazza et al. (2020) demonstrated a 28.6% increase in BSFL weight gain and high material reduction rates when rearing BSFL on chicken manure enriched with a $1\%_{WW}$ solution of companion bacteria isolated from the BSF egg surface (*Kocuria marina*, *Micrococcus luteus*, *Enterococcus faecalis*, *Lysinibacillus boronitolerans*, *Gordonia sihwensis* and *Proteus mirabilis*) and BSFL gut (*Bacillus subtilis*). Somroo et al. (2019) reported a considerable increase in material reduction and biomass conversion, as well as crude protein content of larvae, when co-digesting soybean curd with *Lactobacillus buchneri*. Microbial pre-treatments thus appear promising for achieving increased efficiencies when rearing BSFL on more difficult substrates. However, one has to consider the increased complexity of the treatment, as well as prolonged treatment times in some cases.

Another interesting route could be the use of enzymes, a commonly used pre-treatment for the hydrolysis of cellulose and hemicellulose in the lignocellulosic feedstock used in second-generation bio-ethanol production (Alvira et al., 2010). To date, little work has focused on enzyme treatment prior to BSFL rearing. Lindberg et al. (2022b) found that it was better to add enzymes directly to vegetable waste than an enzymatic pre-treatment prior to the addition of BSFL. With the direct addition of a cellulase enzyme cocktail (SAE0020 Sigma-Aldrich, 1%$_{ww}$), a 22% increase in conversion efficiency was achieved, while no improvements were found for 2–4 days pre-treatment. This could be an advantage, as no additional step would be necessary, lowering the complexity of the pre-treatment. However, the authors conducted the enzyme pre-treatment at 28°C, which appeared to favour microbial degradation of the material prior to BSFL addition. An enzyme pre-treatment step conducted closer to the optimal temperature of enzymatic hydrolysis of the selected enzyme could yield different results. Theron et al. (2023) conducted enzyme hydrolysis using a cellulolytic enzyme cocktail (Cellic® CTec3, 2 FPU g^{-1} DM) of steam-pre-treated wheat straw and sugar cane bagasse at 40°C for 21 h and found a great improvement in conversion efficiency. Despite these promising results, Theron (2022) concluded that the use of wheat straw and sugar cane bagasse as a feedstock for BSFL is not economically viable due to the high cost associated with required pre-treatments.

8 Sustainability issues in substrate use

As discussed, it is evident that BSFL are highly efficient at converting a wide range of organic waste streams, even though the process efficiency is influenced by multiple interconnected parameters that must be taken into account when aiming for a productive process that yields high-quality products (biomass and fertilizer). It is well known that the sustainability of insect production depends on the substrate on which the larvae are reared (Smetana et al., 2016). As discussed in the introduction, many organic waste streams are not yet permitted as feed substrates for insects. Due to legislative constraints, the insect sector in the EU only has access to approximately 30% of the food waste generated within the EU, with a great share of these streams already having 'traditional' destinations (e.g. direct use as animal feed, such as brewery spent grains). This competition for legal waste streams complicates their use by the insect industry and may lead to an added cost for utilizing more promising substrates.

Many insect producers come from an animal feed production perspective, viewing the insects as mini-livestock, with the aim of attaining a high-quality product of predictable composition. They accomplish this by using substrates that can be used directly as food or animal feed (more heterogeneous waste materials) with the complementary use of by-products of added value, such as commodities (e.g. wheat bran or soybean bran) (Rumbos et al., 2020), to balance diets and ensure an efficient bioconversion process (personal communication). This means that insect breeders currently have considerable control over the process, but the environmental sustainability of the process is low (Fig. 1).

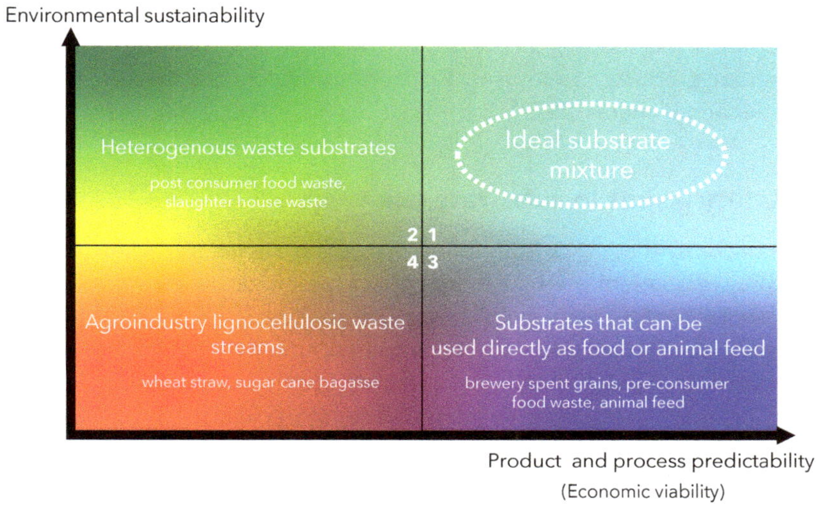

Figure 1 Representation of the equilibrium between environmental sustainability and predictability of the production of BSFL larvae with organic waste streams. Currently, the insect sector is mostly placed in quadrant '3', as only pre-consumer food waste and other high-grade substrates that could be directly used as animal feed are being utilized, resulting in the insect production process being less sustainable but highly predictable in terms of product quality and process performance. The streams represented in quadrant '2', such as post-consumer food waste, have more heterogeneous composition but its use represents lower environmental impact and higher sustainability, while quadrant '4' has streams that are difficult to handle with BSFL due to the high fibre content, which leads to low conversion efficiency. Direct gains in environmental sustainability might represent losses in predictability; however, by mixing feed substrates of varying composition and carrying out the bioconversion process considering all pertinent factors could transfer the industry to higher degrees of sustainability, while maintaining good predictability, represented by quadrant '1'.

Bosch et al. (2019) compared the global warming potential (GWP) of BSFL reared on feed, food or waste substrates and highlighted that when larvae are reared on feeds such as the Gainesville diet (combining water, wheat bran, maize flour and alfalfa), chicken feed or processed wheat, the GWP of larvae meal can be even higher than that of fish and soybean meal. However, the authors assumed low land use values for the latter two and did not consider biodiversity losses in their sustainability assessment, even though both the use of fishmeal and monoculture soy production contributes considerably to biodiversity loss (Green et al., 2019; Ortuño Crespo and Dunn, 2017). Nonetheless, Bosch et al. (2019) highlighted that when using residual streams such as food waste, manures or other animal-derived substrates, the GWP of BSFL production is significantly reduced. In summary, the use of varied waste streams that are currently unauthorized could significantly increase the sustainability of insect production in Europe, even though product and process predictability could be slightly reduced (Fig. 1). Based on the quantities of available food waste in the EU, the insect industry will only thrive, in the sense of having competitive volumes of products, if the use of currently unauthorized wastes is revised (Lalander and Vinnerås, 2022).

9 Safety issues in substrate use

It is noteworthy that the existing barriers to using animal by-products as feed for BSFL were established to reduce the risks of recycling pathogens and other contaminants when recirculating such materials within the food chain (Boqvist et al., 2018). These barriers became even stricter because of the risks associated with these wastes containing prions related to prion disease, which, in turn, have negatively affected the possibility for recirculating food waste into the food chain (Vågsholm et al., 2020b). It is critical to ensure good hygiene in circular systems such as the BSFL rearing systems to avoid recirculating contaminants with substrates. Many studies have assessed the safety potential of rearing BSFL using substrates contaminated with bacteria, fungi, toxins and even pharmaceuticals and pesticides.

BSFL are able to inactivate several microorganisms, such as *Salmonella* spp., *Escherichia coli* (Lopes et al., 2020a; Lalander et al., 2015) and bacteriophages (Lalander et al., 2013), as well as reduce the concentration of some pharmaceuticals and pesticides (Purschke et al., 2017; Lalander et al., 2016) and mycotoxins (Bosch et al., 2017; Purschke et al., 2017) in contaminated feed substrates. No studies on the impact on spore-forming bacteria have been conducted, but Van Looveren et al. (2022) found that *Clostridium perfringens* levels in the BSFL were below the detection limit, even though it was present in their feed substrate. When it comes to more stable biological contaminants, such as helminth eggs, it has been demonstrated that BSFL treatment has no impact on their survival (Lalander et al., 2013, 2015a). Thus, even though BSFL treatment can improve the hygienic quality of the treated substrate, the adoption of post-treatments of final products (larval biomass and frass) is necessary to guarantee the safety of use, as is the case for many other feed products and fertilizers available on the market, including PAPs and manures.

When it comes to more persistent contaminants, such as heavy metals, it has been demonstrated that cadmium and lead are bio-accumulated in BSFL (Purschke et al., 2017), while mercury, for instance, can be bioaccumulated by BSFL but rapidly eliminated from the larvae after a short fasting period (Cardoso et al., 2023). van Der Fels-Klerxet al. (2020) investigated the fate of various chemical contaminants found in bio-wastes of European food waste and found that some of these chemicals were bio-accumulated in the BSFL. However, none accumulated to a level that exceeded the legal thresholds in the EU. The only way to guarantee the quality of products in a circular system is by guaranteeing the quality of the substrate flowing into the system. Substrates with excessive levels of persistent contaminants, such as heavy metals, are not suitable for insect rearing. In this sense, up-stream work is crucial for ensuring sufficiently high quality of the food waste substrates. When it comes to persistent biological contaminants, no studies to date have been conducted on the fate of prions during BSFL rearing. This research is crucial in order for insects to truly unlock the potential of insect-assisted nutrient recycling in Europe. The limitations imposed by the EU and other countries around the world, preventing the use of post-consumer food waste and other animal-containing waste streams as feed substrate for BSFL, are hampering the growth of the insect industry and its competition with less sustainable feed sources, such as fishmeal and soybean meal. Once these substrates are allowed to be used by insect breeders, it will be possible to substantially increase the environmental sustainability of the process, ensuring the provision of high-quality products for the consumer market.

10 Future research trends

The effective bioconversion of different organic waste streams can be achieved by overcoming challenges such as choosing the right feed substrate and understanding how it should be provided to the larvae, as well as adjusting relevant variables within the process (density, feeding rate, substrate nutritional composition, among others). It is worth noting that such variables are always interconnected, making it nearly impossible to separate one from another. Many waste streams are not suitable for larvae alone (e.g. lignocellulosic wastes such as plant parts or brewery spent grains) but can be bioconverted by BSFL if blended with other waste streams or if pre-treatments are carried out, even though economic challenges can arise from the adoption of pre-treatments. Conversely, other waste streams that are more easily degraded by the larvae and are not yet legal as feed substrate in some countries (e.g. post-consumer food waste and slaughterhouse waste), despite their heterogeneity, could benefit the insect sector's environmental sustainability, in particular, if blended with other substrates to ensure product quality (reaching quadrant 1 in Fig. 1). In essence, the holy grail of insect production is to be able to produce insects from waste streams and in that way valorize the waste to improve waste management (Singh and Kumari, 2019), which in turn guarantees that the insect production is sustainable (Parodi et al., 2022), while at the same time producing a high-quality commodity (Lu et al., 2022) that can contribute to improved food security.

11 Conclusion

In this chapter, we have explored available substrates for BSFL rearing and outlined current understanding of how substrates impact both process efficiency and larval composition. Additionally, we have delved into the impact of other critical process parameters, including larval density, larval feed dosage and the depth at which the substrate is provided, on process efficiency and larval composition.

We presented possible strategies for handling nutritionally unbalanced substrates, including blending various waste streams or implementing pre-treatments to enhance nutrient availability for the larvae. Furthermore, we emphasized that substrate properties not only influence process efficiency and larval composition but also have a significant impact on the overall sustainability of the process.

Unfortunately for the European insect sector, currently permitted substrates result in larval biomass with a higher environmental impact compared to BSFL reared on substrates that are not allowed. However, true waste substrates (i.e. post-consumer waste streams) pose challenges due to their heterogeneity, which can lead to unpredictable product quality, and the risk of containing different pollutants.

The current challenge faced by the insect sector is to find ways to safely rear sufficient volumes of larval biomass of predictable quality and with low environmental impact.

12 Where to look for further information

International Platform of Insects for Food and Feed (IPIFF):

- https://ipiff.org/.

13 References

Abduh, M. Y., Nadia, M. H., Syaripudin, S., Manurung, R. and Putra, R. E. 2018. Factors affecting the bioconversion of Philippine Tung seed by black soldier fly larvae for the production of protein and oil-rich biomass. *Journal of Asia-Pacific Entomology* 21(3), 836–842.

Alvira, P., Tomás-Pejó, E., Ballesteros, M. and Negro, M. J. 2010. Pretreatment technologies for an efficient bioethanol production process based on enzymatic hydrolysis: a review. *Bioresource Technology* 101(13), 4851–4861.

Barragan-Fonseca, K. B., Dicke, M. and Van Loon, J. J. A. V. 2017. Nutritional value of the black soldier fly (*Hermetia illucens* L.) and its suitability as animal feed - a review. *Journal of Insects as Food and Feed* 3(2), 105–120.

Barragan-Fonseca, K. B., Dicke, M. and Van Loon, J. J. A. V. 2018. Influence of larval density and dietary nutrient concentration on performance, body protein, and fat contents of black soldier fly larvae (Hermetia illucens). *Entomologia Experimentalis et Applicata* 166(9), 761–770.

Barragan-Fonseca, K. B., Gort, G., Dicke, M. and Van Loon, J. J. A. 2019. Effects of dietary protein and carbohydrate on life-history traits and body protein and fat contents of the black soldier fly Hermetia illucens. *Physiological Entomology* 44(2), 148–159.

Barragan-Fonseca, K. B., Gort, G., Dicke, M. and Van Loon, J. J. A. 2021. Nutritional plasticity of the black soldier fly (Hermetia illucens) in response to artificial diets varying in protein and carbohydrate concentrations. *Journal of Insects as Food and Feed* 7(1), 51–61.

Barrett, M., Chia, S. Y., Fischer, B. and Tomberlin, J. K. 2023. Welfare considerations for farming black soldier flies, Hermetia illucens (Diptera: Stratiomyidae): a model for the insects as food and feed industry. *Journal of Insects as Food and Feed* 9(2), 119–148.

Barroso, F. G., Sánchez-Muros, M. J., Rincón, M. Á., Rodriguez-Rodriguez, M., Fabrikov, D., Morote, E. and Guil-Guerrero, J. L. 2019. Production of n-3-rich insects by bioaccumulation of fishery waste. *Journal of Food Composition and Analysis* 82, 103237.

Bekker, N. S., Heidelbach, S., Vestergaard, S. Z., Nielsen, M. E., Riisgaard-Jensen, M., Zeuner, E. J., Bahrndorff, S. and Eriksen, N. T. 2021. Impact of substrate moisture content on growth and metabolic performance of black soldier fly larvae. *Waste Management* 127, 73–79.

Bellezza Oddon, S., Biasato, I. and Gasco, L. 2022a. Isoenergetic-practical and semi-purified diets for protein requirement determination in Hermetia illucens larvae: consequences on life history traits. *Journal of Animal Science and Biotechnology* 13(1), 17.

Bellezza Oddon, S., Biasato, I., Resconi, A. and Gasco, L. 2022b. Determination of lipid requirements in black soldier fly through semi-purified diets. *Scientific Reports* 12(1), 10922.

Beyers, M., Coudron, C., Ravi, R., Meers, E. and Bruun, S. 2023. Black soldier fly larvae as an alternative feed source and agro-waste disposal route – A life cycle perspective. *Resources, Conservation and Recycling* 192, 106917.

Boqvist, S., Söderqvist, K. and Vågsholm, I. 2018. Food safety challenges and One Health within Europe. *Acta Veterinaria Scandinavica* 60(1), 1.

Bosch, G., Fels-Klerx, H. J. V. D., Rijk, T. C. D. and Oonincx, D. G. A. B. 2017. Aflatoxin b1 Tolerance and Accumulation in Black Soldier Fly Larvae (Hermetia illucens) and Yellow Mealworms (Tenebrio molitor). *Toxins* 9(6), 185.

Bosch, G., Van Zanten, H. H. E., Zamprogna, A., Veenenbos, M., Meijer, N. P., Van Der Fels-Klerx, H. J. and Van Loon, J. J. A. 2019. Conversion of organic resources by black soldier fly larvae: legislation, efficiency and environmental impact. *Journal of Cleaner Production* 222, 355–363.

Brits, D. 2017. Improving feeding efficiencies of black soldier fly larvae, Hermetia illucens (L., 1758) (Diptera: Stratiomyidae: Hermetiinae) through manipulation of feeding conditions for industrial mass rearing. Ph.D. Diss. Stellenbosch University, Stellenbosch.

Cardoso, D. N., Silva, A. R. R., Morgado, R. G., Mostafaie, A., Pereira, A., Pinto, J., Lopes, I. G., Murta, D., Soares, A. M. V. M., Brooks, B. W. and Loureiro, S. 2023. Improving Product Safety for Edible Insects: toxicokinetics of Hg in Tenebrio molitor and Hermetia illucens. *ACS Food Science and Technology* 3(4), 790–798.

Cheng, J. Y. K., Chiu, S. L. H. and Lo, I. M. C. 2017. Effects of moisture content of food waste on residue separation, larval growth and larval survival in black soldier fly bioconversion. *Waste Management* 67, 315–323.

Chia, S. Y., Tanga, C. M., Osuga, I. M., Cheseto, X., Ekesi, S., Dicke, M. and Van Loon, J. J. A. 2020. Nutritional composition of black soldier fly larvae feeding on agro-industrial by-products. *Entomologia Experimentalis et Applicata* 168(6–7), 472–481.

Čičková, H., Newton, G. L., Lacy, R. C. and Kozánek, M. 2015. The use of fly larvae for organic waste treatment. *Waste Management* 35, 68–80.

Dangles, O. and Casas, J. 2019. Ecosystem services provided by insects for achieving sustainable development goals. *Ecosystem Services* 35, 109–115.

Delvendahl, N., Rumpold, B. A. and Langen, N. 2022. Edible insects as food - Insect welfare and ethical aspects from a consumer perspective. *Insects*, 13(2), 121.

Dortmans, B., Egger, J., Diener, S., Verstappen, B. and Zurbrügg, C. 2021. Black soldier fly biowaste processing - A step-by-step guide, 2nd Edition Dübendorf: Eawag – Swiss Federal Institute of Aquatic Science and Technology Department of Sanitation, Water and Solid Waste for Development (Sandec).

Dossey, A. T., Tatum, J. T. and Mcgill, W. L. 2016. Chapter 5. Modern insect-based food industry: current status, insect processing technology, and recommendations moving forward. In: Dossey, A. T., Morales-Ramos, J. A. and Rojas, M. G. (Eds), *Insects as Sustainable Food Ingredients*. San Diego: Academic Press.

Dzepe, D., Nana, P., Fotso, A., Tchuinkam, T. and Djouaka, R. 2020. Influence of larval density, substrate moisture content and feedstock ratio on life history traits of black soldier fly larvae. *Journal of Insects as Food and Feed* 6(2), 133–140.

Ellen Macarthur Foundation 2015. Growth within: a circular economy vision for a competitive Europe. In: Gurewitsch, S. (Ed.). https://ellenmacarthurfoundation.org/growth-within-a-circular-economy-vision-for-a-competitive-europe: Ellen MacArthur Foundation, the McKinsey Center for Business and Environment (sponsered by Stiftungsfonds für Umweltökonomie).

European Commission 2009. Regulation (EC) No 1069/2009 of the European parliament and of the council of 21 October 2009 laying down health rules as regards animal by-products and derived products not intended for human consumption and repealing Regulation (EC) No 1774/2002 (Animal by-Products Regulation). https://eur-lex.europa.eu/eli/reg/2009/1069/oj.

European Commission 2015. Closing the loop - an EU action plan for the circular economy. https://eur-lex.europa.eu/resource.html?uri=cellar:8a8ef5e8-99a0-11e5-b3b7-01aa75ed71a1.0012.02/DOC_1&format=PDF.

European Commission 2020a. Circular economy action plan - the European green deal. https://ec.europa.eu/environment/circular-economy/pdf/new_circular_economy_action_plan.pdf.

European Commission 2020b. Farm to fork strategy - for a fair, healthy and environmentally-friendly food system. https://food.ec.europa.eu/system/files/2020-05/f2f_action-plan_2020_strategy-info_en.pdf.

European Commission 2021a. Commission regulation (EU) 2021/1372 of 17 August 2021 amending Annex IV to Regulation (EC) No 999/2001 of the European Parliament and of the Council as regards the prohibition to feed non-ruminant farmed animals, other than fur animals, with protein derived from animals (Text with EEA relevance). https://eur-lex.europa.eu/legal-content/EN/TXT/?uri=CELEX%3A32021R1372.

European Commission 2021b. Communication from the commission to the European parliament, the council, the European economic and social committee and the committee of the regions contingency plan for ensuring food supply and food security in times of crisis. https://eur-lex.europa.eu/legal-content/EN/TXT/?uri=COM:2021:689:FIN.

European Parliamentary Research Service 2023. EU feed autonomy - Closing the gaps in European food security. https://www.europarl.europa.eu/thinktank/en/document/EPRS_BRI(2023)739328.

Ewald, N., Vidakovic, A., Langeland, M., Kiessling, A., Sampels, S. and Lalander, C. 2020. Fatty acid composition of black soldier fly larvae (*Hermetia illucens*) – possibilities and limitations for modification through diet. *Waste Management* 102, 40–47.

Fowles, T. M. and Nansen, C. 2020. Insect-based bioconversion: value from food waste. In: Närvänen, E., Mesiranta, N., Mattila, M. and Heikkinen, A. (Eds), *Food Waste Management: Solving the Wicked Problem*. Cham: Springer International Publishing.

Fuso, A., Barbi, S., Macavei, L. I., Luparelli, A. V., Maistrello, L., Montorsi, M., Sforza, S. and Caligiani, A. 2021. Effect of the rearing substrate on total protein and amino acid composition in black soldier fly. *Foods*, 10(8).

Gold, M., Cassar, C. M., Zurbrügg, C., Kreuzer, M., Boulos, S., Diener, S. and Mathys, A. 2020. Biowaste treatment with black soldier fly larvae: increasing performance through the formulation of biowastes based on protein and carbohydrates. *Waste Management* 102, 319–329.

Gold, M., Ireri, D., Zurbrugg, C., Fowles, T. and Mathys, A. 2021. Efficient and safe substrates for black soldier fly biowaste treatment along circular economy principles. *Detritus* 16, 31–40.

Gold, M., Tomberlin, J. K., Diener, S., Zurbrügg, C. and Mathys, A. 2018. Decomposition of biowaste macronutrients, microbes, and chemicals in black soldier fly larval treatment: a review. *Waste Management* 82, 302–318.

Green, J. M. H., Croft, S. A., Durán, A. P., Balmford, A. P., Burgess, N. D., Fick, S., Gardner, T. A., Godar, J., Suavet, C., Virah-Sawmy, M., Young, L. E. and West, C. D. 2019. Linking global drivers of agricultural trade to on-the-ground impacts on biodiversity. *Proceedings of the National Academy of Sciences of the United States of America* 116(46), 23202–23208.

Guillaume, J. B., Mezdour, S., Marion-Poll, F., Terrol, C. and Schmidely, P. 2023. Asymptotic estimated digestibility, a new indicator of black soldier fly (Hermetia illucens) conversion efficiency in relation to larval density. *Journal of Insects as Food and Feed* 9(7), 893–906.

Hopkins, I., Newman, L. P., Gill, H. and danaher, J. 2021. The influence of food waste rearing substrates on black soldier fly larvae protein composition: a systematic review. *Insects* 12(7), 608.

Isibika, A., Simha, P., Vinnerås, B., Zurbrügg, C., Kibazohi, O. and Lalander, C. 2023. Food industry waste - an opportunity for black soldier fly larvae protein production in Tanzania. *Science of the Total Environment* 858(3), 159985.

Isibika, A., Vinnerås, B., Kibazohi, O., Zurbrügg, C. and Lalander, C. 2019. Pre-treatment of banana peel to improve composting by black soldier fly (*Hermetia illucens* (L.), Diptera: Stratiomyidae) larvae. *Waste Management* 100, 151–160.

Isibika, A., Vinnerås, B., Kibazohi, O., Zurbrügg, C. and Lalander, C. 2021. Co-composting of banana peel and orange peel waste with fish waste to improve conversion by black soldier fly (*Hermetia illucens* (L.), Diptera: Stratiomyidae) larvae. *Journal of Cleaner Production* 318, 128570.

Jiang, D., Jiang, K., Li, R., Zhao, L., Liu, Z., Xiong, B., Jin, D., Hao, X., Zhu, L., Kang, B. and Bai, L. 2022. Influence of different inoculation densities of black soldier fly larvae

(Hermetia illucens) on heavy metal immobilization in swine manure. *Environmental Science and Pollution Research International* 29(36), 54378–54390.

Lalander, C. H., Diener, S., Magri, M. E., Zurbrügg, C., Lindström, A. and Vinnerås, B. 2013. Faecal sludge management with the larvae of the black soldier fly (Hermetia illucens) – from a hygiene aspect. *Science of the Total Environment* 458–460, 312–318.

Lalander, C. H., Diener, S., Zurbrügg, C. and Vinnerås, B. 2019. Effects of feedstock on larval development and process efficiency in waste treatment with black soldier fly (*Hermetia illucens*). *Journal of Cleaner Production* 208, 211–219.

Lalander, C. H., Ermolaev, E., Wiklicky, V. and Vinnerås, B. 2020. Process efficiency and ventilation requirement in black soldier fly larvae composting of substrates with high water content. *Science of the Total Environment* 729, 138968.

Lalander, C. H., Fidjeland, J., Diener, S., Eriksson, S. and Vinnerås, B. 2015. High waste-to-biomass conversion and efficient *Salmonella* spp. reduction using black soldier fly for waste recycling. *Agronomy for Sustainable Development* 35(1), 261–271.

Lalander, C. H., Senecal, J., Gros Calvo, M., Ahrens, L., Josefsson, S., Wiberg, K. and Vinnerås, B. 2016. Fate of pharmaceuticals and pesticides in fly larvae composting. *Science of the Total Environment* 565, 279–286.

Lalander, C. H. and Vinnerås, B. 2022. Actions needed before insects can contribute to a real closed-loop circular economy in the EU. *Journal of Insects as Food and Feed* 8(4), 337–342.

Lindberg, L., Ermolaev, E., Vinnerås, B. and Lalander, C. 2022a. Process efficiency and greenhouse gas emissions in black soldier fly larvae composting of fruit and vegetable waste with and without pre-treatment. *Journal of Cleaner Production* 338, 130552.

Lindberg, L., Vinnerås, B. and Lalander, C. 2022b. Process efficiency in relation to enzyme pre-treatment duration in black soldier fly larvae composting. *Waste Management* 137, 121–127.

Liu, Z., Minor, M., Morel, P. C. H. and Najar-Rodriguez, A. J. 2018. Bioconversion of three organic wastes by black soldier fly (Diptera: Stratiomyidae) larvae. *Environmental Entomology* 47(6), 1609–1617.

Liu, Z., Morel, P. C. H. and Minor, M. A. 2023. Substrate and moisture content effects on pupation of the black soldier fly (Diptera: Stratiomyidae). *Journal of Insects as Food and Feed* 9(4), 415–425.

Lopes, I. G., Lalander, C., Vidotti, R. M. and Vinnerås, B. 2020a. Reduction of bacteria in relation to feeding regimes when treating aquaculture waste in fly larvae composting. *Frontiers in Microbiology* 11, 1616.

Lopes, I. G., Lalander, C., Vidotti, R. M. and Vinnerås, B. 2020b. Using *Hermetia illucens* larvae to process biowaste from aquaculture production. *Journal of Cleaner Production* 251, 119753.

Lopes, I. G., Wiklicky, V., Ermolaev, E. and Lalander, C. 2023. Dynamics of black soldier fly larvae composting – impact of substrate properties and rearing conditions on process efficiency. *Waste Management* 172, 25–32.

Lopes, I. G., Yong, J. W. H. and Lalander, C. 2022. Frass derived from black soldier fly larvae treatment of biodegradable wastes. A critical review and future perspectives. *Waste Management* 142, 65–76.

Lu, S., Taethaisong, N., Meethip, W., Surakhunthod, J., Sinpru, B., Sroichak, T., Archa, P., Thongpea, S., Paengkoum, S., Purba, R. A. P. and Paengkoum, P. 2022. Nutritional composition of black soldier fly larvae (*Hermetia illucens* L.) and its potential uses as alternative protein sources in animal diets: a review. *Insects* 13(9), 831.

Mazza, L., Xiao, X., Ur Rehman, K., Cai, M., Zhang, D., Fasulo, S., Tomberlin, J. K., Zheng, L., Soomro, A. A., Yu, Z. and Zhang, J. 2020. Management of chicken manure using black soldier fly (Diptera: Stratiomyidae) larvae assisted by companion bacteria. *Waste Management* 102, 312–318.

Miranda, C. D., Cammack, J. A. and Tomberlin, J. K. 2020. Mass production of the black soldier fly, Hermetia illucens (L.), (Diptera: Stratiomyidae) reared on three manure types. *Animals* 10(7).,1243.

Nguyen, T. T. X., Tomberlin, J. K. and Vanlaerhoven, S. 2013. Influence of resources on Hermetia illucens (Diptera: Stratiomyidae) larval development. *Journal of Medical Entomology* 50(4), 898–906.

Ojha, S., Bußler, S. and Schlüter, O. K. 2020. Food waste valorisation and circular economy concepts in insect production and processing. *Waste Management* 118, 600–609.

Opare, L. O., Holm, S. and Esperk, T. 2022. Temperature-modified density effects in the black soldier fly: low larval density leads to large size, short development time and high fat content. *Journal of Insects as Food and Feed* 8(7), 783–802.

Ortuño Crespo, G. and Dunn, D. C. 2017. A review of the impacts of fisheries on open-ocean ecosystems. *ICES Journal of Marine Science* 74(9), 2283–2297.

Parodi, A., Ipema, A. F., Van Zanten, H. H. E., Bolhuis, J. E., Van Loon, J. J. A. and De Boer, I. J. M. 2022. Principles for the responsible use of farmed insects as livestock feed. *Nature Food* 3(12), 972–974.

Parra Paz, A. S., Carrejo, N. S. and Gómez Rodríguez, C. H. 2015. Effects of larval density and feeding rates on the bioconversion of vegetable waste using black soldier fly larvae *Hermetia illucens* (L.), (Diptera: Stratiomyidae). *Waste and Biomass Valorization* 6(6), 1059–1065.

Peguero, D. A., Gold, M., Endara, A., Niu, M., Zurbrügg, C. and Mathys, A. 2023. Evaluation of ammonia pretreatment of four fibrous biowastes and its effect on black soldier fly larvae rearing performance. *Waste Management* 160, 123–134.

Peguero, D. A., Gold, M., Vandeweyer, D., Zurbrügg, C. and Mathys, A. 2022. A review of pretreatment methods to improve Agri-food waste bioconversion by black soldier fly larvae. *Frontiers in Sustainable Food Systems* 5.

Purkayastha, D. and Sarkar, S. 2022. Black soldier fly larvae for treatment and segregation of commingled municipal solid waste at different environmental conditions. *Journal of Environmental Management* 302(A), 114060.

Purschke, B., Scheibelberger, R., Axmann, S., Adler, A. and Jäger, H. 2017. Impact of substrate contamination with mycotoxins, heavy metals and pesticides on the growth performance and composition of black soldier fly larvae (Hermetia illucens) for use in the feed and food value chain. *Food Additives and Contaminants: Part A* 34(8), 1410–1420.

Rumbos, C. I., Karapanagiotidis, I. T., Mente, E., Psofakis, P. and Athanassiou, C. G. 2020. Evaluation of various commodities for the development of the yellow mealworm, Tenebrio molitor. *Scientific Reports* 10(1), 11224.

Sánchez-Bayo, F. and Wyckhuys, K. A. G. 2019. Worldwide decline of the entomofauna: a review of its drivers. *Biological Conservation* 232, 8–27.

Seyedalmoosavi, M. M., Mielenz, M., Veldkamp, T., Daş, G. and Metges, C. C. 2022. Growth efficiency, intestinal biology, and nutrient utilization and requirements of black soldier fly (Hermetia illucens) larvae compared to monogastric livestock species: a review. *Journal of Animal Science and Biotechnology* 13(1), 31.

Singh, A. and Kumari, K. 2019. An inclusive approach for organic waste treatment and valorisation using Black Soldier Fly larvae: a review. *Journal of Environmental Management* 251, 109569.

Smetana, S., Palanisamy, M., Mathys, A. and Heinz, V. 2016. Sustainability of insect use for feed and food: life cycle assessment perspective. *Journal of Cleaner Production* 137, 741–751.

Somroo, A. A., Ur Rehman, K., Zheng, L., Cai, M., Xiao, X., Hu, S., Mathys, A., Gold, M., Yu, Z. and Zhang, J. 2019. Influence of Lactobacillus buchneri on soybean curd residue co-conversion by black soldier fly larvae (Hermetia illucens) for food and feedstock production. *Waste Management* 86, 114–122.

Theron, W. 2022. Steam pretreatment of wheat straw and sugarcane bagasse to produce feed for black soldier fly larvae. Master of Engineering. Stellenbosch University, Stellenbosch.

Theron, W., Teke, G. M., Diedericks, D., Van Rensburg, E. and Görgens, J. F. 2023. Steam pre-treatment of sugarcane bagasse and wheat straw as a cleaner feedstock for black soldier fly larvae rearing. *Journal of Insects as Food and Feed*1-15.

Tokwaro, R., Semiyaga, S., Niwagaba, C. B., Nakagiri, A., Sempewo, J. I., Muoghalu, C. C. and Manga, M. 2023. Application of black soldier fly larvae in decentralized treatment of faecal sludge from pit latrines in informal settlements in Kampala city. *Frontiers in Environmental Science* 11.

United Nations Environment Programme 2021. *Food Waste Index Report 2021*. Nairobi, Kenya.

Vågsholm, I., Arzoomand, N. S. and Boqvist, S. 2020. Food security, safety, and sustainability–getting the trade-offs right. *Frontiers in Sustainable Food Systems* 4.

van Der Fels-Klerx, H. J., Meijer, N., Nijkamp, M. M., Schmitt, E. and Van Loon, J. J. A. 2020. Chemical food safety of using former foodstuffs for rearing black soldier fly larvae (Hermetia illucens) for feed and food use. *Journal of Insects as Food and Feed* 6(5), 475–488.

Van Huis, A. 2020. Insects as food and feed, a new emerging agricultural sector: a review. *Journal of Insects as Food and Feed* 6(1), 27–44.

Van Klink, R., Bowler, D. E., Gongalsky, K. B., Swengel, A. B., Gentile, A. and Chase, J. M. 2020. Meta-analysis reveals declines in terrestrial but increases in freshwater insect abundances. *Science* 368(6489), 417–420.

Van Looveren, N., Vandeweyer, D., Van Schelt, J. and Van Campenhout, L. 2022. Occurrence of *Clostridium perfringens* vegetative cells and spores throughout an industrial production process of black soldier fly larvae (*Hermetia illucens*). *Journal of Insects as Food and Feed* 8(4), 399–407.

Vogel, M., Shah, P. N., Voulgari-Kokota, A., Maistrou, S., Aartsma, Y., Beukeboom, L. W., Salles, J. F., Van Loon, J. J. A., Dicke, M. and Wertheim, B. 2022. Health of the black soldier fly and house fly under mass-rearing conditions: innate immunity and the role of the microbiome. *Journal of Insects as Food and Feed* 8(8), 857–878.

Woelber-Kastner, B. K., Frey, S. D., Howard, D. R. and Hall, C. L. 2021. Insect reproductive behaviors are important mediators of carrion nutrient release into soil. *Scientific Reports* 11(1), 3616.

WWF UK 2021. *Driven to Waste: The Global Impact of Food Loss and Waste on Farms*. WWF-UK.

Yakti, W., Schulz, S., Marten, V., Mewis, I., Padmanabha, M., Hempel, A.-J., Kobelski, A., Streif, S. and Ulrichs, C. 2022. The effect of rearing scale and density on the growth and nutrient composition of Hermetia illucens (L.) (Diptera: Stratiomyidae) larvae. *Sustainability* 14(3), 1772.

Yang, F. and Tomberlin, J. K. 2020. Comparing selected life-history traits of black soldier fly (Diptera: Stratiomyidae) larvae produced in industrial and bench-top-sized containers. *Journal of Insect Science* 20(5).

Yang, L. H. and Gratton, C. 2014. Insects as drivers of ecosystem processes. *Current Opinion in Insect Science* 2, 26–32.

Chapter 2

Pathogens and diseases affecting large-scale production of black soldier fly

Roberto M. Pereira, University of Florida, USA

1 Introduction

Black soldier flies (BSF) (*Hermetia illucens*) have been produced on a large scale using various production methods and food substrates. Depending on cultural methods and conditions, food substrate, and other characteristics of the different production facilities, several microorganisms that could be considered potential pathogens have been identified from either the culture environment or the insects themselves (Joosten et al. 2020, Gorrens et al. 2021a, Klüber et al. 2022). However, many of the organisms listed in some publications have never been observed affecting the soldier fly in large-scale production systems. Some authors doubt that BSF are immune to pathogens and indicate that the lack of reports on pathogen outbreaks does not indicate that these do not occur (Barrett et al. 2023).

During BSF production, several antimicrobial peptides (AMPs) are released onto the production medium. These peptides are important in the prevention of diseases and the maintenance of healthy fly colonies (Xia et al. 2021). The use of this AMP-rich medium in feeding other animals transfers this chemical immunity to the animals feeding on the BSF larvae and its growth medium. This transfer of immunological compounds occurs not only from the nutritional content of the fly larvae but also from the immunological content left by the insects in their growth medium.

http://dx.doi.org/10.19103/AS.2024.0139.03

Understanding the organisms that can affect BSF in large-scale production facilities, as well as the immunological properties present in the insect bodies and added to the insect growth medium, are important elements in evaluating production systems. This evaluation must consider methods that can prevent widespread pathological phenomena with the potential to decrease the efficiency of fly production. These methods should promote the production of immunological products that protect not only the insect culture but also benefit other animals and humans that may consume the insects and insect products.

It is important to remember that the vast majority of microorganisms that have been isolated from or somewhat associated with BSF are not pathogenic and would not have any deleterious effect on the BSF production. On the contrary, the majority of these microorganisms have a beneficial effect on the fly organism or culture, either by helping in the digestion of products or serving the insect host in another beneficial role. Therefore, a general look at the many organisms that have been associated with BSF and its culture would not be pertinent if we aim to understand the organisms that can interfere negatively with the normal development and survival of BSF in its different life stages and under different culture conditions. The vestigial presence of genetic material, which could be associated with viruses and other potential pathogens, is an indication of past interactions between the organisms but not necessarily an indication of potential pathogenic interference that could be of concern to practitioners interested in the small-to-large scale culture of the BSF.

Another important factor to be considered is that BSF larvae are reported to produce several products (Choi et al. 2012, Zdybicka-Barabas et al. 2017, Gorrens et al. 2021b, Xia et al. 2021, Van Moll et al. 2022) that act as anti-microbial agents and are released into the medium in which the fly larvae develop. A deeper analysis of these anti-microbial factors produced by BSF immatures is outside the scope of this work. However, BSF seems to have an adaptive immune system that allows this insect to live and thrive on substrates and ambients that have a very high microbial contamination (Klüber et al. 2022).

Disease outbreaks are common in many farmed invertebrate populations and can result in significant mortality, as well as symptom-related suffering and loss of natural behaviors before death. Apparent symptoms of insect diseases vary based on the pathogen but may include behavioral changes, lethargy, sepsis, swollen abdomens, feeding cessation, and other symptoms (Joosten et al. 2020, Vega and Kaya 2012).

Although BSF is certainly not immune to pathogens, this insect may be less affected by many microorganisms due to BSF's ecological function as a decomposer, which means it is exposed to temperatures that may be higher than many potential natural enemies could withstand. The decomposing matter that BSF normally uses as a development medium reaches temperatures that eliminate some potential pathogens. A second level of protection comes from deposits of lauric acid, a predominant fatty acid deposited in BSF, which has been shown to have antimicrobial activity both *in vitro* and *in vivo* (Koutsos et al. 2022). Products added by larval stage to the growth medium provide a high level of protection to BSF larvae against fungal species (Kuznetsova et al. 2022) and potentially other pathogens.

Despite a fairly large volume of information on different systems for large-scale production of BSF, an interested person looking for details on the potential pathogens and diseases that may affect BSF production will find little immediately available. This lack of information on pathogens may be a bit misleading and may lead potentially interested parties to underestimate the potential for serious disruptions in the production of BSF due to diseases. Although this insect seems to be relatively resistant to the many pathogens that readily decimate other insect production systems, a false sense of security may prove to be disastrous if disease outbreaks do occur.

Beyond the technical aspects related to the success of the insect production, including those related to prevention and control of potential diseases, concerns have been raised about farmed animal welfare in any large-scale exploitation of animals for consumption or other purposes, and the farming of BSF is not exempt from these concerns (Birch 2020), although they are less intense than those raised about the farming of higher animals. When considering the presence of pathogens and the possibility that the environment imposed on these flies may facilitate disease outbreaks, the focus on welfare and health may need to be considered beyond the economic aspects related to the production and utilization of the flies. As we learn more about the relationships between disease occurrence and the conditions imposed on large populations of farmed insects, new research and development may lead to changes in insect rearing to satisfy society's commitment to animal welfare and the humane treatment of the animals we chose to use and produce for various purposes in modern society.

Modern genetic techniques allow us to determine the presence of infinitely small amounts of genetic material that may be related to a pathogenic entity. These observations are extremely valuable in understanding the biological relationships between BSF and other entities in nature, but they are not sufficient to allow an understanding of the potential dangers that pathogens may represent to the insects and those involved in the large-scale production of these insects. The listing of potential pathogens without a clear demonstration of a pathogenic relationship between the agent and the BSF host has little value, especially because BSF larvae develop in substrates that harbor many microorganisms that fall within groups that include previously described insect pathogens. This chapter will be restricted to those organisms that are known to be involved in true pathogenic relationships with the BSF (not aided by artificial experimental methods), which could cause negative effects on the insect culture. Despite warnings from some authors about the potentially devastating effects of pathogens in BSF production systems (Jensen and Lecocq 2023), presently, these organisms do not seem to represent a substantial threat to the production of BSF larvae as this insect has developed morphological and immunological defenses that work efficiently against pathogens (Tourtois et al. 2017).

2 Pathogens: viruses, bacteria, fungi, protists, and nematodes

The current list of organisms that can cause disease in the BSF population is not very extensive, despite relatively high industrial interest in this insect. Existing books on insect pathology and microbial control of insects (Lacey 2012, Vega and Kaya 2012, Poinar and Thomas 1984) do not mention BSF. It is reasonable to expect that, as interest and economic activity around BSF production and use grows, more pathogens will be described as causing disease and potentially large outbreaks in BSF production facilities. The most relevant information available on pathogens and disease outbreaks currently available is presented here, but future readers are directed to search for other information because the science on the pathology of BSF is in its initial stages. Known pathogens of *H. illucens* are listed in Table 1.

2.1 Viruses

An exogenous totivirus (*H. illucens* totivirus 1, HiTV1) infects BSF but has not been observed to cause any defects or physiological changes in the fly (Pienaar et al. 2022). Although there is genetic evidence of interactions between the BSF and viral entities, no other virus has been found to affect this fly species, and there are no reports of viral disease outbreaks in large-scale production facilities (Joosten et al. 2020). Because of the cryptic nature of the larvae of BSF, viral infections may likely go unnoticed even in laboratory conditions, especially if the viruses do not cause massive mortality. Mild infections that lead to lower productivity may not be recognized under small- and large-scale production systems if there is little impact on the overall status of the colony.

Given the significant interest in BSF, combined with the lack of reports on the other viruses and any widespread disease in the BSF populations of unknown causes, it seems unlikely that a virus of great consequences for the production of BSF exists. However, it is impossible to predict whether such a virus might exist and which could cause great damage to commercial BSF production. BSF producers must be on alert for potential signs that may indicate the presence of a virus in the insect population. Also, close attention to optimum parameters for the production of the flies may avoid catastrophic effects of any virus in the fly

Table 1 Pathogens reported infecting the black soldier fly

Pathogen group	Species	Specific stage	References	Note
Viruses	Totivirus (*Hermetia illucens* totivirus 1, HiTV1		Pienaar et al. 2022	
	Virus ??		Yang, 2017	90% mortality
	Penaeus merguiensis densovirus (PmergDNV)		Fauce and Owens, 2008	
Bacteria	Xanthomonadales order		Zheng et al. 2013	Genetic study
	Lysobacter spp.		Zheng et al. 2013	Genetic study
	Burkholderiales order		Zheng et al. 2013	Genetic study
	Bacteroides spp.		Zheng et al. 2013	Genetic study
	Clostridia spp.		Zheng et al. 2013	Genetic study
	Bacillus spp.		Zheng et al. 2013	Genetic study
	Paenibacillus thiaminolyticus		She et al. 2023	
	Alcaligenes faecalis		Klüber et al. 2022	
	Diutina rugosa		Klüber et al. 2022	
	Klebsiella pneumoniae		Klüber et al. 2022	
	Microbacterium thalassium		Klüber et al. 2022	
	Pseudomonas aeruginosa		Klüber et al. 2022	
	Serratia marcescens		Joosten et al. 2020	
Fungi	*Isaria* spp.	Adult	Leatherdale 1970	Observation
	Beauveria bassiana	Adult	Lecocq et al. 2021	Picture

Pathogen group	Species	Specific stage	References	Note
	Enthomophthora sp.		Balazy 1993	Undetermined sp.
			Eilenberg 2000	
	Unidentified fungus	Pupae	Barrett et al. 2023	
	Paecilomyces suffultus	on fly pupae (Psychodidae, Tipulidae, and Stratiomyidae)	Ellis 1957	Cited as Cylindrodendrum suffultum
	Fusarium solani		Klüber et al. 2022	
Nematodes	*Steinernema carpocapsae*	Young stages	Tourtois et al. 2017	
	Steinernema feltiae	Young stages	Tourtois et al. 2017	
	Steinernema riobrave	Young stages	Tourtois et al. 2017	
	Heterorhabditis bacteriophora	Young stages	Tourtois et al. 2017	
Protists	NONE reported			

population, since viral diseases are known to flare up at times when the insect populations are under greater stress and suboptimal conditions.

Although no virus has been observed in eliminating BSF populations, some viruses that have a broad host spectrum could cause problems in BSF production facilities after moving over from other species and locations. One of these viruses is the *Penaeus merguiensis* densovirus (PmergDNV), which can be transmitted to other invertebrates outside of its original crustacean host (Fauce and Owens 2008). This is especially true if products of subproducts of different invertebrate rearing facilities are reused, so this utilization must be monitored and precautions such as temperature or chemical sterilizations must be used to prevent the establishment of undesirable viruses in the insect population. This is especially important because potential viruses could go undetectable for a long time in BSF production facilities before environmental changes lead to a flare-up of the virus.

Another important factor to consider is the possibility that the interaction of viruses with other pathogens (Patterson et al. 2020), including viruses that can be transferred from insects to humans (Ferreira et al. 2020). The possibility that mass-reared BSF could serve as a link in this chain that would eventually be detrimental for humans involved in the rearing of BSF is a remote one, but must be considered when safety measures related to the mass rearing and utilization of the flies are implemented.

Constant vigilance, perhaps using kits that employ modern molecular diagnosis tools for potential specific viruses and other pathogens, may help the BSF industry and other insect-related facilities prevent the occurrence of catastrophic viral outbreaks in rearing. Additionally, precise control of temperature, medium moisture, and other factors that affect the growth of the BSF larvae may be the best way to prevent the occurrence of viral outbreaks that could decimate the BSF population. Despite the belief expressed by some that BSF may be less susceptible to disease outbreaks than other insect populations (Joosten et al. 2020), this comparative advantage may quickly disappear when BSF becomes the focus of an industrial endeavor that seeks to lower the cost of production and maximize the profits.

2.2 Bacteria

Many bacteria have been found in media occupied by *H. illucens* (Klüber et al. 2022, Shumo et al. 2021), although no true bacterial pathogen had been identified affecting this insect until recently. Several bacteria (*Alcaligenes faecalis, Diutina rugosa, Klebsiella pneumoniae, Microbacterium thalassium, Pseudomonas aeruginosa*) were considered pathogenic to BSF when injected into fly larvae (Klüber et al. 2022), but these organisms do not seem to be problematic in large-scale production systems where the carriers to bacterial entry into the fly larvae bodies are not compromised.

A bacterium causing soft rot on BSF larvae has been identified as affecting BSF production in China (She et al. 2023). The causative agent was the bacterium *Paenibacillus thiaminolyticus*, which is an environmental bacterium that has also been isolated causing an infection in humans affecting catheters used for hemodialysis. The bacterial infection caused the swelling of the BSF midgut area and destroyed the peritrophic membrane and microvilli, allowing the growth and spread of other bacterial species in the insect gut. Infected larvae turn dark and lose appetite. After death, infected larvae turn black and have a putrid smell. The body-covering skin becomes very fragile, tearing easily, and facilitating the spread of the bacterium in the rearing environment. Because the effect of the bacterium can be stronger under higher temperatures, careful control of the rearing facility may prevent drastic consequences if this bacterium is affecting the BSF colony.

H. illucens, like many other insects, can be affected by *Serratia marcescens*, but no outbreaks of this bacterium have been reported (Joosten et al. 2020). Some research indicates that immune resistance to potential bacterial pathogens (Bruno et al. 2021) responds quickly, with both phagocytosis and encapsulation acting rapidly after a bacterial challenge, followed by robust phagocytosis and encapsulation functions (von Bredow et al. 2022). Because *Serratia* infection can be very common in insects living in crowded conditions, due to the possibility that the insect cuticle can be damaged and the bacteria can gain access to the hemolymph and reproduce rapidly in the insect, it would be expected that *S. marcescens* would appear in soldier fly population in the wild. However, the dark insect larval skin, the hidden nature of the larval stage, and the black adult cuticle may hide any obvious disease symptoms, and the bacteria-rich environment where the insect develops may hide any minor occurrence of this bacterium in the developing fly population.

Given the medium in which BSF larvae develop and its high bacterial content, strong immunological protection against bacterial infection would be expected in this insect (Joosten et al. 2020). *H. illucens* larvae can trigger both phagocytosis and encapsulation rapidly when exposed to bacteria and other potentially pathogenic microorganisms. That process is followed by humoral components that reinforce the insect's defenses against potentially infective bacteria (Bruno et al. 2021). High levels of lysozyme and phenoloxidase, two key elements in the insect immune system, are responsible for the robust systemic attack on infecting bacteria, providing BSF with a robust protection against potential disease-causing bacteria and other organisms. Thus, despite genetic material related to many bacterial species and groups being detected on the surface of BSF eggs (Heussler et al. 2022), this bacterial load does not result in high mortality of larvae because many of these bacteria are likely either neutral or beneficial to the development of the BSF larvae, not representing a risk to the normal development of the insect.

2.3 Fungi

Several entomopathogenic fungi have been described as affecting BSF individuals and populations (Joosten et al. 2020), but massive epizootics have not been reported affecting mass-production systems. There is no mention of fungal strains derived from BSF in the USDA-ARS Catalog of fungal strains (USDA-ARSef 2020), although several are designated as originating from the Stratiomyidae family. The temperatures of the medium where BSF larvae normally develop do not seem to favor the development of fungi, and only special strains adapted to higher temperatures would be able to survive and thrive in populations of BSF. Indeed, the fungal species that dominate populations found on the soldier fly larvae are in the *Candida* genus (Vitenberg and Opatovsky 2022), which presents no threat to the BSF and serves as a nutritional source for fly development.

Although *Beauveria bassiana* can be infective to adult flies under laboratory conditions (Lecocq et al. 2021), this fungus has not been observed in the wild attacking BSF or in colonies under artificial rearing conditions. Under laboratory conditions, the fungus was able to affect egg production and cause mortality of the fly adults, but similar observations have not been made in the field or under normal conditions in production systems. *B. bassiana* could be a threat both to the larvae and the adult flies, but it does not seem to be observed with frequency in large-scale production systems, therefore its status as a threat to the colonies under artificial conditions seems to be minimum.

Several Entomophthorales fungi have been observed on Stratiomyidae flies and must be considered as potential pathogens (Balazy 1993, Eilenberg 2000), along with other fungal species (Leatherdale 1970). However, their occurrence of these fungi on BSF larvae and adults seems to be at the level of a mild moderator of the fly population rather than a very serious mortality factor in most occasions. Of course, this status of a mild pathogen can be changed considerably when large-scale production systems are in place. Interestingly, the inoculation of *Fusarium solani*, a known plant pathogen, can cause high mortality in BSF larvae (Klüber et al. 2022), which serves as a reminder that, when large-scale production of the BSF becomes widespread, unknown pathogens may become more important and widespread.

The fungal population associated with BSF rearing has been described in detail (Boccazzi et al. 2017). However, this research demonstrates that the fungal species associated with BSF rearing are mostly restricted to yeast species that have an important role in food substrate decomposition, and not pathogenic effects on the insect larvae. Furthermore, fungal species associated with BSF production are dominated by substrate degradation species, rather than insect pathogenic species that could harm the insect population.

2.4 Protists

A recent review of pathogens affecting BSF (Joosten et al. 2020) and other publications (Barrett et al. 2023, Eilenberg et al. 2015) have not referred to any protist affecting the BSF. Although the number of protist pathogens in insects, in general, is not very high, it is not unusual for well-studied insects to have some protist pathogen listed for the species. For a relatively widely used species, a protist absence in BSF is surprising, especially when we consider the fact that protist pathogens have been a consistent headache in many other insect-rearing systems.

2.5 Nematodes

Steinernema spp. and *Heterorhabditis bacteriophora* are common insect-pathogenic nematodes capable of penetrating and colonizing many insects from various orders. These nematodes can colonize BSF larvae, but *Steinernema* spp. were not attracted to BSF larvae, possibly due to the absence of attracting cues (Tourtois et al. 2017). Despite this initial lack of attraction, the nematodes completed their development in the BSF larvae. However, the number of second-generation infective juveniles produced within the host was substantially lower than what could have been expected given the size of the host.

This lower nematode production demonstrates a lower suitability of the BSF as an entomopathogenic nematode host, which may provide some protection to this insect in large BSF-rearing facilities. However, if nematodes are present in a BSF-rearing facility, it is doubtful that this low production of second-generation nematodes in the BSF body would be sufficient to protect the facility from considerable damage if a widespread nematode infestation was present.

3 Managing rearing conditions to prevent disease outbreaks

Optimal control of rearing conditions is the best defense against any disease outbreaks. Insect populations reared under optimal conditions develop faster, experience lower levels of stress, and therefore are less susceptible to diseases. It is not uncommon in insect-rearing facilities, as well as in other animal-rearing facilities, for temperature fluctuations and other destabilizing conditions to slow growth, increase stress in the insect population, and trigger disease outbreaks.

Although BSF can survive temperatures close to 0°C, the ideal rearing temperature for the insect to consume food and develop at a steady pace is around 35°C. This optimum development temperature, though not extremely high, is sufficiently high to at least serve as a deterrent to some potential pathogens. BSF can survive very low temperatures, down to close to 0°C, but the minimum temperature for development is well above that minimum, at 10°C. This range of temperature allows the manager of a BSF facility a wide temperature zone to work with, in case an increase or a decrease in temperature becomes an option in disease management. Of course, the rearing medium must be maintained as much as possible within the optimum range for BSF development (25–30°C) as long as there is a temperature control system in place that allows for both lowering and increasing the temperature of the rearing medium.

Another factor that must be regulated very closely is the water content of the medium. Many facilities keep the water content of the medium used for BSF rearing at 70–80%, but control of moisture content, is not always very easy to maintain, depending on the insect rearing facilities and the ability to control medium condition during the fly development. Moisture levels are especially important for the presence and performance of the microbial community (Bekker et al. 2021), which has a strong effect on the development and well-being of the BSF population and, therefore, a strong influence on the maintenance of a healthy population in the rearing facility.

Balancing rearing conditions to obtain optimal BSF development and minimize disease occurrence may require some level of experimentation by the BSF facility operator in order to balance the well-being of the BSF population and the need to prevent disease outbreaks that can be caused by abnormal fluctuations of the growth medium and air moisture content and temperatures. Depending on the air-handling equipment that may be available in the BSF production facility, a proper adjustment may be necessary to compensate for normal daily variations in temperature and relative humidity.

4 Disease prevention and control measures

When dealing with the mass production of insects, disease prevention should be foremost in the planning and execution of all tasks. Due to the density of the organisms in rearing facilities and the cryptic nature of many insects, the presence of diseases may not be noticeable until a large number of insects are exposed to pathogens. Additionally, due to the high population densities typically found in insect production systems, the chances of spread of an infectious disease are extremely high. Despite normally developing in waste and other unsanitary material, BSFs are still susceptible to microbial diseases that can flourish at times when conditions put the population under stress. Sanitation practices are usually not practical during the short larval development of the soldier fly, so sanitary disease prevention practices must be implemented before the addition of BSF eggs to the development medium.

The first line of defense against diseases in an insect control facility relies on the defenses that the insect species produces naturally. The BSF larvae are known to produce several AMPs (Elhag et al. 2017) and are capable of reducing bacterial loads in different growth media and substrates (Erickson et al. 2004). These natural defenses are the result of survival and adaptation in media that tend to be rich in microbial life and are essential elements in the utilization of the BSF larvae. One of the elements of this chemical barrier against microorganisms includes cecropins that have a broad spectrum of activities not only against bacteria but also fungi, viruses, and other parasites (Elhag et al. 2017). Because the gut microbiome varies with diet (Chen et al. 2023), it is possible to enhance the natural disease resistance of the BSF larvae by manipulating the diet composition.

The immune response to bacterial challenge in BSF larvae consists of rapid cellular response represented by phagocytosis and encapsulation of any invading organism, most of which are removed from the hemolymph in a short time, followed by humoral responses by a lysozyme and phenoloxidase that targets the remaining products of circulating pathogens (Bruno et al. 2021). This special adaptation, which is certainly necessary for an insect that lives in the microbe-rich environment inhabited by BSF larvae, allows the insect to respond to pathogen challenges quickly and effectively.

The cellular processes involved in the immune response that results in the fast elimination of invaders from *H. illucens* larvae consist of nodulation, along with phenoloxidase activity, and other antimicrobial peptides present in the hemolymph (Zdybicka-Barabas et al. 2017). However, any immune system can be overwhelmed by pathogens if the microbial load is sufficiently large, therefore, the manager of BSF-producing facilities must prevent the presence of excessive loads of pathogenic organisms that could overwhelm the BSF immune system. With that in mind, procedures must be implemented that minimize the pathogen inoculum including partial decontamination of the growth medium, elimination of procedures that may result in larger pathogen loads, and tight control of environmental conditions that favor rapid and vigorous growth of the BSF larvae, including the addition of supplements that may improve the quality of the growth medium.

Physical and temporal barriers that separate different populations of BSF in a production facility, and prevent the flow of flies from wild populations from invading and contaminating the captive BSF population, are useful in limiting the possibility of disease spread within a facility. The strict control of the production system and maintenance of the growing insect population within strict environmental limits can be used in maintaining the whole population within a narrow range of developmental stage, which diminish the exposure of the population to pathogens, especially those that require longer incubation periods.

5 Conclusion and future trends

Serious disease outbreaks in BSF populations seem to be very rare since they have not been reported by producers (Eilenberg et al. 2015) or documented in scientific or popular literature. This seems to be the case for the different geographic locations where BSF is reared in large production facilities. Whether this absence of disease outbreak reports is due to a natural resistance of BSF populations to these widespread outbreaks, due to natural resistance of the insect to specific pathogens, or to the environment in which the insect develops, is not entirely settled yet. However, likely, disease outbreaks in large-scale BSF production will not be as serious or as constant as observed in other reared insect populations.

Despite past observations indicating a lower susceptibility of BSF to common pathogens observed in insect-rearing facilities, this should not be taken as a reason for relaxed hygienic conditions and lacking controls designed to prevent and control disease outbreaks in insect-rearing facilities. Given the evidence presented here and reported in the scientific literature, the practitioner involved in BSF production can breathe easier but should maintain constant vigilance and adopt procedures that prevent or minimize the chances for catastrophic disease outbreaks. This is especially important given that there is still much research that must be done on the subject of diseases that can affect *H. illucens*.

However, it is possible that with greater BSF-rearing activity, serious disease outbreaks may become more common than they seem to be, based on the current literature. It is possible that pathogens that do not seem to cause serious problems now, could become a greater problem in the future due to, either changes in the BSF, or the pathogen populations, along with potential changes in the production practices that may cause greater exposure or susceptibility of the insect to pathogens.

There are some warning signals in the scientific literature that deserve attention, such as reports of greater than 90% mortality in populations exposed to viral infection (Yang 2017) and reports on colony losses (Miranda et al. 2021). As BSF facilities continue to scale up their production, disease outbreaks and observations of novel pathogens in the industry are more likely (Joosten et al. 2020).

Like any other organisms in nature, the BSF, in all its stages, serves as a host to several described pathogens and likely others that have not been reported in the literature. For an insect that has been the focus of much interest in recent years as an alternative source of nutritional material for both humans and other animals, it is surprising that more information is not available on the pathology of this organism and the potential remedies that can be utilized in case these become a problem in the various production systems proposed for this insect.

This is probably due, at least partially, to the natural resistance of this insect to common pathogens that affect several other insect populations. However, some of the immunity in the BSF to many potential pathogens is also related to the inhospitable environment where these fly larvae develop, where high temperatures and the presence of by-products, some of which are released in the medium by the developing fly maggot, make the ambient less adequate for many potential pathogens and other natural enemies.

Despite these considerations, the reader should be aware of the adaptive capacity of many potential pathogens and the opportunistic nature of microorganisms and other parasites that could adapt to and take advantage of the rearing conditions humans create to favor the economic and efficient production of BSF. With increased protection afforded to captive BSF populations in rearing facilities, these populations may start to lose some of the mechanisms that protect the individuals from occasional pathogen attacks because their exposure to these potential pathogens is diminished (Jensen and Lecocq 2023). Over time, captive populations could experience decreased resistance to pathogen attacks. As BSF production increases around the world, we will likely encounter more challenges related to pathogens and the diseases they cause in *H. illucens*.

6 Where to look for further information

Further information on diseases that affect insects can be found in the textbook by Vega and Kaya listed in the References. For information on techniques that may be useful in determining the presence of diseases in insect populations, consult the textbook by Lacey.

7 References

Balazy, S. (1993), *Entomophthorales*. Polish Academy of Sciences, Krakow, Poland.

Barrett, M., Chia, S. Y., Fischer, B., Tomberlin, J. (2023). Welfare considerations for farmingblack soldier flies, *Hermetia illucens* (Diptera: Stratiomyidae): a model for the insects as food and feed industry. *J. Ins. Food Feed* 9 (2):119–148. https://doi.org/10.3920/JIFF2022.0041

Bekker, N. S., Heidelbach, S., Vestergaard, S. Z., Nielsen, M. E., Riisgaard-Jensen, M., Zeuner, E. J., Bahrndorff, S. and Eriksen, N. T. (2021), Impact of substrate moisture content on growth and metabolic performance of black soldier fly larvae. *Waste Managem*. 127(15): 73–79 https://doi.org/10.1016/j.wasman.2021.04.028.

Birch, J. (2020), The search for invertebrate consciousness. *Noûs*. https://doi.org/10.1111/nous.12351.

Boccazzi, i. V., Ottoboni, M., Martin, E., Comandatore, F., Vallone, L., Spranghers, T., Eeckhout5, M., Mereghetti, V., Pinotti, L., Epis, S. (2017) A survey of the mycobiota associated with larvae of the black soldier fly (*Hermetia illucens*) reared for feed production. *PLoS ONE* 12(8): e0182533. https://doi.org/10.1371/journal.pone.0182533.

Bruno, D., Montali, A., Mastore M., Brivio, M. F., Mohamed, A., Tian, L. Grimaldi, A., Casartelli, M. and Tettamanti, G. (2021), Insights into the immune response of the black soldier fly larvae to bacteria. *Comp. Immunol.* 12. https://doi.org/10.3389/fimmu.2021.745160.

Chen, Y., Huang, M., Li, J., Zhang, Y., Wang, H., Qi, G., Shi, Q., Zhang, Z., Yang, M.,Chen, H., Pang, R. (2023). The dynamics of the microbial community in fall armyworm *Spodoptera frugiperda* during a life cycle. Entomol. Exp. Applic. 171: 502–513. https://doi.org/10.1111/eea.13282.

Choi, W.-H., Yun, J.-H., Chu, J.-P. and Chu, K.-B. (2012). Antibacterial effect of extracts of *Hermetia illucens (Diptera: Stratiomyidae) larvae against gram-negative bacteria. Ent. Res. 42:* 219–226.

Elhag, O., Zhou, D., Song, Q., Soomro, A. A., Cai, M., Zheng L., Yu, Z., Zhang, J. (2017), Screening, expression, purification and functional characterization of novel antimicrobial peptide genes from *Hermetia illucens* (L.). *PLoS One.* 12(1): e0169582. https://doi.org/10.1371/journal.pone.0169582.

Eilenberg, J. (2000), Entomophthorales on diptera. In: L. Papp and B. Darvas (Eds.), *Contributions to a manual of palaearctic diptera (with special reference to flies of economic importance): general and applied dipterology,* Vol. 1, pp. 519–534. Science Herald, Budapest.

Eilenberg, J., Vlak, J.M., Nielsen-LeRoux, C., Cappellozza, S. and Jensen, A.B. (2015), Diseases in insects produced for food and feed. *J. Insects Food Feed.* 1(2): 87–102. https://doi.org/10.3920/JIFF2014.0022.

Ellis, E.A. 1957. Entomogenous fungi in Norfolk. Trans. Norf. Norw. Nat. Soc. 18: 23-28.

Erickson, M.C., Islam, M., Sheppard, C., Liao, J. and Doyle, M.P. (2004), Reduction *of Escherichia coli* O157:H7 and *Salmonella enterica* serovar *Enteritidis* in chicken manure by larvae of the black soldier fly. *J. Food Prot.* 67(4): 685–890. https://doi.org/10.4315/0362-028x-67.4.685.

Fauce, K.A. and Owens, L., 2008. The use of insects as a bioassay for *Penaeus merguiensis* densovirus (*Pmerg*DNV). J. Inverteb. Pathol. 98: 1–6.

Ferreira, R. S., Cruz, L.C.T.A., Souza, V.J., Neves, N.A.S., Souza. V.C., Franco, L.C., Lemos, P.S., Lima, C.P.S., Naveca, F.G., Atanaka, M., Nunes, M.R.T. and Slhessarenko, R.D. (2020), Insect-specific viruses and arboviruses in adult male culicids from Midwestern Brazil. *Inf. Gen. Evol.* 85, 104561. https://www.sciencedirect.com/journal/infection-genetics-and-evolution/vol/85/suppl/C.

Gorrens, E., Van Moll, L., Frooninckx, L., De Smet, J. and Van Campenhout, L. (2021a), Isolation and identification of dominant bacteria from black soldier fly larvae (*Hermetia illucens*) envisaging practical applications. *Front. Microbiol.* 12: 665546. https://www.frontiersin.org/articles/10.3389/fmicb.2021.665546.

Gorrens, E., Van Looveren, N., Van Moll, L., Vandeweyer, D., Lachi, D., De Smet, J. and Van Campenhout, L. (2021b), *Staphylococcus aureus* in substrates for Black Soldier Fly Larvae (*Hermetia illucens*) and its dynamics during rearing. *L. Microbiol Spectr.* 9(3): e0218321.

Heussler, C.D., Klammsteiner, T., Stonig, K.T., Insam, H., Schlick-Steiner, B.C. and Steiner, F.M. (2022), bioRxiv preprint. https://doi.org/10.1101/2022.12.22.520758.

Jensen, A.B. and Lecocq, A. (2023), Diseases of black soldier flies *Hermetia illucens* L. a future challenge for production? *J. Ins. Food Feed.* 9: 1–4. https://www.wageningenacademic.com/doi/epdf/10.3920/JIFF2023.0030?role=tab.

Joosten, L., Lecocq, A., Jensen, A. B., Haenen, O., Schmitt, E. and Eilenberg, J. (2020), Review of insect pathogen risks for the black soldier fly (*Hermetia illucens*) and guidelines for reliable production. *Entomol. Exper. Applic.* 168: 432–447.

Klüber, P., Müller, S., Schmidt, J., Zorn, H. and Rühl, M. (2022), Isolation of bacterial and fungal microbiota associated with *Hermetia illucens* larvae reveals novel insights into entomopathogenicity. *Microorganisms*, 10(2): 319. https://doi.org/10.3390/microorganisms10020319.

Koutsos, E., Modica, B. and Freel, T. (2022), Immunomodulatory potential of black soldier fly larvae: applications beyond nutrition in animal feeding programs. *Translat. Ani. Sci.* 6(3): 1–9. https://doi.org/10.1093/tas/txac084.

Kuznetsova, T.A., Vecherskii, M.V., Khayrullin, D.R., Stepankov, A.A., Maximova, I.A., Kachalkin, A.V. and Ushakova, N.A. (2022), Dramatic effect of black soldier fly larvae on fungal community in a compost. *J. Sci. Food Agric.* 102(6): 2598–2603. https://doi.org/10.1002/jsfa.11601.

Lacey, L.A. (2012), *Manual of Techniques in Invertebrate Pathology*, 2nd ed. Academic Press, San Diego.

Leatherdale, D. (1970), The arthropod hosts of entomogenous fungi in Britain. *Entomophaga*. 15: 419–435.

Lecocq, A., Joosten, L., Schmitt, E., Eilenberg, J. and Jensen, A.B. (2021), *Hermetia illucens* adults are susceptible to infection by the fungus *Beauveria bassiana* in laboratory experiments. *J. Ins. Food Feed*. 7: 63–68.

Miranda, C.D., Cammack, J.A. and Tomberlin, J.K. (2021). Mass production of the black soldier fly, *Hermetia illucens* (L.), (Diptera: Stratiomyidae) reared on three manure types. *Animals*. 10(7): 1243. https://doi.org/10.3390/ani10071243.

Patterson, E.I., Villinger, J., Muthoni, J.N., Dobel-Ober, L. and Hughes, G.L. (2020), Exploiting insect-specific viruses as a novel strategy to control vector-borne disease. *Cur. Opin. Ins. Sci.* 39: 50–56.

Pienaar, R.D., Gilbert, C., Belliardo, C., Herrero, S. and Herniou, E.A. (2022), First evidence of past and present interactions between viruses and the black soldier fly, *Hermetia illucens. Viruses*. 14(6): 1274. https://doi.org/10.3390/v14061274.

Poinar, G.O. and Thomas, G.M. (1984), *Laboratory Guide to Insect Pathogens and Parasites*. Plenum Press, New York and London.

She, W., Xiao, Q., Meng, Y., Zhao, P., Wu, C. Huang, F., Cai, M., Zhang, J., Yu. Z., Rehman, K., Peng, D., Zheng, L. (2023), Isolated and identified pathogenic bacteria from black soldier fly larvae with "soft rot" reared in mass production facilities and its incidence characteristics. *Waste Manag*. 163: 85–95.

Shumo, M., Khamis, F.M., Ombura, F.L., Tanga, C.M., Fiaboe, K.K.M., Subramanian, S., Ekesi, S.Schlüter, O.K., van Huis, A., Borgemeister, C. 2021. A Molecular Survey of Bacterial Species in the Guts of Black Soldier Fly Larvae (*Hermetia illucens*) Reared on Two Urban Organic Waste Streams in Kenya. Front. *Microbiol.* 12:687103. doi: 10.3389/fmicb.2021.687103.

Tourtois, J., Ali, J.G. and Grieshop, M.J. (2017), Susceptibility of wounded and intact black soldier fly *Hermetia illucens* (L.) (Diptera: Stratiomyidae) to entomopathogenic nematodes. *J. Invertebr. Pathol.* 150: 127–129. https://doi.org/10.1016/j.jip.2017.10.002.

USDA-ARSef-. (2020), ARS collection of entomopathogenic fungal cultures – isolates from diptera. https://www.ars.usda.gov/ARSUserFiles/80620520/Diptera_June2020.pdf.

Van Moll, L., De Smet, J., Paas, P, Tegtmeier, D., Vilcinskas, A., Cos, P. and Van Campenhout, L. (2022), *In vitro* evaluation of antimicrobial peptides from the Black Soldier Fly (*Hermetia Illucens*) against a selection of human pathogens. *Microbiol. Spectr.* 10(1): e0166421.

Vega, F.E. and Kaya, H.K. (2012). *Insect Pathology*, 2nd ed., Elsevier Inc., San Diego.

Vitenberg, T. and Opatovsky, I. (2022). Assessing fungal diversity and abundance in the black soldier fly and its environment. *J. Ins. Sci.* 22(6): 1–6. https://doi.org/10.1093/jisesa/ieac066.

von Bredow, Y.M., Müller, A., Popp, P.F., Iliasov, D. and von Bredow, C.-R. (2022). Characterization and mode of action analysis of black soldier fly (*Hermetia illucens*) larva-derived hemocytes. *Ins. Sci.* 29: 1071–1095. https://doi.org/10.1111/1744-7917.12977.

Xia, J., Ge, C. and Yao, H. (2021), Antimicrobial peptides from black soldier fly (*Hermetia illucens*) as potential antimicrobial factors representing an alternative to antibiotics in livestock farming. *Animals.* 11: 1937. https://doi.org/10.3390/ani11071937.

Yang, S. 2017. Intensive black soldier fly farming. https://symtonbsf.com/blogs/blog/intensive-black-soldier-fly-farming.

Zdybicka-Barabas, A., Bulak, P., Polakowski, C., Bieganowski, A., Waśko, A. and Cytryńska, M. (2017). Immune response in the larvae of the black soldier fly *Hermetia illucens*. *Invert. Surv. J.* 14: 9–17. https://www.isj.unimore.it/index.php/ISJ/article/view/19/323.

Zheng, L., Crippen, T.L., Singh, B., Tarone, A.M., Dowd, S., Yu, Z., Wood, T.K. and Tomberlin, J.K. (2013), A survey of bacterial diversity from successive life stages of black soldier fly (Diptera: Stratiomyidae) by using 16S rDNA pyrosequencing. *J. Med. Entomol.* 50: 647–658.

Chapter 3

Use of protein from black soldier flies in poultry feed

Arya Rezaei Far, Somaya Naser El Deen and Teun Veldkamp, Wageningen University and Research, The Netherlands

1 Introduction

Insects have been proposed as a promising alternative protein source to solve the global challenge of sustainable protein production for human food and animal feed. Among insect species, black soldier fly (*Hermetia illucens*) larvae (BSFL) are most commonly processed into BSFL protein meal (BSFLM) and BSFL oil. The nutritional value of defatted BSFLM is similar to or higher than conventional protein sources such as soybean meal (SBM) and other insect species, offering the possibility of completely replacing SBM in the diet of laying hens and broilers (van der Heide et al., 2021).

However, there are inconsistencies in studies of the impacts of including BSFM in diets on the performance of broilers and laying hens. Similar to conventional ingredients such as SBM (Ibáñez et al., 2020; Frikha et al., 2012; Ravindran et al., 2014), there are variations in the nutritional value of BSFLM, which can cause inconsistencies in the results (Kierończyk et al., 2022). This

http://dx.doi.org/10.19103/AS.2024.0139.04

means that there is currently no consensus on the optimum inclusion levels for BSFLM in the diets of broilers and laying hens.

Successful inclusion of BSFLM in the diets of broilers and laying hens requires a precise determination of its nutritional value. Replacing SBM with BSFLM in poultry diets is complicated by challenges such as accurate determination of crude protein content, establishing an appropriate conversion factor for nitrogen, and assessing chitin content and its effects on nutritional value. These issues are addressed in the first part of this chapter, along with current knowledge on amino acid content, amino acid digestibility and metabolizable energy. The chapter proceeds with an overview of the impact of dietary BSFLM inclusion on the performance and health of broilers and laying hens. Finally, the chapter concludes with an assessment of the current use of insects as feed and the future prospects for these promising alternative protein sources in poultry nutrition.

2 Nutritional value of black soldier fly larvae meal in poultry diets: crude protein content

BSFLM is incorporated to the diet of poultry as a protein source. This requires an accurate quantification of its crude protein content. The most common method for determining protein content in the feed sector is based on assessing the nitrogen content using the Kjeldahl or Dumas methods. The total nitrogen content is then multiplied by the protein conversion factor (k_p) of 6.25 to calculate the crude protein content. However, feed ingredients often contain non-protein nitrogen (NPN) and both the acid digestion step of the Kjeldahl method and the combustion in Dumas can release this nitrogen. Consequently, the higher the content of NPN is, the more the crude protein content will be overestimated. Notably, chitin in BSFL contains nitrogen that is considered indigestible or biologically unavailable to chickens, emphasizing the importance of determining an accurate conversion factor.

In contrast, true protein is measured as the sum of all the amino acids. However, compared to crude protein analysis, amino acids analysis is more expensive and less commonly performed on a batch-to-batch basis for feed formulation in the feed sector. In addition, crude protein content is analysed to assess crude protein digestibility, which is crucial to determine the nutritional value of BSFLM.

As insect products gain more attention, some studies have sought to determine an accurate k_p factor of BSFL and BSFLM (Eggink & Dalsgaard, 2023; Smets et al., 2021; Jansen et al., 2017). Jansen et al. (2017) determined the k_p of BSFL and its protein extract to be 4.67 and 5.62, respectively. Using the sum of the anhydrous amino acids to determine the k_p factor in BSF, Eggink and Dalsgaard (2023) suggested the k_p factors between 3.36 and 4.19 for larvae or prepupae being harvested from the same rearing substrate between 12 and 24 days of age. Smets et al. (2021) determined the k_P factors between 3.98 and 4.76 in BSFL reared on different substrates and harvested between 14 and 21 days of age. These outcomes highlight the deviation of the k_P factor of BSFL from the standard value of 6.25, and the existing variations related to the nutritional composition of the rearing substrates, larval stage, and the harvesting age. Considering the effects of processing on the nutritional value of proteins in BSFLM, it is necessary to determine the k_P factor of the meal and understand the factors contributing to these variations.

Following the accurate quantification of BSFLM crude protein, the digestibility of this protein needs to be determined in poultry to determine the nutritional value of BSFLM as an alternative ingredient. Several factors, such as the nutritional composition of the rearing substrate, processing methods, and fibre content, could affect the protein digestibility of BSFLM (Do et al., 2020). Smets et al. (2021) observed no correlation between the k_P factors of BSFL and chitin, indicating that chitin content might not be the main indicator for predicting protein quality.

In addition to the nutritional composition of larvae, processing can affect BSFLM protein digestibility. Devitalization of BSFL by freezing promotes enzymatic browning of meal through endogenous phenoloxidase, which affects organoleptic characteristics and protein digestibility, as well as the content of digestible amino acids, particularly lysine and cysteine. Blanching prevents this effect resulting in BSFLM with improved amino acids and protein digestibility (Rodríguez-Rodríguez et al., 2022). Moreover, exposure to heat during the processing of BSFL into meals can affect digestibility as well. Processing introduces additional variability in the nutritional value of BSFLM. It is necessary to identify indicators which can accurately predict protein quality and nutritional value of BSFLM for poultry. Feed manufacturers would apply these indicators across different BSFLM batches to formulate precise diets for poultry.

It is well known that reducing the amount of undigested protein in poultry has beneficial effects on both bird health and the environment (Strifler et al., 2023). Although undigested protein bound to chitin along with other undigested nitrogen of BSFLM could increase the undigested nitrogen content at the end of the small intestine (impairing gut functionality by enhancing the bacterial fermentation of protein), this has not been consistently observed. For instance, impaired performance of broilers fed a diet with 10% BSFLM in comparison to 5% or a control diet (0% BSFLM) was not associated with negative effects on the histomorphometry of the small intestine, and no significant differences were observed in the content of biogenic amines in the caecal digesta (Hartinger et al., 2021). Further understanding the effects of BSFLM on the gut microbiome of poultry could help clarify these findings.

In conclusion, BSFLM is a high-quality source of protein with a high nutritional value for poultry. Despite the existing variabilities in the protein and amino acid composition of BSFLM, it has a comparable or superior protein digestibility compared to conventional ingredients such as SBM. There is an urgent need to identify the indicators for assessing the protein quality and the nutritional value of BSFLM to facilitate its large-scale application in poultry feed.

3 Chitin and fibre content of black soldier fly larvae meal

The exoskeleton of BSFL contains chitin, a polysaccharide of glucosamine and a component of the scleroprotein matrix. Compared to other animal-derived protein sources, BSFLM has a higher carbohydrate content (13.6 ± 0.16% as fed) (Li & Wu, 2020). With an average chitin content of 7% in BSFLM, chitin represents a significant fraction of carbohydrates in BSFLM. Although further research is needed to understand the potential effects of endogenous chitinase in poultry on the degradation of BSFLM chitin fraction, the general idea is that the chitin nitrogen is not digestible or absorbable in the gastrointestinal tract of monogastric animals. Therefore, the presence of chitin in BSFLM reduces protein digestibility and affects the accuracy of crude protein quantification. It seems that the functional impact of chitin on the host depends on its intake level. High dietary chitin intake hampers nutrient digestibility while lower levels provide health-promoting effects (Gasco et al., 2021, 2023). Therefore an accurate quantification of chitin is essential to determine the optimum inclusion level and intake of chitin in poultry.

Various methods have been used to estimate the chitin content of BSFLM. Commonly, the content of acid detergent fibre (ADF), neutral detergent fibre (NDF) or crude fibre (CF) has been used widely to estimate the chitin content of insects in the studies on the application of BSFLM in poultry feed (Facey et al., 2023; Fruci et al., 2023; Hartinger et al., 2021; de Marco et al., 2015). Correcting for the amino acid content of ADF increases the accuracy of the estimation (Eggink & Dalsgaard, 2023; Finke, 2007). Other methods are based on chitin hydrolysis and quantification of glucosamine via spectrophotometry or chromatography (Han & Heinonen, 2021), however, are not commonly executed in commercial animal nutrition laboratories.

Recently, Eggink and Dalsgaard (2023) assessed the chitin content in BSFL throughout its entire life comparing six different methodologies. They concluded that quantification of glucosamine by liquid chromatography method is the most accurate compared to gravimetrical methods. Gravimetrical methods, in particular NDF and CF, tend to overestimate the chitin content in BSFLM samples. Although ADF corrected for amino acids could be an effective alternative to chitin assessment, its high detection limit (1.0%) significantly affects accuracy, particularly in samples with low chitin content.

Estimating BSFLM chitin content by assessing fibre fraction with gravimetrical methods is likely to remain the most common method in the poultry feed sector. However, this method presents certain limitations, such as a high detection limit and the overestimation of chitin related to the presence of residual fibre from the gastrointestinal tract of BSFL and substrate in the meal.

In general, the chitin content of BSFLM in the literature ranges between 3% and 8% (Fruci et al., 2023; Cutrignelli et al., 2018; Marono et al., 2015; Mwaniki et al., 2019). Methods based on glucosamine titration measured chitin content values 4.65 and 7.33 (% in DM). The highest and the lowest chitin contents were observed in BSFL reared on four different substrates and harvested at 21 days of age (Smets et al., 2021). Eggink and Dalsgaard (2023) reported chitin content of 5.4, 4.7 and 11.3 (% as DM) in BSFL on days 12, 15 and 24, respectively. Therefore, harvesting age as well as the rearing substrates contribute to variations in BSFL chitin content. Although the results from the few studies using more accurate methods, compared to fibre-based methods, are not drastically different, they indicate existing variations in the BSFLM chitin content.

Despite the expected health-promoting effects of BSFLM chitin (Gasco et al., 2021), it can negatively affect protein digestibility (Marono et al., 2015). Additionally, it has been suggested that excessive chitin intake can reduce feed intake and body weight gain (BWG) in poultry (Cutrignelli et al., 2018). Therefore, it is crucial to accurately assess the chitin content of BSFLM and to better understand its interactions with the digestive tract of chickens. An optimum chitin intake prevents negative effects on nutrient digestibility while enabling its health-promoting effects.

4 Amino acid digestibility and metabolizable energy

Formulating poultry feed based on digestible amino acids is more accurate than using crude protein or amino acid content. The digestibility coefficients of amino acids are determined by *in vitro* models or *in vivo* studies in poultry. Few studies have assessed the nutrient digestibility of BSFLM (Marono et al., 2015; Schiavone et al., 2017; Mwaniki & Kiarie, 2019; Matin et al., 2021). The results of these studies have been used to formulate diets to evaluate the effect of different inclusion levels of BSFLM in poultry feed on performance parameters.

Table 1 presents the ileal digestibility coefficients of amino acids from different studies. Although the methodological variations in the digestibility studies limit the direct comparison, the results still indicate the extent of variation across different batches of BSFLM. For example, in the study of Matin et al. (2021), different BSFLM products were assessed with the same methodology

Table 1 Apparent ileal digestibility of amino acids % (unless mentioned), and the concentration of crude protein (CP) and crude fat (CF) (g/kg as DM) in BSFLM from different studies

CP	Fat	Arg	His	Isol	Leu	Lys	Met	Phe	Thr	Trp	Val	Ala	Asp	Cys	Gly	Glu	Pro	Ser	Tyr	
369	343	0.83	0.81	0.45	0.76	0.46	0.42	0.63	0.75	ND	0.62	0.86	0.61	0.82	0.67	0.74	0.89	0.82	0.43	de Marco et al., 2015
553	180	0.79	0.64	0.83	0.84	0.8	0.83	0.82	0.73	ND	0.9	0.92	0.82	0.44	0.066	0.81	0.65	0.71	0.92	Schiavone et al., 2017
655	46	0.80	0.63	0.87	0.89	0.8	0.78	0.86	0.77	ND	0.91	0.99	0.8	0.45	0.65	0.85	0.82	0.77	0.95	
575	70	0.88	0.55	0.86	0.86	0.84	0.85	0.87	0.78	ND	0.85	0.86	0.78	0.57	0.85	0.75	0.87	0.81	0.87	Mwaniki & Kiari, 2018
377	344	0.94	0.91	0.92	0.94	0.91	0.93	0.92	0.92	0.94	0.85	0.94	0.93	0.79	0.78	0.9	0.91	0.9	0.93	Do et al., 2020
575	207	0.95	0.90	0.92	0.93	0.90	0.95	0.93	0.92	0.97	0.81	0.92	0.93	0.80	ND	0.91	0.92	0.91	0.94	Matin et al., 2021[1]
581	192	0.93	0.89	0.90	0.91	0.88	0.92	0.91	0.88	0.97	0.77	0.91	0.90	0.73	ND	0.89	0.90	0.87	0.91	
452	190	0.91	0.84	0.90	0.90	0.79	0.90	0.90	0.87	0.66	0.76	0.89	0.88	0.66	ND	0.88	0..86	0.86	0.92	
567	122	0.91	0.85	0.86	0.86	0.89	0.92	0.88	0.84	0.97	0.60	0.84	0.88	0.69	ND	0.87	0.83	0.82	0.88	

[1]Standardized ileal digestibility coefficient. ND = not determined.

and meaningful variations were observed in the digestibility of certain amino acids. The digestibility coefficient of lysine in full-fat BSFLM was between 90% and 79% and resulted in a digestible lysine content of 39.5 and 26.6 g/kg (in DM). Given a 20% inclusion of BSFLM in broiler diets, this difference can lead to approximately 30% deviation in digestible lysine content, which could adversely affect the broilers' performance.

Despite the limited number of studies on the amino acid digestibility assessment of BSFLM, the results indicate comparable or even higher digestibility of amino acids compared with SBM, except cysteine and valine. Researchers have discussed the clinical effects of a lower intake of digestible cysteine (Brede et al., 2018) and valine (Mwaniki et al., 2020) when replacing SBM with BSFLM, indicating negative impacts on the broilers' growth performance. Therefore, it is necessary to accurately predict the BSFLM digestible amino acid content to ensure that the requirements of poultry are met.

The nutritional composition of BSFLM, particularly the dietary fibre and chitin fraction, may influence the digestibility of amino acids. The result of *in vitro* protein and amino acid digestibility assessments of six different BSFLM, varying in protein and fat content, indicated the protein digestibility coefficient ranging from 66.0% to 68.7%. The study found that acid-detergent fibre and chitin content negatively impacted the amino acids digestibility coefficients (Marono et al., 2015). The higher NDF content (32.7% vs 12.4 and 15.3%) in BSFLM was associated with a lower digestibility of amino acids, particularly lysine, valine and cysteine (Matin et al., 2021).

Mwaniki and Kiarie (2018) determined the nutrients digestibility in BSFLM for broilers. Despite the relatively high amino acids digestibility compared to conventional ingredients, they observed low retention of crude protein (47%), gross energy (64.5%) and apparent metabolizable energy (AME) content 2902 kcal kg^{-1}DM. The authors attributed these findings to the negative impacts of chitin content on nutrient digestibility. Although the chitin content has been suggested as an important factor affecting protein digestibility of BSFLM in chickens, none of these studies precisely quantified or characterized the chitin content of BSFLM. The amino acid, crude protein, and nitrogen content in BSFLM is presented in Table 2 and the energy content, crude protein, and crude fat content in BSFLM is presented in Table 3 to show the variation between studies. Further research is needed to better understand the causative effects of the nutritional composition of BSFL on amino acid digestibility and to assess the effects of the processing methods on the nutritional value of amino acids in BSFLM.

4.1 Metabolizable energy

In addition to the variations in BSFLM products, different methods were used to assess the apparent metabolisable energy of BSFLM in studies. Therefore,

Table 2 Amino acid content (g/16N), crude protein and nitrogen (g/kg as DM), sum of the analysed amino acids (SAA g/kg DM) and the ratio of total amino acids to the total nitrogen of BSFLM from different studies (SAA:N)

		1	2	3	4			5		6	7	8	9		10	11
Arg	3.9	4.1	4.7	4.8	4.8	4.7	4.9	4.8	5.0	5.0	4.2	4.6	4.4	4.6	3.4	5.3
His	2.2	2.5	2.1	9.8	3.2	2.8	3.4	3.2	3.3	3.1	2.8	2.8	2.8	2.9	2.9	3.1
Iso	3.4	3.7	4.1	4.2	4.4	4.6	4.4	4.7	4.7	3.6	4.2	4.3	7.2	4.1	3.3	4.7
Le	5.2	5.6	6.9	6.8	6.8	6.8	6.6	7.2	7.2	6.5	7.4	6.8	6.5	6.4	5.3	6.5
Lys	3.8	3.8	5.9	5.7	7.1	7.4	7.7	7.4	6.8	5.0	5.4	5.6	6.0	5.2	4.2	6.0
Met	1.2	1.3	1.3	1.6	1.8	1.8	1.6	1.5	1.6	1.5	1.4	1.8	1.4	1.8	1.4	2.5
Phe	3.0	3.3	4.0	3.8	4.4	4.5	4.2	4.4	4.3	5.0	3.6	4.2	4.0	4.2	3.5	3.9
Th	3.1	3.3	3.7	4.0	4.0	3.9	3.9	4.2	4.1	3.5	3.4	3.9	3.6	3.6	3.6	4.1
Val	4.9	5.3	6.2	6.0	7.7	7.2	8.1	9.0	13.1	5.7	5.6	5.8	5.3	6.0	5.0	6.0
Try	ND	ND	ND	ND	1.5	1.5	1.4	1.5	1.4	1.3	1.4	ND	1.1	ND	ND	ND
Ala	6.2	6.7	ND	6.8	7.2	6.4	7.2	7.1	6.8	6.1	6.3	6.9	7.5	5.7	4.9	8.2
Asp	6.7	7.5	ND	9.1	9.0	9.1	8.6	9.6	9.2	8.9	7.7	8.2	8.0	8.4	7.1	8.7
Cys	0.0	0.0	0.9	0.7	1.0	1.0	1.2	1.3	1.1	0.9	0.8	ND	0.7	0.6	0.7	3.7
Gly	4.3	4.6	5.8	5.3	5.2	5.3	6.0	7.2	7.8	9.9	5.1	5.0	5.2	5.5	4.7	5.2
Glu	8.8	9.7	ND	11.9	10.1	9.8	10.0	11.9	11.4	4.2	9.5	10.9	9.8	9.4	9.1	10.4
Pro	5.5	5.0	ND	6.0	6.3	4.7	5.3	6.5	7.0	6.4	5.2	5.2	5.4	5.0	4.6	10.1
Ser	3.7	4.1	4.1	4.5	4.1	3.5	3.6	4.1	4.3	4.3	3.9	4.0	3.9	3.7	3.3	5.0
Tyr	4.8	5.2	5.7	4.9	6.6	6.0	6.0	6.2	6.4	3.2	ND	5.2	6.1	6.1	5.0	5.9
SAA	391	496	ND	553	359	523	546	461	600	464	ND	341.7	455	434	315	366
CP	553	655	478	575	377	575	581	452	567	551	712	401.0	509	522	438	369
N	88.5	104.8	76.4	92.1	60.4	92.0	93.0	72.3	90.7	88.2	113.9	64.2	81.5	83.5	70.0	59.0
SAA:N	4.4	4.7	ND	6.0	6.0	5.7	5.9	6.4	6.6	5.3	ND	5.3	5.6	5.2	4.5	6.2

[1]Schiavone et al., 2017; [2]Patterson et al. 2021; [3]Mwaniki et al., 2018; [4]Do et al., 2021; [5]Matin et al., 2021; [6]Facey et al 2023; [7]van der Heide et al., 2021; [8]de Souza Viella et al., 2021; [9]Zawisza et al., 2023; [10]Petkov et al., 2022; [11]de Marco et al., 2015.

Table 4 only provides an overview of the available energy content in BSFLM. Due to variations in BSFLM products and the differences in the methods used to assess the AME across studies, the results presented in Table 4 are not directly comparable. Nevertheless, examining the nutritional composition of BSFLM in each study alongside the corresponding AME values provides insight into the factors contributing to the variations in AME values.

For instance, one of the lowest AME values reported is 2902 kcal/kg, with a gross energy retention of 64.5%. This was observed in a highly defatted BSFLM (6.8%) having a high NDF content (38.7%) (Mwaniki & Kiarie, 2019). In another example, two BSFLM samples with similar crude fat content (19%) but highly different NDF content (15% vs 33%) showed an 11% difference in true metabolizable energy (Matin et al., 2021). The higher the NDF content, the lower the true metabolizable energy of the BSFLM. These findings highlight the significant role of fibre content, particularly the chitin fraction, in contributing to variations in AME values.

5 Including black soldier fly larvae meal in the diet of broilers

The effects of including BSFLM as an alternative protein source in broiler diets have been the subject of research aiming to find the highest inclusion level in which broilers maintain feed intake and growth performance. A recent review suggests that an inclusion level of up to 10% can enhance feed intake and BWG, while higher inclusion levels impair growth performance (Dörper et al., 2020). Meta-analyses on the BSFLM inclusion in broiler diets indicated that levels of 10% (Nayohan et al., 2022) or 15% (Martinez Marin et al., 2023) do not negatively affect the growth performance of broilers. The negative effects of BSFLM chitin on protein digestibility in broilers may impair growth performance (Facey et al., 2023; Murawska et al., 2021; Hartinger et al., 2021; Dabbou et al., 2018). However, some studies indicate that broilers maintain growth performance even when the inclusion of BSFLM exceeded 10% or higher in the case of the complete substitution of SBM with BSFLM (Heuel et al., 2021; Brede et al., 2018; Neumann et al., 2018a). Research on BSFLM inclusion in poultry diets remains limited, and further research is needed to precisely determine the optimal inclusion level for broilers.

BSFLM can serve as the main or the sole protein source in broiler feed without compromising growth performance. De Souza Vilela et al. (2021) suggested 15–20% as the optimal BSFLM levels in broiler diets. The highest BWG and lowest feed conversion ratio were observed when SBM was partially replaced with 10% full-fat BSFLM in the starter and 20% in the grower and finisher diets. These inclusion levels did not affect crude protein digestibility in broilers. Some studies showed that supplementing diets with the extra amount of

Table 3 Results of the energy availability (kcal/kg) of different black soldier fly larvae meal including the crude protein (CP %), crude fat (CF %) and gross energy (kcal/kg) from different studies

Strain	Age (days)	CP	CF	GE	Method	parameter	Results (kcal/kg)	Reference
Ross 708 male	32–35	36.9	34.4	5688	Basal diet and total excreta collection	ME Men	4156 3965	de Marco et al., 2015
Ross 708 male	17–21	56.1	6.8	4973	Basal diet and total excreta collection	AMEn	2902	Mwaniki & Kiarie, 2018
Ross 308 male	32–35	55.2	18	5828	Total excreta collection and semi-purified diet	AME AMEn	3881 3552	Schiavone et al., 2017
		65.6	4.6	5064			2759 2357	
WLR (1)	NA	57.5		5410	Precision-fed rooster assay	TMEn	4250	Matin et al., 2021
		58.1	19.2	5325			3998	
		45.2	19	5159			3561	
		56.7	12.1	5110			3561	
cobb 500	16-21	55.1	14.8	5135	Basal diet and total excreta collection	AME AMEn	4729 4228	Osunbami & Adeola, 2022

[1]WLR = Single Comb White Leghorn roosters; AMEn = nitrogen-corrected Apparent Metabolizable Energy; TMEn = nitrogen-corrected True Metabolizable Energy; GE = Gross Energy.

synthetic amino acids prevents the negative effects on the growth performance of broilers when the BSFLM inclusion level is high (Heuel et al., 2021; Brede et al., 2018; Velten 2018a, 2018b). Therefore, BSFLM can even completely replace SBM without negatively impacting performance parameters, provided that amino acid content, particularly the sulphuric amino acids, is sufficient.

The potential imbalances or deficiencies in amino acid intake associated with BSFLM inclusion may impair broiler performance (Fruci et al., 2023 Facey et al., 2022, Hartinger et al., 2021). Both the deviation in amino acid content in BSFLM from the expected values and the reduction in the amino acids digestibility of the diet can reduce the digestible amino acids content of the final feed. Therefore, increasing the supplementation of synthetic amino acids in diets with high BSFL levels, compared to a control diet, could help broilers to maintain or improve growth performance. Further research on the content of digestible amino acids and their balance is needed to determine the optimal inclusion level of BSFLM and to optimise the digestible amino acids balance in broiler diets.

In addition to the amino acid deficiency or intake imbalance, several factors have been suggested as the reason for the impaired performance of broilers when the inclusion level of BSFLM is high. Some common suggestions include a decline in feed intake due to increased dietary fibre content, and the dark colour of BSFLM and its negative impacts on palatability (Marono et al., 2017; Heuel et al., 2022). Another factor is reduced protein or fat utilization due to a higher content of chitin (Hartinger et al., 2021; Cutrignelli et al., 2018). Additionally, nutrient absorption may be reduced due to the histomorphological changes in the mucosal surface of the small intestine (Dabbou et al., 2018; Velten et al., 2018a). The chitin content of BSFLM and its negative effects on protein digestibility are particularly highlighted in literature as key reasons for impaired performance observed with increasing levels of BSFLM in broiler diets.

Considering the negative impact of chitin on broiler performance, it is important to determine the threshold for chitin intake in broilers fed with BSFLM. In the starter phase, up to 11.4 g/kg chitin in feed (262 mg chitin/day/broiler) showed no effects, while feed intake decreased at a chitin content of 17.7 g/kg (Facey et al., 2023). Another study found that up to 15.7 g/kg chitin in feed (332 mg/day/broiler) showed no negative effects in the starter phase, but feed intake declined when the chitin content was 23.5 g/kg feed (Fruci et al., 2023). Some studies indicate that chitin may not be the sole factor for impaired growth performance in broilers fed high levels of BSFLM. For instance, Heuel et al. (2021) observed that including 20% BSFLM in the diet of slow-growing broilers increased the chitin content to 22.5 g/kg feed without negatively affecting the growth performance. In fact, the feed intake even slightly increased (Heuel et al., 2021). Conversely, using a different BSFLM product with lower chitin content (187 vs 224 g/kg as is) was associated with

Table 4 Effects of replacing soybean meal with full-fat or defatted black soldier fly larvae meal in broiler diets on the feed intake (FI) in the starter phase, and other performance parameters including body weight gain (BWG), feed conversion ratio (FCR) and average daily weight gain (ADG) in recent studies

BSFLM (a)	Replacement (b)	BSF inclusion, %	Feed intake in the starter phase (c)	Results in relation with BSF inclusion level (c)	References
D	P & C (1)	Starter: 0, 4, 8, 16, 25 Grower: 3, 7, 14, 21 Finisher: 3, 5, 10, 16	No effects up to 8% Reduced at 16% & 25%	• In the starter phase, up to 8% improved BWG and FCR, while 16% and 25% reduced FI, BWG and increased FCR. • Same trend continues over the experimental period.	Facey et al., 2023
D	P	Starter: 0, 5, 10, 20, 29 Grower: 0, 5, 9, 18, 27 Finisher: 0, 4, 8, 15, 23	No effects up to 20% Reduced at 29%	• Up to 10% improved BWG and FCR; 20% had no effect on BWG but impaired FCR; 29% reduced BWG and impaired FCR in the Starter phase • In grower and finisher, different effects • Overall, negative effect on BWG in 20% and 29% inclusion levels. Slightly improved FCR at the lowest inclusion level	Fruci et al., 2023
D	P (2)	Starter: 10 Grower: 9	No effects	• No effect on growth performance	van der Heide et al., 2021
F	P	5%		• No effect on growth performance	Elangovan et al., 2021
F	P (3)	Starter: 0, 2.5, 5.0, 7.5, 10.0 Grower: 0, 5, 10, 15, 20 Finisher: 0, 5, 10, 15, 20	No effects	• FCR improved (quadratically) • BWG improved linearly	de Souza Vilela et al., 2021
D	P	Starter: 0, 5, 10 Grower: 0, 4.4, 8.7 Finisher: 0, 4.1, 8.1	Increased	• In the starter phase, higher FI was not associated with higher BWG and BWG was reduced in 10% • FCR impaired by including BSFLM • Overall, 5% did not affect performance parameters while 10% impaired FI and BWG.	Hartinger et al., 2021

BSFLM (a)	Replacement (b)	BSF inclusion, %	Feed intake in the starter phase (c)	Results in relation with BSF inclusion level (c)	References

F		Starter: 0, 20, 30, 40
		Grower: 0, 17, 25, 34
		Finisher: 0, 13, 20, 27
			Reduced	• Including BSFLM:
				  ◦ Reduced FI in all phases
				  ◦ Reduced ADG in starter and finisher while increased ADG linearly in finisher.
				• Improved FCR linearly in grower and finisher.
					Murawska et al., 2021

D	P (4)	4, 8, 12	Increased at 8% and no effects at 4% and 12%	• Overall, at 4% BSFLM broilers gained higher BW than the control and other levels. While 8% and 12% impaired performance.	Mat et al., 2021
D	P (5)	10%	No effects	• Including 10% BSFLM improved BWG and FCR in the grower and finisher phases	Lalev et al., 2022
D	C (6)	Starter: 23 Grower: 21	Reduced	• Total replacement of SBM with BSFLM reached the same body weight gain with lower FI and FCR when Met:Cys ratio of the final feed was 50:50	Brede et al., 2018
D		0, 5, 10, 15	Increased up to 10% inclusion level	• Best growth performance at 10% inclusion, while performance impaired at 15% inclusion	Dabbou et al., 2018
P	P	14.5 and 11.9%		• Supplementing amino acids to the diet containing BSFLM prevented impaired performance	Velten et al., 2018a
P	P	14.5 and 11.9%	No effects	• Supplementing higher levels of amino acids to the diet containing BSFLM improved BWG and FCR	Velten et al. 2018b

(Continued)

Table 4 (*Continued*)

BSFLM (a)	Replacement (b)	BSF inclusion, %	Feed intake in the starter phase (c)	Results in relation with BSF inclusion level (c)	References
P	C	Starter: 26 Grower: 22	No effects	• Extra Supplementation of most limiting amino acids to the diet containing BSFLM prevented impaired performance	Neumann et al., 2018a
P	C (7)	20%	No effects in meal 1 and impaired in meal 2	• Including BSFLM in diets deficient in Lys and Met+Cys resulted in ADG similar to a non-deficient diet depending on the choice of BSFLM	Heuel et al., 2022

(a) Defatted (D) indicates products with crude protein content of 47-71% and crude fat content of 6-14%; full-fat (F) indicates products with crude protein content of 40-50% and crude fat content of 21-32% 6-14% crude fat; full fat indicates products with 21-35% crude fat.
(b) Partial or complete replacement of SBM as the main protein source in diets (unless else mentioned). The protein content of grains or the supplemented amino acids was not considered as the protein source of diets. Diets in each growth phase were formulated as iso-caloric and iso-nitrogenous indicating similar content of crude protein and the most limiting essential amino acids unless else mentioned. *Diets include (1) 4-6% pork meal; (2) rapeseed meal; (3) meat and bone meal; (4) fish meal; (5) 5-7% SFSM; (6) diets with different Met:Cys ratio; (7) lysine and Met+cys-deficient diet.
(c) Only significant differences ($P < 0.05$) are mentioned. Tendency indicates $P < 0.1$.

- Male Ross 308 broilers were used in the studies except the studies used mixed genders of Ross 308 (Hartinger et al., 2021), Ross 708 (Facey et al., 2023; Fruci et al., 2023) and Cobb (Mat et al., 2021).
- Production period was 35 or 49 days.

impaired growth performance. An impaired feed intake has been observed in broilers fed diets with lower BSFLM inclusion levels contained 8.6 g/kg chitin (Kim et al., 2021). The available data are limited and do not provide sufficient evidence to draw a definitive conclusion regarding the broilers threshold for chitin intake. Further research is needed to explore the potential interaction between the birds genetics and the effects of BSFLM chitin on performance. In addition, substituting SBM with BSFLM is associated with other changes in feed characteristics, which could affect feed intake and growth performance.

The existing variations in the nutritional value of BSFLM can lead to overestimation or underestimation of its overall nutritional value, not just in terms of amino acids. When the nutritional value of BSFLM is overestimated, diets might not meet the broilers' requirements, potentially leading to increased feed intake or impaired growth performance. However, when the nutritional value of BSFLM is underestimated, it could reduce the feed intake and the intake of other essential nutrients. For example, there was a negative linear relationship between the inclusion level of BSFLM in broiler diets (from 13% to 40%) and average feed intake. However, the relative weight of the liver and abdominal fat was higher in broilers fed high levels of BSFLM (Murawska et al., 2020) which could indicate an underestimation of the available metabolizable energy. The number of studies on nutrients digestibility in BSFLM, particularly amino acids and the metabolizable energy, is limited. In addition, there are variations in the nutritional value of different BSFLM products. Therefore, the deviations of BSFLM nutritional values from the expected values cannot be ruled out as an important cause of inconsistencies in the outcome of studies on the effects of replacing SBM with BSFLM on broilers performance.

When determining the optimal inclusion level of BSFLM in broiler diets, feed intake and performance are critical factors to consider. However, other important considerations include the impact on feed and broiler products, the flock health, the environmental impact of broiler production and ensuring consistent supply of the ingredient (BSFLM) to the market with reliable nutritional value. While it is possible to completely replace conventional protein sources with BSFLM in broiler diets, there is still a tendency to include BSFLM at a lower level.

Observing positive effects on broiler performance has supported the use of BSFLM at lower inclusion levels or as a supplementary ingredient with potential health benefits (Hartinger et al., 2022). Limiting BSFLM inclusion levels in diets could also mitigate the negative impacts of variations in the nutritional value on broiler performance. For instance, adding 4.1% or 8.1% BSFLM to the diet of broilers without coccidiostat supplementation resulted in a higher body weight gain (BWG) during the starter phase (0–10 days) compared to the control diet without coccidiostat, and was even comparable to the control diet supplemented with coccidiostat and growth promoter antibiotic (Facey et al.,

2023). In another study, including 5% or 10% BSFLM in the diet of broilers up to 14 days of age resulted in a higher BWG and feed intake compared to the control diet. In addition, the results were similar to broilers fed the those of the control diet supplemented with a growth promotor antibiotic (Fruci et al., 2023). Hartinger et al. (2022) also observed positive effects of a low inclusion level of BSFLM (5.2%) on BWG and FCR compared to SBM-based control diet, particularly during the starter phase. While we cannot explain why these positive effects were limited to broilers at early age and were not observed after the 1 or 2 weeks, in older broilers, it highlights the health-promoting effects of BSFLM as an alternative protein source.

These results support the idea of using BSFLM as a supplement in inclusion levels lower than 10% during the early stages of growth to enhance performance and health, even though the exact mechanism remains unclear. In fact, limiting the inclusion of BSFLM is a reasonable approach to avoid potential negative effects on broiler performance due to variations in the nutritional value of feed.

Currently, there is no common agreement on the effects of including BSFLM in broiler diets on growth performance. Variations in study designs, in terms of feed formulation and test ingredients, make it challenging to compare results across studies. Meanwhile, the tendency to include BSFLM at lower levels has emerged which aim to only partially replace SBM or to use BSFLM as a functional ingredient and supplement in poultry diets. This approach is favored because higher inclusion levels may reduce feed intake and impair performance even though these negative impacts have not been constantly observed. Therefore, to establish the optimum inclusion level of BSFLM in poultry diets, it is necessary to find the reasons for reduced feed intake and performance at higher inclusion levels of BSFLM.

6 Including black soldier fly larvae meal in the diet of laying hens

In comparison to SBM, BSFLM has a comparable or higher nutritional value of protein and a higher calcium (Ca) content. This makes it a valuable alternative to both SBM and fish meal (FM) in the diet of laying hens. Due to the lower protein requirement in the diet of laying hens compared to broilers, it is possible to completely replace SBM with a low inclusion level of BSFLM.

Recent studies have investigated the effects of BSFLM inclusion in the diet of laying hens (Table 5). A review by Dorper et al. (2020) suggested that inclusion levels above 10% may reduce production performance. However, the limited number of studies and variations in the methodologies in terms of laying hen strains, age, nutritional value of BSFLM and feed formulation prevent drawing definitive conclusions at this time.

Table 5 Effects of replacing conventional protein sources with full-fat or defatted BSFLM in the diet of laying hens on the feed intake (FI), body weight (BW), egg weight (EW), egg mass (EM), laying rate (LR %), the feed conversion ratio (FCR) egg quality and performance parameters in recent studies

Strain	Age (week)	BSF product*	Target replacement,[a] inclusion (%)	Effects on feed intake[b]	Effects on egg quality[b]	Results in terms of the inclusion level of BSFLM (a, b)	References
Lohmann Brown Classic	28–36	D	C[1], 15%	No effect	No effects on eggshell characteristics, egg composition, albumen composition Yolk colour became more red and yellow	No effect on FI, BW, LR Tendency to improve feed efficiency Higher EW	Heuel et al., 2021
Hy-line Brown	26–33	D	P, 2% and 4%	No effect	No effect on EW, eggshell characteristics Yolk colour increased No effects on the haugh unit and chemical composition of the eggs Tendency for a lower cholesterol content	At 4% BSFLM LR slightly reduced but no effects on EM At 2% no effects on other performance parameters	Park et al., 2021
Bovans Brown	25–37	D	P, 5% 10% 15%	No effect	Lower cholesterol in yolks	No effects on performance Higher blood Ca and uric acid	Zawisza et al., 2023
Lohmann Brown Classic	24–45	D	C[2], 17%	Lower than the control treatment	Lower average EW and rise in the number of small- and medium-sized eggs as well as the XL eggs	Reduced BW, LR, EW, EM Improved FCR linearly increased blood Ca	Marono et al., 2017

(Continued)

Table 5 *(Continued)*

Strain	Age (week)	BSF product*	Target replacement,[a] inclusion (%)	Effects on feed intake[b]	Effects on egg quality[b]	Results in terms of the inclusion level of BSFLM (a, b)	References
Hy-line brown	20–40	D	P, 7 & 15%	No effects	No effects on the relative weight of yolk and albumen Lower relative weight of eggshell in higher inclusion of BSFLM Linear reduction of the relative weight and thickness of the eggshell	Quadratic effects on BW with BSFLM-14.6% the highest BW No effects on EW, FCR Increased LR and EM	Bovera et al., 2018
Lohmann Brown Classic breeder	57–63	D & F	P[3], 5%	Lower than the control treatment	NA	Reduced FI while LR was higher in BSFLM-fed layers	Petkov et al., 2022
Shave White	19–27	D	P, 5 & 7.5%	No effects at BSFLM 5% but Increased in BSFLM 7.5%	Increased yolk colour and uniformity of the egg weight Linearly improved shell thickness but the effects on breaking strength was only significant in 5% BSF	Quadratic relationship between the inclusion level of BSFLM and LR, EW, EM, while linear relationship with FI and FCR. At 5%, FCR impaired as LR and EW reduced while no effect on FI. At 7.5% FCR impaired as FI increased while there was no effect on LR and EW. BSFLM resulted in a higher BW.	Mwaniki et al., 2018

Strain	Age (week)	BSF product*	Target replacement,[a] inclusion (%)	Effects on feed intake[b]	Effects on egg quality[b]	Results in terms of the inclusion level of BSFLM (a, b)	References
Shaver white	28–43	D	P & C, 10& 15%	No effect	Linearly increased yolk colour and shell BS, no effects on HU and shell thickness	No effects on LR, reduced EW, a tendency to reduce EM, increased BW and FCR.	Mwaniki et al., 2020
White leghorn	51–55	D	P & C, 8 & 16 & 24%	Lower at 24% than other treatments	No effects on EW, albumen height, haugh unit, blood and meat spots Linear increase in yolk colour Quadratic effects on the egg specific gravity as at 8% BSFLM the lowest and at 24% BSFLM the highest	Up to 16% had no effects on performance parameters, while at 24% reduced FI, BW, and LR.	Patterson et al., 2021
Lohmann Select Leghorn classic	64–74 (5)	D	P & C[4], 12&24%	No effect	No effects	No effects on performance parameters	Maurer wt al., 2016
Black-bone chickens	56–64	F	P, 1&3&5%	Increase	No effects on the breaking strength of the eggshell At 3% and 5% lower eggshell thickness, yolk index, albumen height BSFLM affect the fatty acid and amino acid content of the eggs as well as the antioxidant activity	At 3% and 5% had higher EW and LR and at 3% the highest EW At 1% no effects on performance At 5% best FCR and the highest LR	Liu et al., 2021

(Continued)

Table 5 (*Continued*)

Strain	Age (week)	BSF product*	Target replacement,[a] inclusion (%)	Effects on feed intake[b]	Effects on egg quality[b]	Results in terms of the inclusion level of BSFLM (a, b)	References
Julia	24–29	F	P, 10%	No effects	Higher EW, albumen weight, eggshell thickness and albumen height	No effects on LR & BWLower level of plasma Ca	Kawasaki et al., 2019

*Defatted (D) includes products with CP between 46% and 61% and crude fat between 5% and 17% (except one product in the study of Heuel et al., 2021 with 30% crude fat); full fat (F) includes products with crude protein of 37% and 44 % and crude fat of 31–42%.
(a) P stands for partial replacement of the soybean meal (SBM) and C stands for a complete replacement of SBM (else other protein source indicated) while diets were iso-caloric and with similar content of crude protein and the most limiting essential amino acids. The protein content of grains or the supplemented amino acids was not considered as the protein source of diets. Feed and water were provided ad libitum unless otherwise noted: (1) diets including SFSM (6%) and without amino acids supplementation, (2) diets including 5% sunflower seed meal, (3) diets including 13% sunflower seed meal; (4) diets including 8% sunflower seed meal without amino acids supplementation; (5) four experimental round of 3 weeks.
(b) Only significant differences ($P < .05$) are mentioned. Tendency indicates $P < 0.1$.

Some studies indicate that higher levels of BSFLM, even beyond 10%, can partially or completely replace SBM without negatively affecting production performance. For example, Zawisza et al. (2023) replaced SBM with BSFLM at levels ranging from 5% to 15% reporting no adverse effects on feed intake and production performance. Similarly, Heuel et al. (2021) found that completely replacing SBM with 15% BSFLM, without amino acid supplementation, slightly improved egg weight, egg mass and feed efficiency. However, another study by Mwaniki et al. (2020) reported that while feed intake, laying rate and feed efficiency were not affected, egg weight decreased when SBM was completely replaced with 15% BSFLM. This reduction in egg weight was likely due to a decrease in limiting amino acids and linoleic acid in the diet when BSFLM replaced SBM (Mwaniki et al., 2020). Despite supplementing the diet with synthetic amino acids, replacing SBM with BSFLM led to about 13% lower amino acids such as lysine and Met+Cys, and about 25% lower Arg in the diet (Mwaniki et al., 2020). Further research has confirmed similar trends. Patterson et al. (2021) found that replacing SBM with 16% BSFLM did not affect performance parameters. However, at 24% inclusion, there were reductions in feed intake, egg production and body weight. These negative effects were attributed to lower levels of the most limiting amino acids, such as lysine, methionine, arginine, tryptophan, and isoleucine, in diets with higher BSFLM content. In particular, the dietary concentrations of arginine and tryptophan at 24% BSFLM inclusion level were lower than the breeding company's recommendations.

In summary, these findings suggest that BSFLM can potentially replace SBM in laying hen diets without compromising production performance and egg quality, as long as amino acids and linoleic acid requirements are met according to the specific strain and age of laying hens. While promising, further research is needed to fully understand the long-term effects of higher BSFLM inclusion levels.

In contrast to these outcomes, Marono et al. (2017) observed negative effects when SBM was completely replaced with 17% BSFLM in the diet of laying hens. Feed intake immediately decreased and remained lower than in the control group for the rest of the 21 weeks of the experimental period. Lower feed intake (108 vs 125 g/day/hen) was associated with reduced egg production and egg weight resulting in a 5.5% decrease in egg mass. Further investigation of the laying hens in this study indicated that feeding laying hens with 17% BSFLM compared to the control group resulted in a lower apparent ileal digestibility of crude protein (67.7 vs 78.3%), a tendency for a lower digestibility of fat (73.0 vs 77.3%), and reduced villus height in the jejunum and ileum (Cutrignelli et al., 2017). The cause of the impaired feed intake is not clear, but the dark colour of BSFLM may have reduced feed palatability (Marono et al., 2017). The dark colour of BSFLM may also indicate potential heat damage during processing which could reduce the availability of nutrients, in particular amino acids, and impair protein digestibility and even damage the small intestine.

Chitin may also negatively affect protein and lipid digestibility (Marono et al., 2015). In the study by Marono et al. (2017), they found that the chitin content of 9.2 g/kg in the diet led to an average intake of about 1 g/day, which was associated with reduced feed intake. However, in another study, the chitin content of 9.7 g/kg in the diet by including 15% BSFLM led to an average chitin intake of 1 g/day with no negative effects on the feed intake (Bovera et al., 2018). Despite this, the apparent ileal digestibility of crude protein decreased (76.1 vs 86.2%) (Bovera et al., 2018).

In contrast, laying hens maintained feed intake with a higher chitin content of 11.2 g/kg (from 15% BSFLM), resulting in a daily chitin intake of about 1.4 g/day. Moreover, the digestibility of nitrogen and energy slightly increased compared to the control diet (Heuel et al., 2021). Although the dietary chitin content in these studies (9.2–11.2 g/kg) was much higher than in other laying hen studies (Park et al., 2021; Kawasaki et al., 2019), the absence of negative effects in the study by Heuel et al. (2021) suggests that chitin might not be the primary cause of impaired feed intake. Nonetheless, its negative impact on protein digestibility can exacerbate the deficiency of certain amino acids present in marginal concentrations.

All the results assessing chitin content of BSFLM suffer from the fact that the value is based on indirect measurement of chitin content which could affect the conclusions (Eggink & Dalsgaard, 2023). Further study is necessary to quantify the chitin content and variations among different BSFLM products and to better understand the effects of the chitin content of BSFLM on feed intake and nutrient digestibility as well as on the microbiome and intestinal functionality in laying hens. Determining an optimum intake of chitin for laying hens could facilitate reaching a consistent outcome when SBM is completely replaced with BSFLM.

Including BSFLM in the diet of laying hens can enhance Ca metabolism and eggshell quality (Kawasaki et al., 2019; Mwaniki et al., 2018, 2020). This positive effect was observed as higher blood Ca levels associated with better eggshell quality in laying hens fed BSFLM (Kawasaki et al., 2019). For example, laying hens fed a diet with 17% BSFLM showed higher blood Ca levels compared to the control group (0% BSFLM), despite similar dietary Ca intakes (5.36 and 5.33 g Ca/day) (Marono et al., 2017; Secci et al., 2018). Additionally, including 5–15% BSFLM in the diet increased blood Ca levels in laying hens. The level of blood Ca was significantly higher (26.9 vs 25.0 mg/dL) in laying hens fed 15% BSFLM than that of laying hens in the control group even when the dietary Ca intake was 13% lower (Zawisza et al., 2023). Similar effects were observed in broilers, where partial and complete replacement of SBM with BSFLM increased blood

Ca levels despite reduced feed intake and dietary Ca (Tykałowski et al., 2023). Unpublished data from our team further support these findings, showing that laying hens maintained egg production and eggshell quality with 10% BSFLM in the diet, despite reduced feed and Ca intake.

The main hypothesis for the improved Ca metabolism in laying hens fed diets containing BSFLM is that the chitin content in BSFLM enhances fermentation in the hindgut. Fermentation increases the content of short-chain fatty acids (SCFA) which can increase the intestinal absorption of Ca and other minerals (Metzler-Zebeli et al., 2010). This effect on Ca absorption was observed by the increase in caecal SCFA of laying hens associated with a higher plasma Ca level when SBM was completely replaced by BSFLM (Cutrignelli et al., 2018). However, these results should be interpreted cautiously, as changes in egg production rate and egg weight affect eggshell quality in laying hens. In addition, several organs are involved in Ca metabolism in laying hens and further research is required to investigate the positive effects of BSFLM on Ca metabolism.

In addition to the effects on the eggshell, replacing SBM with BSFLM also affected the internal quality of eggs. Improvements in yolk colour and reduction in cholesterol content of the yolk have been common findings (Zawisza et al., 2023; Park et al., 2021; Secci et al., 2018). So far, no study observed a rise in the cholesterol content of the eggs when SBM is replaced with BSFLM. However, lower blood cholestrol levels have been observed in hens, explaining the decline in eggs cholesterol content (Bovera et al., 2018; Marono et al., 2017). The chitin content of BSFLM and its interaction with fat digestion might affect fat metabolism and reduce cholesterol levels in laying hens.

Overall, it is possible to replace SBM with BSFLM in the diet of laying hens without negative effects on performance parameters as long as the nutritional requirements of the flock are met. Special attention needs to be given to providing sufficient digestible amino acids with a correct balance of amino acids and meeting recommendations for essential fatty acids. When SBM is completely replaced with BSFM, the digestible concentration of limiting amino acids may fall below the breeding company recommendations. Therefore, it is necessary to increase supplementation of limiting amino acids in the diet. In addition, replacing SBM and soybean oil with BSFM reduces the linoleic acid content in laying hen diets (Zawisza et al., 2023; Kawasaki et al., 2019) while its deficiency reduces egg weight. For example, a complete replacement of soy oil (1.48% in the control diet) by including 10% or 15% defatted BSFLM resulted in a 35% reduction in linoleic acid associated with a reduction in egg weight (57.8 vs 56.9 and 56.5 g) (Mwaniki et al., 2020). Understanding the existing variations in the nutritional value of BSFLM for laying hens remains limited and further research is needed.

7 Effects of black soldier fly larvae meal on poultry health

In addition to its high nutritional value, BSFLM contains bioactive compounds, such as chitin, lauric acid, and short peptides, which may enhance animal health. These compounds can exert direct antimicrobial and immune modulatory effects, or indirectly promote a healthier gut microbiota (Veldkamp et al., 2022; Gasco et al., 2021). Health-promoting effects of BSFLM have been observed in studies *in vitro*, *in vivo* and with pathogen challenge models. A review of *in vitro* studies suggests that BSFLM can inhibit the replication of pathogenic bacteria such as *Salmonella typhimurium*, *Escherichia Coli* and *Pseudomonas aeruginosa* (Koutsos et al., 2022). Moreover, BSFLM in feed can suppress the replication of *Clostridium perfringens* and *Clostridium difficile* (Dong et al., 2021a, 2021b). These findings emphasize the role of BSFLM as a sustainable alternative ingredient poultry diets, offering both nutritional and health benefits.

Studies in broilers and laying hens further demonstrate the effects of replacing SBM with BSFLM on gut health, influencing intestinal histomorphometry, SCFA production and microbiota composition (Hartinger et al., 2021; Biasato et al., 2020; Moniello et al., 2019; Cutrignelli et al., 2018, Borrelli et al., 2017). Even at low inclusion levels (2% and 4%), BSFLM was associated with increased SCFA production in the caeca (Park et al., 2021) likely due to chitin fermentation. However, some studies reported negative effects on gut functionality with higher BSFLM inclusion (Hartinger et al., 2021; Biasato et al., 2020) which align with results indicate reductions in nutrient digestibility. Further investigation into the mechanism behind these positive and negative effects is necessary. This research could help determine the optimal inclusion level of BSFLM to maximize both its nutritional and health benefits.

The immunomodulatory effects of BSFLM in poultry, potentially mediated directly or indirectly via the effects on gut microbiome, could be a mode of action explaining its health-promoting effects. Higher blood globulin levels (2.74 vs 2.12 g/dL) by including 17% (Marono et al., 2017) or 15% (3.41 vs 2.66 g/dL) BSFLM in the diet promoted a better systemic immune response in laying hens (Bovera et al., 2018). In broilers, including 2.5% to 20% BSFLM in the diet was linked to a linear reduction in white blood cells in the blood and intraepithelial T lymphocytes in the jejunum. These were associated with improved BWG and feed efficiency (De Souza Vilela et al., 2021). Unpublished data of our team support these findings, showing improved feed efficiency and increased natural killer cells counts in laying hens fed BSFLM. These results suggest that BSFLM may enhance the innate immune function, which can reduce the need for stimulation of specific adaptive immune responses reducing their catabolic cost.

In addition, BSFLM has been shown to positively modulate mucosal immune responses. Including BSFLM (3–9%) in the diet linearly increased the level of secretory IgA in the ileum associated with a linear effect on feed efficiency (Chu et al., 2020). Similarly, low levels of BSFLM (1–3%) increased lymphocytes proliferation in the spleen and the number of T helper cells in the blood, alongside elevated serum lysozyme levels reflecting enhanced macrophage activity (Lee et al., 2018). Although the results of studies on the effects of including BSFLM on the immune response of poultry are not consistent, several studies have reported changes in the innate immune response and T lymphocytes population in different lymphoid tissues and peripheral blood suggesting the immune modulatory effects in poultry.

Studies assessing the effects of including BSFLM on the health of poultry during pathogenic challenges are limited but promising (Table 6). Trials involving pathogenic *Salmonella*, infectious bronchitis virus and coccidiosis show reduced pathogen replication in broilers fed diets containing BSFLM (Lee et al., 2018; Zhang et al., 2021; Liman et al., 2022). For example, feeding broilers with 1–3% BSFLM from day old and challenged with *S. gallinarum*

Table 6 Impacts of including BSFLM in the diet on the response of broilers to pathogenic challenges

Challenge model	BSF inclusion	Results	References
Salmonella gallinarum via oral route on 21 days of age	1%, 2%, 3%	Positive linear relationship between BSF inclusion level and T helper Cells count in the spleen Spleen lymphocytes proliferation during a T-cell challenge Serum lysozyme concentration Survivability of broilers in *Salmonella gallinarum* infection Including BSFL in the diet reduced the concentration of viable bacteria in the liver, spleen, bursa of Fabricius and caecum	Lee et al., 2018
QX-like infectious bronchitis virus via eye-drop on 10 days of age	1%, 5%, 10%	Reduced IBV viral loads in the trachea and kidneys Increased CD8 T cells in the blood Increased IFN-gamma gene expression and protein expression in the spleen	Zhang et al., 2021
Coccidiosis by oral inoculation of *E. acervulina* and *E. tenella* on 10 days of age	0.05%	Decreased the excretion of oocytes by 1.4-fold Improved live bodyweight gain in challenged broilers	Liman et al., 2022

at 21 days of age showed increased survivability and reduced replication of pathogenic bacteria in target tissues. Survivability in the control treatment (BSFLM-free diet) was 50% while it increased to 85% in broilers fed 3% BSFLM (Lee et al., 2018). Additionally, broilers fed 0.05% BSFLM in the diet and challenged with *E. acervulina* and *E. tenella* on day 10 exhibited reduced oocysts excretion compared to infected broilers fed a diet without BSFLM. The impaired BWG in the coccidiosis challenge was partially recovered up to 20 days of age (Liman et al., 2022). Replacing SBM with different levels of full-fat BSFLM, 1–10%, reduced morbidity and mortality in an infectious bronchitis virus challenge in broilers associated with a decreased viral replication in target tissues, trachea and kidney (Zhang et al., 2021). These finding align with studies indicating enhanced activity of the innate immune cells and T lymphocytes led to improved feed efficiency. Despite the limited number of challenge studies, the reduction in pathogen replication and improvement in survival rates observed with BSFLM inclusion is promising and warrants further research to elaborate its underlying mechanisms.

8 Current use of insects as feed and future trends

The range of studies reviewed in this chapter indicate the potential of BSFLM as alternative, more sustainable protein ingredient with a nutritional value comparable to conventional ingredients such as SBM. Besides, its function as a protein ingredient in the diet of poultry, BSFLM contains bioactive compounds such as antimicrobial peptides, lauric acid, chitin and chitosan, with potential health-promoting effects. Evidence indicating direct or indirect antimicrobial and immunomodulatory effects requires further research to understand the potential modes of action involved.

Given this potential, global production of insects for use as petfood and animal feed is expanding rapidly. Companies in the European Union (EU) initially focused on the petfood market. Since the EU's approval of insect-derived processed animal proteins (PAPs) in aquaculture in July 2017, the aquafeed market has taken over as the primary animal feed market. In 2021, the use of insect PAPs was also authorized for poultry and pig nutrition. IPIFF (2021) conducted a survey among insect feed operators in Europe. Figure 1 indicates the expected growth of the main markets by 2025 and 2030.

In 2025, the petfood sector is expected to remain the main sector for insect PAPs. The authorization of insect PAPs in aquaculture will increase the share of this sector to 25–35%. After the approval of insect PAPs for poultry and pigs in 2021, these markets will be the next markets for insects as feed. The proportion of insect PAPs in the animal feed market is anticipated to fluctuate in the future due to changes in the production, availability and subsequently price of traditional animal and aquafeed ingredients, such as FM and SBM.

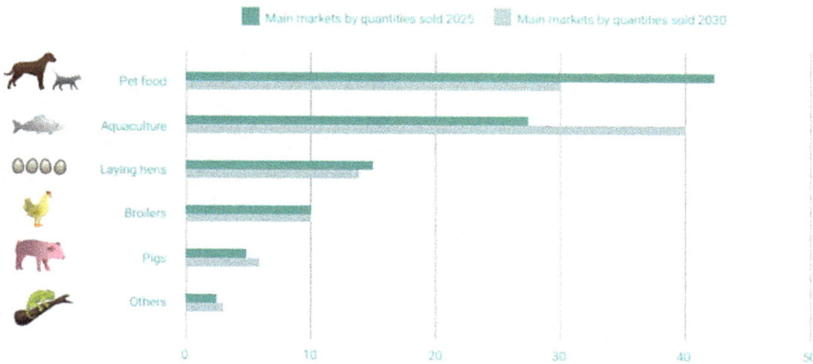

Figure 1 Expected main markets for insect PAPs in 2025 and 2030 based on a survey conducted by IPIFF (2021).

The accessibility, quality and consistency of supply of insect PAPs are key factors that might influence insect-based products being regularly incorporated in feed formulations. The application of insect PAPs in poultry and pig feed will also depend on costs and extra benefits the farmer can gain when insect PAPs are included in animal feed. It is expected that further upscaling of the insect PAPs volumes produced and broadening the list of permitted substrates for growing insects will decrease the costs. The use of new low-value biowaste sources as substrate to grow insects will also further improve sustainability. Furthermore, the potential additional health and welfare effects of including insect PAPs and insect oil in animal feed may provide an extra driver to include insect products in animal feed.

9 Conclusion

BSFLM offers comparable or higher nutritional value than SBM, which is the most common protein source in poultry feed. Rearing larvae on agri-food by-products and using BSFLM in feed helps recycle nutrients and reduces environmental impact. BSFLM contains bioactive compounds, such as chitin, lauric acid and antimicrobial peptides, which can enhance poultry health. Although BSFLM is a promising sustainable alternative to SBM, study outcomes on its use in broiler and laying hen diets are inconsistent. Differences in study design, variations in the nutritional value of BSFLM for poultry, influenced by the nutritional composition of larvae and processing methods, are key factors contributing to inconsistent results. Application of BSFLM as a primary protein source in poultry feed requires a precise prediction of its digestible amino acids and metabolisable energy content as well as reliable quality indicators.

There are several constraints to the commercial application of BSFLM as an alternative protein source in poultry feed. In addition to higher purchasing costs and limited production capacity, inconsistencies in its nutritional value pose significant challenges. These limitations, coupled with the growing evidence of BSFLM's health-promoting effects in poultry, have led to a tendency to use it at low inclusion levels as a supplemental ingredient rather than as a primary one. Incorporating BSFLM as a supplement mitigates the impact of nutritional variability on the production performance of poultry. Therefore, addressing these constraints through collaboration between the insect rearing sector and the feed industry is essential to facilitate the broader application of BSFLM as the primary ingredient or supplemental health promoting ingredient in poultry feed.

10 Where to look for further information

Due to the novelty of using BSFL in poultry feed, the research landscape is rapidly evolving. It is important to closely follow recent issues of relevant journals in the fields of insects as food and feed, as well as poultry nutrition and health, in order to stay updated. In addition to the references provided in this chapter, further resources are suggested below:

10.1 Further reading

- Dörper, A., Veldkamp, T. & Dicke, M. (2021). Use of black soldier fly and house fly in feed to promote sustainable poultry production. *Journal of Insects as Food and Feed, 7*, 761–780.
- Gasco, L., Oddon, S. B., Vandenberg, G. W., Veldkamp, T. & Biasato, I. (2023). Factors affecting the decision-making process of using insect-based products in animal feed formulations. *Journal of Insects as Food and Feed*, 1–12. https://doi.org/10.3920/jiff2022.0164
- Sogari, G., Amato, M., Biasato, I., Chiesa, S. & Gasco, L. (2019). The potential role of insects as feed: a multi-perspective review. *Animals*, *9*(4), 119.
- Veldkamp, T., Claeys, J., Haenen, O. L. M., Van Loon, J. J. A. & Spranghers, T. (Eds.). (2021). *The basics of edible insect rearing: handbook for the production chain*. Wageningen Academic Publishers.
- Veldkamp, T. & Gasco, L. (2023). Insects as global opportunity. *Animal Frontiers*, *13*(4), 3–5.
- Veldkamp, T., Schiavone, A. & Gasco, L. (2022). Introducing the special issue 'Insects on the monogastric menu'. *Journal of Insects as Food and Feed*, *8*(9), 951–952.
- The latest list of publications on the website of Study Commision Insects at https://insects.eaap.org/.

10.2 Key Journals and Conferences

- Proceedings of EAAP. Each year a growing number of sessions are allocated to topics related to insects as feed. https://brill.com/display/serial/EAAB?rskey=e7MZ0l&result=2.

- Journal of Insects as Food and Feed. https://brill.com/view/journals/jiff/jiff-overview.xml?rskey=BNRhBC&result=1.
- Major European and international research projects and organisations. The European Federation for Animal Science (EAAP): Study Commission Insects. https://insects.eaap.org/.
- Sustainable Insect Chain SUSINCHAIN project has received funding from the European Union's Horizon 2020 research and innovation programme under grant agreement n° 861976.It aimed to overcome the barriers to increasing the economic viability of the insect value chain by contributing to novel protein provision for feed and food in Europe. Project outcomes are accessible at https://susinchain.eu/International.
- Platform of Insects for Food and Feed (IPIFF) is an EU non-profit organisation which represents the interests of the insect production sector towards EU policy makers, European stakeholders and citizens. https://ipiff.org/publications-position-papers/.
- Insect as food and feed dossier in Wageningen University and Research. https://www.wur.nl/en/research-results/dossiers/file/insects-food-and-feed.htm.
- InsectFeed programme aimed to generate fundamental knowledge on using Black Soldier Fly Larvae in poultry feed. The list of publications is available on https://insectfeed.nl/.
- Network for Insect Knowledge – NIK – https://nfik.nl/. ADVAGROMED – ADVanced AGROecological approaches based on the integration of insect farming with local field practices in MEDiterranean countries. Funded by the PRIMA foundation. Publications on the application of insects in poultry feed. https://advagromed.com/poultry.

11 References

Biasato, I., Ferrocino, I., Dabbou, S., Evangelista, R., Gai, F., Gasco, L. & Schiavone, A. (2020). Black soldier fly and gut health in broiler chickens: insights into the relationship between cecal microbiota and intestinal mucin composition. *Journal of Animal Science and Biotechnology*, *11*, 1–12.

Borrelli, L., Coretti, L., Dipineto, L., Bovera, F., Menna, F., Chiariotti, L. & Fioretti, A. (2017). Insect-based diet, a promising nutritional source, modulates gut microbiota composition and SCFAs production in laying hens. *Scientific Reports*, *7*(1), 1–11.

Bovera, F., Loponte, R., Pero, M. E., Cutrignelli, M. I., Calabrò, S., Musco, N. & Moniello, G. (2018). Laying performance, blood profiles, nutrient digestibility and inner organs traits of hens fed an insect meal from *Hermetia illucens* larvae. *Research in Veterinary Science*, *120*, 86–93.

Cutrignelli, M. I., Messina, M., Tulli, F., Randazzo, B., Olivotto, I., Gasco, L. & Bovera, F. (2018). Evaluation of an insect meal of the Black Soldier Fly (*Hermetia illucens*) as soybean substitute: intestinal morphometry, enzymatic and microbial activity in laying hens. *Research in Veterinary Science*, *117*, 209–215.

Chu, X., Li, M., Wang, G., Wang, K., Shang, R., Wang, Z. & Li, L. (2020). Evaluation of the low inclusion of full-fatted *Hermetia illucens* larvae meal for layer chickens: growth performance, nutrient digestibility, and gut health. *Frontiers in Veterinary Science, 7*, 585843.

Dabbou, S., Gai, F., Biasato, I., Capucchio, M. T., Biasibetti, E., Dezzutto, D. & Schiavone, A. (2018). Black soldier fly defatted meal as a dietary protein source for broiler chickens: effects on growth performance, blood traits, gut morphology and histological features. *Journal of Animal Science and biotechnology, 9*, 1–10.

De Marco, M., Martínez, S., Hernandez, F., Madrid, J., Gai, F., Rotolo, L., Belforti, M., Bergero, D., Katze, H., Dabbou, S., Kovitvadhi, A., Zoccarato, I., Gasco, L., Schiavone, A. (2015). Nutritional value of two insect larval meals (*Tenebrio molitor* and *Hermetia illucens*) for broiler chickens: Apparent nutrient digestibility, apparent ileal amino acid digestibility and apparent metabolizable energy. *Animal Feed Science Technology*, 209: 211–218.

Do, S., Koutsos, L., Utterback, P. L., Parsons, C. M., De Godoy, M. R. & Swanson, K. S. (2020). Nutrient and AA digestibility of black soldier fly larvae differing in age using the precision-fed cecectomized rooster assay. *Journal of Animal Science, 98*(1), skz363.

Dong, L., Ariëns, R. M., America, A. H., Paul, A., Veldkamp, T., Mes, J. J. & Govers, C. (2021a). *Clostridium perfringens* suppressing activity in black soldier fly protein preparations. *LWT, 149*, 111806.

Dong, L., Tomassen, M. M., Ariëns, R. M., Oosterink, E., Wichers, H. J., Veldkamp, T. & Govers, C. (2021b). Clostridioides difficile toxin A-mediated Caco-2 cell barrier damage was attenuated by insect-derived fractions and corresponded to increased gene transcription of cell junctional and proliferation proteins. *Food & Function, 12*(19), 9248–9260.

Dörper, A., Veldkamp, T. & Dicke, M. (2021). Use of black soldier fly and house fly in feed to promote sustainable poultry production. *Journal of Insects as Food and Feed, 7*(5), 761–780.

Eggink, K. M. & Dalsgaard, J. (2023). Chitin contents in different black soldier fly (*Hermetia illucens*) life stages. *Journal of Insects as Food and Feed*, 1–10.

Elangovan, A. V., Udayakumar, A., Saravanakumar, M., Awachat, V. B., Mohan, M., Yandigeri, M. S. & Bhatta, R. (2021). Effect of black soldier fly, *Hermetia illucens* (Linnaeus) prepupae meal on growth performance and gut development in broiler chicken. *International Journal of Tropical Insect Science, 41*, 2077–2082.

Facey, H., Kithama, M., Mohammadigheisar, M., Huber, L. A., Shoveller, A. K. & Kiarie, E. G. (2023). Complete replacement of soybean meal with black soldier fly larvae meal in feeding program for broiler chickens from placement through to 49 days of age reduced growth performance and altered organs morphology. *Poultry Science, 102*(1), 102293.

Finke, M. D. (2007). Estimate of chitin in raw whole insects. *Zoo Biology: published in affiliation with the American zoo and aquarium association, 26*(2), 105–115.

Frikha, M., Serrano, M. P., Valencia, D. G., Rebollar, P. G., Fickler, J. & Mateos, G. G. (2012). Correlation between ileal digestibility of amino acids and chemical composition of soybean meals in broilers at 21 days of age. *Animal Feed Science and Technology, 178*(1–2), 103–114.

Fruci, M., Kithama, M., Kiarie, E. G., Shao, S., Liu, H., Topp, E. & Diarra, M. S. (2023). Effects of partial or complete replacement of soybean meal with commercial black soldier fly larvae (*Hermetia illucens*) meal on growth performance, cecal short chain fatty acids, and excreta metabolome of broiler chickens. *Poultry Science*, *102*(4), 102463.

Gasco, L., Józefiak, A. & Henry, M. (2021). Beyond the protein concept: health aspects of using edible insects on animals. *Journal of Insects as Food and Feed*, *7*(5), 715–741.

Gasco, L., Oddon, S. B., Vandenberg, G. W., Veldkamp, T. & Biasato, I. (2023). Factors affecting the decision-making process of using insect-based products in animal feed formulations. *Journal of Insects as Food and Feed*, 1–12.

Han, X. & Heinonen, M. (2021). Development of ultra-high performance liquid chromatographic and fluorescent method for the analysis of insect chitin. *Food Chemistry*, *334*, 127577.

Hartinger, K., Fröschl, K., Ebbing, M. A., Bruschek-Pfleger, B., Schedle, K., Schwarz, C. & Gierus, M. (2022). Suitability of *Hermetia illucens* larvae meal and fat in broiler diets: effects on animal performance, apparent ileal digestibility, gut histology, and microbial metabolites. *Journal of Animal Science and Biotechnology*, *13*(1), 1–16.

Hartinger, K., Greinix, J., Thaler, N., Ebbing, M. A., Yacoubi, N., Schedle, K. & Gierus, M. (2021). Effect of graded substitution of soybean meal by *Hermetia illucens* larvae meal on animal performance, apparent ileal digestibility, gut histology and microbial metabolites of broilers. *Animals*, *11*(6), 1628.

van der Heide, M. E., Nørgaard, J. V. & Engberg, R. M. (2021). Performance, nutrient digestibility and selected gut health parameters of broilers fed with black soldier fly, lesser mealworm and yellow mealworm. *Journal of Insects as Food and Feed*, *7*(6), 1011–1022.

Heuel, M., Sandrock, C., Leiber, F., Mathys, A., Gold, M., Zurbrüegg, C. & Terranova, M. (2022). Black soldier fly larvae meal and fat as a replacement for soybeans in organic broiler diets: effects on performance, body N retention, carcase and meat quality. *British Poultry Science*, *63*(5), 650–661.

Heuel, M., Sandrock, C., Leiber, F., Mathys, A., Gold, M., Zurbrügg, C. & Terranova, M. (2021). Black soldier fly larvae meal and fat can completely replace soybean cake and oil in diets for laying hens. *Poultry Science*, *100*(4), 101034.

Heuel, M., Kreuzer, M., Sandrock, C., Leiber, F., Mathys, A., Guggenbühl, B. & Terranova, M. (2022). Feeding value of black soldier fly larvae compared to soybean in methionine-and lysine-deficient laying hen diets. *Journal of Insects as Food and Feed*, *8*(9), 989–999.

Ibáñez, M. A., De Blas, C., Cámara, L. & Mateos, G. G. (2020). Chemical composition, protein quality and nutritive value of commercial soybean meals produced from beans from different countries: a meta-analytical study. *Animal Feed Science and Technology*, *267*, 114531.

International, Platform, of, Insects, for, Food, and, and Feed, The European insect sector today: I, I and regulatory landscape. IPIFF vision paper on the future of the insect sector towards 2030. 2018: Brussels, Belgium.

IPIFF. (2021). An overview of the European market of insects as feed. Factsheet. https://ipiff.org/wp-content/uploads/2021/04/Apr-27-2021-IPIFF_The-European-market-of-insects-as-feed.pdf

Janssen, R. H., Vincken, J. P., Arts, N. J., Fogliano, V. & Lakemond, C. M. (2019). Effect of endogenous phenoloxidase on protein solubility and digestibility after processing of *Tenebrio molitor*, *Alphitobius diaperinus* and *Hermetia illucens*. *Food Research International*, *121*, 684–690.

Janssen, R. H., Vincken, J. P., van den Broek, L. A., Fogliano, V. & Lakemond, C. M. (2017). Nitrogen-to-protein conversion factors for three edible insects: *Tenebrio molitor*, *Alphitobius diaperinus*, and *Hermetia illucens*. *Journal of Agricultural and Food Chemistry*, *65*(11), 2275–2278.

Kawasaki, K., Hashimoto, Y., Hori, A., Kawasaki, T., Hirayasu, H., Iwase, S. I. & Fujitani, Y. (2019). Evaluation of black soldier fly (*Hermetia illucens*) larvae and pre-pupae raised on household organic waste, as potential ingredients for poultry feed. *Animals*, *9*(3), 98.

Koutsos, E., Modica, B. & Freel, T. (2022). Immunomodulatory potential of black soldier fly larvae: applications beyond nutrition in animal feeding programs. *Translational Animal Science*, *6*(3), txac084.

Lalev, M., Hristakieva, P., Mincheva, N., Oblakova, M. & Ivanova, I. (2022). Insect meal as alternative protein ingredient in broiler feed. *Bulgarian Journal of Agricultural Science*, *28*(4), 743–751.

Lee, J., Kim, Y. M., Park, Y. K., Yang, Y. C., Jung, B. G. & Lee, B. J. (2018). Black soldier fly (*Hermetia illucens*) larvae enhances immune activities and increases survivability of broiler chicks against experimental infection of *Salmonella Gallinarum*. *Journal of Veterinary Medical Science*, *80*(5), 736–740.

Liman, S., Lebedev, V., Simonova, E., Biryukov, I. & Ushakova, N. (2022). Bioactive feed additive based on the larvae of Black Soldier Fly (*Hermetia Illucens*) in diets for broilers with experimental Coccidiosis. In A. Muratov & S. Ignateva (Eds.), *Fundamental and Applied Scientific Research in the Development of Agriculture in the Far East (AFE-2021) Agricultural Innovation Systems, Volume 2* (pp. 1008–1014). Springer International Publishing.

Liu, X., Liu, X., Yao, Y., Qu, X., Chen, J., Xie, K. & He, C. (2021). Effects of different levels of *Hermetia illucens* larvae meal on performance, egg quality, yolk fatty acid composition and oxidative status of laying hens. *Italian Journal of Animal Science*, *20*(1), 256–266.

Lokaewmanee, K., Suttibak, S., Sukthanapirat, R., Sriyoha, R., Chanasakhatana, N., Baotong, S. & Trithalen, U. (2023). Laying hen performance, feed economy, egg quality and yolk fatty acid profiles from laying hens fed live black soldier fly larvae. *Czech Journal of Animal Science*, *68*(4), 169–177.

Marono, S., Piccolo, G., Loponte, R., Di Meo, C., Attia, Y. A., Nizza, A. & Bovera, F. (2015). *In vitro* crude protein digestibility of *Tenebrio molitor* and *Hermetia illucens* insect meals and its correlation with chemical composition traits. *Italian Journal of Animal Science*, *14*(3), 3889.

Marono, S., Loponte, R., Lombardi, P., Vassalotti, G., Pero, M. E., Russo, F. & Bovera, F. (2017). Productive performance and blood profiles of laying hens fed *Hermetia illucens* larvae meal as total replacement of soybean meal from 24 to 45 weeks of age. *Poultry Science*, *96*(6), 1783–1790.

Mat, K., Kari, Z. A., Rusli, N. D., Rahman, M. M., Harun, H. C., Al-Amsyar, S. M. & Hassan, A. M. (2022). Effects of the inclusion of black soldier fly larvae (*Hermetia illucens*) meal on growth performance and blood plasma constituents in broiler chicken

(*Gallus gallus domesticus*) production. *Saudi Journal of Biological Sciences*, *29*(2), 809–815.

Matin, N., Utterback, P. & Parsons, C. M. (2021). True metabolizable energy and amino acid digestibility in black soldier fly larvae meals, cricket meal, and mealworms using a precision-fed rooster assay. *Poultry Science*, *100*(7), 101146.

Maurer, V., Holinger, M., Amsler, Z., Früh, B., Wohlfahrt, J., Stamer, A. & Leiber, F. (2016). Replacement of soybean cake by *Hermetia illucens* meal in diets for layers. *Journal of Insects as Food and Feed*, *2*(2), 83–90.

Moniello, G., Ariano, A., Panettieri, V., Tulli, F., Olivotto, I., Messina, M. & Bovera, F. (2019). Intestinal morphometry, enzymatic and microbial activity in laying hens fed different levels of a *Hermetia illucens* larvae meal and toxic elements content of the insect meal and diets. *Animals*, *9*(3), 86.

Moula, N. & Detilleux, J. (2019). A meta-analysis of the effects of insects in feed on poultry growth performances. *Animals*, *9*(5), 201.

Murawska, D., Daszkiewicz, T., Sobotka, W., Gesek, M., Witkowska, D., Matusevičius, P. & Bakuła, T. (2021). Partial and total replacement of soybean meal with full-fat black soldier fly (*Hermetia illucens* L.) Larvae meal in broiler chicken diets: impact on growth performance, carcass quality and meat quality. *Animals*, *11*(9), 2715.

Mwaniki, Z. N. & Kiarie, E. (2018). Standardized ileal digestible amino acids and apparent metabolizable energy content in defatted black soldier fly larvae meal fed to broiler chickens. *Canadian Journal of Animal Science*, *99*(2), 211–217.

Mwaniki, Z., Neijat, M. & Kiarie, E. (2018). Egg production and quality responses of adding up to 7.5% defatted black soldier fly larvae meal in a corn–soybean meal diet fed to Shaver White Leghorns from wk 19 to 27 of age. *Poultry Science*, *97*(8), 2829–2835.

Mwaniki, Z., Shoveller, A. K., Huber, L. A. & Kiarie, E. G. (2020). Complete replacement of soybean meal with defatted black soldier fly larvae meal in Shaver White hens feeding program (28–43 wks of age): impact on egg production, egg quality, organ weight, and apparent retention of components. *Poultry Science*, *99*(2), 959–965.

Neumann, C., Velten, S. & Liebert, F. (2018). The graded inclusion of algae (*Spirulina platensis*) or insect (*Hermetia illucens*) meal as a soybean meal substitute in meat type chicken diets impacts on growth, nutrient deposition and dietary protein quality depending on the extent of amino acid supplementation. *Open Journal of Animal Sciences*, *8*(2), 163–183.

Onsongo, V., Osuga, I., Gachuiri, C., Wachira, A., Miano, D., Tanga, C., Ekesi, S., Nakimbugwe, D. & Fiaboe, K. (2018). Insects for income generation through animal feed: effect of dietary replacement of soybean and fish meal with black soldier fly meal on broiler growth and economic performance. *Journal of Economic Entomology*, *111*(4), 1966–1973.

Osunbami, O. T. & Adeola, O. (2022). Regression method-derived digestible and metabolizable energy concentrations of partially defatted black soldier fly larvae meal for broiler chickens and pigs. *Livestock Science*, *264*, 105042.

Park, S. H., Kim, H. R., Baek, Y. C., Ryu, C. H., Ji, S. Y., Jeong, J. Y. & Kim, B. (2021). Effects of dietary inclusion level of microwave-dried and press-defatted Black Soldier Fly (*Hermetia illucens*) larvae meal on productive performance, cecal volatile fatty acid profile, and egg quality in laying hens. *Animals*, *11*(6), 1486.

Patterson, P. H., Acar, N., Ferguson, A. D., Trimble, L. D., Sciubba, H. B. & Koutsos, E. A. (2021). The impact of dietary Black Soldier Fly larvae oil and meal on laying hen performance and egg quality. *Poultry Science*, *100*(8), 101272.

Petkov, E., Ignatova, M. & Popova, T. (2022). Layers' performance and egg hatchability as affected by the dietary inclusion of two meals of black soldier fly (*Hermetia illucens*). *Journal of Insects as Food and Feed*, *8*(10), 1077–1084.

Ravindran, V., Abdollahi, M. R. & Bootwalla, S. M. (2014). Nutrient analysis, metabolizable energy, and digestible amino acids of soybean meals of different origins for broilers. *Poultry Science*, *93*(10), 2567–2577.

Rodríguez-Rodríguez, M., Barroso, F. G., Fabrikov, D. & Sánchez-Muros, M. J. (2022). *In vitro* crude protein digestibility of insects: a review. *Insects*, *13*(8), 682.

Schiavone, A., de Marco, M., Martínez, S., Dabbou, S., Renna, M., Madrid, J. & Gasco, L. (2017). Nutritional value of a partially defatted and a highly defatted black soldier fly larvae (*Hermetia illucens* L.) meal for broiler chickens: apparent nutrient digestibility, apparent metabolizable energy and apparent ileal amino acid digestibility. *Journal of Animal Science and Biotechnology*, *8*, 1–9.

Secci, G., Bovera, F., Nizza, S., Baronti, N., Gasco, L., Conte, G. & Parisi, G. (2018). Quality of eggs from Lohmann Brown Classic laying hens fed black soldier fly meal as substitute for soya bean. *Animal*, *12*(10), 2191–2197.

de Souza Vilela, J., Andronicos, N. M., Kolakshyapati, M., Hilliar, M., Sibanda, T. Z., Andrew, N. R. & Ruhnke, I. (2021). Black soldier fly larvae in broiler diets improve broiler performance and modulate the immune system. *Animal Nutrition*, *7*(3), 695–706.

Smets, R., Claes, J. & Van Der Borght, M. (2021). On the nitrogen content and a robust nitrogen-to-protein conversion factor of black soldier fly larvae (*Hermetia illucens*). *Analytical and Bioanalytical Chemistry*, *413*(25), 6365–6377.

Strifler, P., Horváth, B., Such, N., Farkas, V., Wágner, L., Dublecz, K. & Pál, L. (2023). Effects of feeding low protein diets with different energy-to-protein ratios on performance, carcass characteristics, and nitrogen excretion of broilers. *Animals*, *13*(9), 1476.

Tykałowski, B., Koncicki, A., Kowalczyk, J., Śmiałek, M., Bakuła, T., Murawska, D., Sobotka, W. & Stenzel, T. (2023). The impact of full-fat Hermetia illucens larvae meal on the health and immune system function of broiler chickens. *Journal of Veterinary Research*, *0*(0). https://doi.org/10.2478/jvetres-2023-0022

Veldkamp, T. & Bosch, G. (2015). Insects: a protein-rich feed ingredient in pig and poultry diets. *Animal Frontiers*, *5*(2), 45–50.

Veldkamp, T., Dong, L., Paul, A. & Govers, C. C. F. M. (2022). Bioactive properties of insect products for monogastric animals–a review. *Journal of Insects as Food and Feed*, *8*(9), 1027–1040.

Velten, S., Neumann, C., Bleyer, M., Gruber-Dujardin, E., Hanuszewska, M., Przybylska-Gornowicz, B. & Liebert, F. (2018a). Effects of 50 percent substitution of soybean meal by alternative proteins from *Hermetia illucens* or *Spirulina platensis* in meat-type chicken diets with graded amino acid supply. *Open Journal of Animal Sciences*, *8*(2), 119.

Velten, S., Neumann, C., Schäfer, J. & Liebert, F. (2018b). Effects of the partial replacement of soybean meal by insect or algae meal in chicken diets with graded amino acid supply on parameters of gut microbiology and dietary protein quality. *Open Journal of Animal Sciences*, *8*(3), 259–279.

Zawisza, P., Szymczyk, B., Arczewska-Włosek, A. & Szczepanik, K. (2023). Effects of partial replacement of soybean meal with defatted *Hermetia illucens* meal in the diet of laying hens on performance, dietary egg quality, and serum biochemical and redox indices. *Animals*, *13*(3), 527.

Zhao, J., Kawasaki, K., Miyawaki, H., Hirayasu, H., Izumo, A., Iwase, S. I. & Kasai, K. (2022). Egg quality and laying performance of Julia laying hens fed with black soldier fly (*Hermetia illucens*) larvae meal as a long-term substitute for fish meal. *Poultry Science*, *101*(8), 101986.

Zhang, Y., Yang, C. Y., Li, C., Xu, Z., Peng, P., Xue, C. & Cao, Y. (2021). Black soldier fly (*Hermetia illucens* L.) larval diet improves CD8+ lymphocytes proliferation to eliminate chicken coronavirus at an early infection stage. *Veterinary Microbiology*, *260*, 109151.

Chapter 4

Black soldier fly meal: an alternative protein source for pigs

S. Struthers, and J. G. M. Houdijk, Scotland's Rural College (SRUC), UK; and H. N. Hall, Anpario plc, UK

1 Introduction

The rise in the global human population, in conjunction with higher standards of living, has increased the demand for animal protein for human consumption. The overall demand for protein annually is estimated to be 202 million tonnes but is expected to reach 435 million tonnes by 2050 (Boland et al., 2013; Henchion et al., 2017). It is predicted that a large proportion of this demand will be for pork and poultry products (Kim et al., 2019), given that these are the more efficient land-based, farmed animals for converting feed protein into animal protein, but this would result in increased pressure on sustainable protein sourcing of their nutrition.

Pork is currently the second most consumed protein (globally), behind poultry. It is projected to grow by 11% to 129 million tonnes by 2032 (OECD/FAO, 2023), and as a versatile and lean meat, pork can be highly efficient to produce. Currently, most of the protein requirement in pig diets is fulfilled by plant sources, predominantly soya bean meal (SBM) (Boland et al., 2013; Kim et al., 2019; Alagappan et al., 2022). However, using soya products in

http://dx.doi.org/10.19103/AS.2024.0139.05

livestock feed is largely considered unsustainable due to increasing demand for SBM without an increase in supply, which drives prices upwards, but also the significant climate impact of land use change especially for meeting this additional SBM demand, as well as the carbon footprint of the associated transport distances (Ffoulkes et al., 2021; Gupta et al., 2021). Fish meal is no longer routinely included in pig diets due to unfavourable raw material costs, as the majority of FM produced is used in aquaculture, but it is known to be a valuable and well balanced protein for young pigs especially in creep and post-weaning diets. In summary, the increased protein requirements for human consumption drives a need for alternative sustainable, protein-rich, raw materials to feed livestock.

Insects, especially as it can be argued that using pigs as bio-converters to produce high-value protein products from nutritional resources that are not suitable for direct consumption by humans, will assist in the overall sustainability of animal protein production. Since humans can readily utilise soya as a source of protein, following appropriate food preparation, more protein sources are required which do not compete with human edible sources. These alternative proteins should be suitable for feeding, cost effective and have improved sustainability credentials compared to traditional sources such as SBM.

Black soldier fly (BSF; *Hermetia illucens*) is a promising alternative protein for livestock diets. BSF can be reared on a variety of co-product streams from the worldwide food and drink industries, as well as from household and retail food surplus. This allows for bioconversion of a wide array of organic substrates, converting this material from a relatively low protein value, in terms of variable quality, into a source of much less variable, high-quality nutrients for human consumption (Spranghers et al., 2017). BSF reared on organic substrates is rich in crude protein and fats and has a nutritional profile comparable to that of SBM and FM in terms of amino acid profile. Therefore, it offers excellent soya-replacement potential (Spranghers et al., 2017; Heuel et al., 2022).

2 Nutritional composition of black soldier fly

The nutritional composition and dietary value of BSF have been studied and reviewed in depth in several papers (Barragan-Fonseca et al., 2017; Spranghers et al., 2017; Veldkamp and Vernooij, 2021; Hawkey and Hall, 2023). However, papers reviewing the commercial replacement of traditional protein sources in pig diets in a commercial setting are still relatively lacking. As protein makes up the largest component of BSF on a dry matter basis, followed by fat (Barragan-Fonseca et al., 2017), it would also require post-harvest processing to best fit the commercial diet of pigs and more closely resemble current protein sources; this would then enable the most straightforward nutritional comparison and exchange of the tradition protein source for BSF in the diet. The average crude protein and fat content in BSF are 40% and 26% on a dry matter basis, respectively; however, these values can vary based on the feed substrate and insect developmental stage at the time of harvest (Barragan-Fonseca et al., 2017; Hawkey and Hall, 2023).

BSF's amino acid profile is comparable to that of SBM and fishmeal (FM) (Hawkey and Hall, 2023). The amino acid content of BSF also does not vary much between studies; however, specific amino acid levels can differ due to feed substrate (Barragan-Fonseca et al., 2017; Spranghers et al., 2017). BSF protein contains 10 essential amino acids and has higher levels of histidine, methionine, and tryptophan when compared to SBM (Hawkey and Hall, 2023).

Saturated fatty acids comprise most of the total fat content in BSF (Barragan-Fonseca et al., 2017), with the major fatty acids being palmitic, oleic, and linoleic acids. BSF also has a high level of lauric acid, a medium-chain saturated fatty acid, which has been reported to have anti-microbial and anti-inflammatory properties (Spranghers et al., 2017).

The mineral content of BSF is generally higher than other insect species used as feed (Finke, 2013). BSF larvae, in particular, have high calcium content due to their mineralised exoskeleton (Finke, 2013); however, some of this calcium is bound to chitin, the structural carbohydrate found in insects and shellfish, making its bioavailability to the pig limited as, like most animals, it does not produce chitinases endogenously (Rathore and Gupta, 2015).

Chitin is a long-chain polysaccharide found within the exoskeleton of insects and shellfish, including prawns and lobsters (Komi et al., 2018). Chitin is resistant to degradation by the animal in the digestive tract; however, it has been shown to promote hindgut fermentation of short-chain fatty acids and have immunomodulatory effects (Komi et al., 2018; Yu et al., 2020b). This might suggest that chitin can exert prebiotic properties. Chitin has been shown to have value once extracted from BSF for use in industrial processes (bioplastics) and animal diets but at lower levels. When used to modulate gut health, it has the potential to bind endotoxins and mycotoxins.

When estimating the crude protein content of insect products, it is worthwhile noting that chitin affects this calculation due to its high non-protein nitrogen (N) content. Therefore, N should be multiplied by a factor of 5.60 for BSF protein and 4.76 for whole larvae rather than the usual 6.25 factor commonly used (Janssen et al., 2017; Ewald et al., 2020). As chitin content varies with insect life stage, it is also important to note that comparisons between sources may be affected differently. Evaluating insects on a true protein (sum of amino acids) basis over a crude protein basis is recommended to avoid these inaccuracies (Hawkey et al., 2021). This further allows better appreciation of soya replacement potential, the basis of which is digestible lysine, and not crude protein.

3 Production performance of pigs fed diets containing black soldier fly larvae

The existing literature demonstrates that BSF larvae are suitable to be included as a feed ingredient in pig diets (Newton et al., 1977; Spranghers et al., 2018; Biasato et al., 2019; Zhu et al., 2022; van Heugten et al., 2022). Research has shown that BSF's nutritional value, digestibility, and palatability (particularly larvae) are comparable to SBM and FM, highlighting its potential to partially replace these feed ingredients without adversely affecting performance and improving sustainability.

3.1 Growth performance of pigs feddiets containing protein from black soldier fly

In nursing piglets, partial replacement (3.5%) of FM in the creep diet with BSF larvae did not affect growth performance (Driemeyer, 2016) (Table 1).

This has also been observed in weaned piglets. Biasato et al. (2019) found that the inclusion of up to 10% partially defatted BSF larvae meal (replacing 60% of the SBM inclusion) in weaned piglet diets had no overall effect on growth performance. Average daily feed intake (ADFI) showed a linear response to increasing BSF larvae meal levels from 24 to 61 days post-weaning, with piglets fed 10% BSF larvae meal having the highest ADFI. This increase was attributed to increased palatability of the diet by the inclusion of partially defatted BSF. Spranghers et al. (2018) also reported no effect of full-fat (4% and 8% inclusion) and defatted (5.4% inclusion) BSF prepupae meal on the growth performance of weaned piglets when replacing whole toasted soya beans. However, the authors did note that, despite the lack of statistical differences between the treatment groups, piglets fed full-fat BSF prepupae meal at 4% or 8% inclusion gained less weight compared to control piglets fed toasted soya beans and piglets fed defatted BSF prepupae meal. Piglets fed full-fat BSF prepupae meal at 8% inclusion also had lower ADFI. They suggested this may have been caused by reduced palatability due to the large amount of free medium-chain fatty acids present in the feed containing full-fat BSF prepupae (Dierick et al., 2002; Spranghers et al., 2018). However, Newton et al. (1977) reported no differences in palatability between piglets fed BSF prepupae or SBM diets.

More recently, Boontiam et al. (2022) studied the effect of supplementing 6% or 12% full-fat BSF larvae on the growth performance of weaned piglets under poor hygiene conditions. Compared to the negative control (poor hygiene and 0% BSF larvae), pigs fed BSF larvae had higher body weights and ADFI from 1 to 28 days post-weaning. Dietary inclusion of BSF also improved body weight and ADFI from 15 to 28 days post-weaning compared to the positive control group (good hygiene and 0% BSF larvae). Over the entire experimental period (1 to 28 days post-weaning), no differences in growth

Table [illegible] Performance of pigs fed protein from black soldier fly (BSF).

Pig growth stage	Age at weaning	Experimental period	BSF life stage	BSF feed form	BSF inclusion (%)	Target replacement	Replacement (%)	Results	Reference
Nursing		10–28 d of age	Larvae	Meal	0, 3.5	Fishmeal	3.5	Growth performance is not affected. Haematological and biochemical parameters are not affected.	Driemeyer (2016)
Weaned piglets	20 d	1–61 d post-wean	Larvae	Partially defatted meal	0, 5, 10	Soybean meal	0, 31, 62	Growth performance is not affected. Haematological parameters are not affected. Gut morphology and histology are not affected. Trend for higher FI with 10% BSF inclusion.	Biasato et al. (2019)
Weaned piglets	28 d	1–28 d post-wean	Larvae	Full-fat meal	0, 6, 12	Soybean meal	0, 20, 40	Growth performance is not affected. Diarrhoea rate was reduced in pigs fed 6 or 12% BSF.	Boontiam et al. (2022)
	21 d	0–42 d post-wean	Larvae	Full-fat meal	0.5–15 (depending on phase)	Dried whey, fishmeal, blood meal, blood plasma	0, 25, 50	Growth performance is not affected.	Crosbie et al. (2021)
	32 d	0–27 d post-wean	Larvae	Meal	0, 5, 10, 20	Fishmeal	0, 25, 50, 100	Growth performance is not affected except for reduced ADG with 5% BSF inclusion.	Håkenåsen et al. (2021)
	28 d	1–28 d post-wean	Larvae	Meal	0, 4.5, 9.1, 13.7, 18.3	Soybean meal	0, 25, 50, 75, 100	ADG and ADFI were improved with 4.5 and 9.1% BSF inclusion. FCR was not affected.	Liu et al. (2023)

(Continued)

Table 1 *(Continued)*

Pig growth stage	Age at weaning	Experimental period	BSF life stage	BSF feed form	BSF inclusion (%)	Target replacement	Replacement (%)	Results	Reference
	21 d	1–15 d post-wean	Prepupae	Full-fat meal	0, 4, 8	Toasted soybean	0, 50, 100 72	Growth performance is not affected. Gut morphology and histology are not affected.	Spranghers et al. (2018)
	21 d	1–15 d post-wean	Prepupae	Defatted meal	5.4	Toasted soybean	0, 50, 100 72	Growth performance is not affected. Gut morphology and histology are not affected.	Spranghers et al. (2018)
	21 d	1–28 d post-wean	Larvae	Full-fat meal	0 1, 2, 4	Fishmeal	0, 25, 50, 100	Growth performance is not affected. Serum total protein and globulin levels were highest while urea and triglyceride levels were lowest at 2% BSF inclusion. Jejunal villi height was the longest with 2% BSF inclusion. Inclusion of 2% BSF affected specific gut microbial populations, metabolic profiles, and mucosal immune gene expression.	(Yu et al. (2020a,b)

Pig growth stage	Age at weaning	Experimental period	BSF life stage	BSF feed form	BSF inclusion (%)	Target replacement	Replacement (%)	Results	Reference
Grower		9 wk	Larvae	Full-fat meal	0, 9, 12, 14.5, 18.5	Fishmeal	0, 25, 50, 75, 100	Growth performance is not affected. FCR was higher at 14.5 and 18.5% inclusion. Neutrophil count was higher and platelet counts were lower at 14.5 and 18.5% BSF inclusion. No other haematological parameters were affected.	Chia et al. (2019)
Grower		4 wk	Larvae		0, 3	Poultry offal	100	Growth and intake were not affected though FCR was increased.	Go et al. (2022)
Finisher		46 d	Larvae	Dried powder meal	0, 4, 8	Soybean meal	0, 18, 36	Final BW and ADG were the highest and FCR lowest at 4% inclusion. BSF inclusion positively impacted meat quality. BSF inclusion impacted microbial populations, metabolic profiles, and immune gene expression in the gut.	Yu et al. (2019a, 2019b)
Finisher		14 wk	Larvae	Full-fat meal	0, 6, 9, 12, 14	Fishmeal	0, 25, 50, 75, 100	ADFI was not affected; the Inclusion of 9 to 14% BSF improved ADG and reduced FCR.	Chia et al. (2021)

ADG = average daily gain; ADFI = average daily feed intake; BW = body weight; FCR = feed conversion ratio (feed over gain); FI = feed intake performance were found between the treatment groups; however, diarrhoea rate was significantly reduced in piglets fed BSF larvae (Boontiam et al., 2022). This suggests that feeding BSF can help improve piglet health and mitigate the effects of housing pigs in poor hygiene conditions (lighter body weight, increased diarrhoea, decreased nutrient utilisation). This is likely due to the antimicrobial properties of the fatty acid profile in the full-fat BSF, though prebiotic benefits arising from chitin cannot be excluded.

Furthermore, the inclusion of full-fat BSF larvae meal to partially replace (25% or 50%) animal protein sources (FM, spray-dried blood meal, and blood plasma) in nursery pig diets supported growth performance suggesting that dietary animal protein sources could be replaced by as much as 50% with full-fat BSF larvae meal (inclusion of up to approximately 15% in the diet) without negative consequences (Crosbie et al., 2021).

Partially replacing FM with 2% full-fat BSF larvae meal positively impacted ADG and FCR (F:G) in weaned piglets during the first two weeks post-weaning, with ADG being highest and FCR lowest in piglets fed 2% BSF larvae meal (Yu et al., 2020a). Over the 28-day post-weaning experimental period, no differences were found between the treatment groups fed 0% to 4% full-fat BSF larvae meal in replacement of 0% to 100% of the FM inclusion (Yu et al., 2020a). This is similar to what was reported by Håkenåsen et al. (2021) except the BSF inclusion rate was higher (0%, 5%, 10%, 20% of the diet), the BSF still replaced FM at 0%, 25%, 50%, 75%, and 100%. Interestingly, the only effect was seen for ADG over the 27-day post-weaning experimental period with piglets fed 5% full-fat BSF larvae meal having lower ADG compared to the control diet with the 10% and 20% BSF inclusion groups being intermediate (Håkenåsen et al., 2021). As this is the lowest BSF inclusion studied in this chapter, it may be that a higher inclusion and replacement of traditional protein sources are needed for the accurate evaluation of novel proteins in the diet.

Liu et al. (2023) investigated the effect of total replacement of SBM with increasing levels of BSF larvae meal (0% to 18.3%). The authors found that ADG and ADFI improved in piglets fed 4.5 and 9.1% BSF larvae meal (replacing 25% and 50% SBM, respectively), although FCR was unaffected.

In grower pigs, replacing up to 100% of fishmeal in the diet with full-fat BSF larvae meal (18.5% inclusion) did not negatively impact growth performance (Chia et al., 2019). Pigs fed 9% BSF larvae meal had better FCR than those fed 14.5 or 18.5% BSF larvae meal; however, all three groups were not different from the fishmeal control diet (Chia et al., 2019). Haematological parameters were generally unaffected, except for neutrophil and platelet counts, which were higher in the 14.5 and 18.5% BSF larvae meal inclusion groups (representing 75% and 100% fishmeal replacement, respectively).

Go et al. (2022) found that ADG was reduced during the first 2 weeks of feeding 3% BSF larvae compared to pigs fed a control diet containing poultry offal (100% replacement). However, by the end of the 4-week experimental period, no differences in ADG were noted. Pigs fed the 3% BSF larvae diet had lower FCR (G:F) than controls over the 4-week experiment. These minor differences in growth performance were attributed to anti-nutritional factors (ANFs) in the BSF, such as chitin. Chitin is a component of the insect's exoskeleton, and if not digested, it can bind to proteins and reduce digestibility (Wang and Shelomi, 2017).

In finisher pigs, partially replacing soybean meal with 4% dried BSF larvae meal (18% replacement) improved body weight and ADG and reduced FCR (F:G) compared to the control and 8% BSF inclusion diets (Yu et al., 2019a). This agrees with what has been reported in more recent studies. Chia et al. (2021) found replacing up to 100% fishmeal with up to 14% full-fat BSF larvae meal did not affect body weight or ADFI; however, the higher levels of BSF inclusion (9% to 14%) improved ADG and lowered FCR (F:G). Yu et al. (2019a) suggested that the differences between their treatment groups could be related to the upregulated expression of genes related to lipogenic potential and muscle fibre composition. They also attributed the lack of effect of the 8% BSF inclusion to higher chitin levels in the BSF diet. The improved growth performance in pigs fed increasing levels of BSF larvae meal in the study by Chia et al. (2021) was attributed to increased palatability.

The aforementioned studies were all carried out with the idea of using BSF as an alternative protein source. One study though looked at the use of BSF oil. Thus, when BSF larvae were included in the diet of weaned piglets as an extracted insect oil to replace corn oil, body weight, and ADG linearly increased with increasing BSF oil inclusion (0%, 2%, 4%, 6%). Serum cholesterol and platelet count also increased linearly with increasing BSF oil inclusion (van Heugten et al., 2022). This suggests that BSF larvae oil is at least as palatable as corn oil suitable to be considered as an alternative energy source.

Overall, these studies suggest that the inclusion of BSF in the diets of piglets both immediately post-weaning and at later ages does not adversely affect growth performance. The number of studies carried out in growing and finisher pigs is rather limited and so more work would be required to confirm the benefits of replacing SBM with BSF in these diets.

3.2 Nutrient digestibility

From the 12 studies reviewed, that reported pig performance when replacing traditional protein sources (one study included reviewed oil replacement), only three of these also reported nutrient digestibility. Commercial scale studies are still required to further evaluate the nutrient digestibility of BSF inclusion in pig diets, however we have summarised the current findings below.

Nutrient digestibility was not influenced by the inclusion of partially defatted 10% BSF larvae meal in the diet of weaned piglets (Biasato et al., 2019).

Dietary supplementation of 6% or 12% full-fat BSF larvae meal improved dry matter, crude protein, and ether extract digestibility in weaned piglets when compared to the negative control group (poor hygiene and 0% BSF); however, the BSF inclusion groups did not differ from the positive control (good hygiene and 0% BSF) (Boontiam et al., 2022). This improved digestibility is likely due to the improved gut health of piglets fed full-fat BSF larvae, which would help reduce unwanted inflammation and can result in increased energy partitioning to growth.

Including 5.4% defatted BSF larvae meal in weaned piglet diets resulted in equal or improved ileal nutrient digestibility compared to the control diet containing toasted soybean (Spranghers et al., 2018). Including full-fat BSF larvae meal at 4% or 8% reduced ileal energy digestibility; however, 4% BSF larvae meal inclusion increased ileal crude protein digestibility (Spranghers et al., 2018).

The dietary BSF inclusion in these studies is relatively low. At higher levels of BSF larvae inclusion (33%) in a study from almost 50 years ago, Newton et al. (1977) found reduced dry matter, increased ether extract, and similar crude protein digestibility in weaned piglets compared to those fed soybean meal. These differences may be due to changes in gut microbiome following the feeding of a novel feed ingredient.

3.3 Gut health, haematological parameters, and meat quality

Driemeyer (2016) found that including 3.5% BSF larvae meal in nursing piglet diets (replacing 100% fishmeal inclusion) did not affect haematological or biochemical parameters. The authors did observe increasing haemoglobin and haematocrit levels over the experimental period with BSF inclusion. Although the differences were not statistically significant, the authors suggested that they may be biologically important as higher haemoglobin and haematocrit levels can be indicators of immunological stress; however, the authors also noted that none of the animals showed physical signs of distress (Driemeyer, 2016).

Biasato et al. (2019) found that BSF inclusion in the diet of weaned piglets did not affect haematological parameters except for monocytes and neutrophils, which showed linear and quadratic responses to increasing BSF inclusion, respectively. Regardless, all of the haematological parameters recorded fell within the normal physiological range for pigs, suggesting that BSF inclusion did not negatively impact the piglet's health (Biasato et al., 2019). Inclusion of up to 10% partially defatted BSF larvae meal also did not influence gut morphology or histology (Biasato et al., 2019).

Serum cholesterol linearly increased in weaned piglets fed increasing levels of BSF larvae oil (van Heugten et al., 2022). Platelet count also tended to increase linearly with increasing BSF oil inclusion. BSF larvae oil inclusion affected no other haematological or serum parameters (van Heugten et al., 2022).

Yu et al. (2020a, 2020b) reported that weaned piglets fed 2% full-fat BSF larvae meal had higher serum total protein and globulin levels than piglets fed FM. Gut morphology was also influenced, with piglets fed 2% BSF larvae meal having longer jejunal villi heights than piglets fed fishmeal or 1% BSF larvae meal (Yu et al., 2020a). Including 2% BSF larvae meal also affected specific microbial populations, their metabolic profiles, and mucosal immune gene expressions in the gut (Yu et al., 2020b).

In grower pigs, neutrophil counts were increased, and platelet counts were reduced in pigs fed 14.5 or 18.5% full-fat BSF larvae meal instead of fishmeal (Chia et al., 2019), which coincided with greater (more detrimental) FCR. Since no other haematological were affected by BSF inclusion in this study, an elevated neutrophil level might be an indication of a stress response (Widowski et al, 1989), in accord with as suggested by Driemeyer (2016).

In finisher pigs, Yu et al. (2019a) found that including dried BSF larvae meal (4% or 8%) positively influenced carcass traits and meat quality as loin eye area and marbling scores were higher with BSF inclusion. Similar to what was found in the weaned piglets, Yu et al. (2019b) reported that BSF inclusion in finisher pig diets altered the microbial populations, their metabolic profiles, and mucosal immune gene expression in the gut.

4 Benefits of using black soldier fly

4.1 Reducing the use of traditional protein sources, soya bean meal, and fishmeal

This chapter evaluates peer-reviewed papers where BSF has been utilised to replace mainly SBM or FM in pig diets and these clearly indicate that while there is evidence to support the use of BSF as a suitable alternative, there is still a significant degree of variability regarding the upper limit and dose-dependency of this approach. For most studies, the largest inclusion of BSF used, ranging from 5% to 20%, did not impact growth performance. However, in some studies, positive effects on performance were observed at levels below the maximum tested. This supports that, as is the case with any ingredient, there are constraints to consider, which in the case of BSF are likely coming from the level of post-harvest processing regarding protein and oil extraction and chitin levels, which can be a result of variation in growth stage and processing. As such, owing to its positive effect on protein concentration, defatting BSF would be expected to increase its suitability in the commercial pig diet and therefore its soya replacement potential, though it should be noted that in the inclusion levels of full-fat, partially defatted, or defatted BSF reported to date, an upper limit on SBM replacement in young pigs seems not to have been demonstrated yet. There are also trade-offs to be considered between benefits and constraints from both BSF and the replaced ingredients. For the BSF, these trade-offs may come from the level of chitin, which at low inclusion level might be providing prebiotic benefits but over a threshold might result in fibre-based constraints. As such, replacing SBM with BSF will inevitably reduce the levels of fibre in the diet. Furthermore, if the replaced ingredient is arguably of greater digestibility, replacement on a digestible amino acid basis would be expected to detriment feed efficiency, as observed for BSF inclusion at the expense of poultry offal (Go et al., 2022). Finally, it should also be noted that whilst in some studies all the SBM was safely replaced, i.e. without impacting performance, this will be sensitive to the basal level of SBM in the first place. The lower the latter, the more likely it can be completely replaced with an alternative.

The vast majority of growth performance studies using BSF have been on peri-weaning pigs, whilst only a few have been carried out on grower and finisher pigs. Whilst arguably the level of SBM in pig rations reduces as pigs move through the different feeding phases, the greater feed intake of grower and particularly finisher pigs compared to peri-weaning pigs means that most SBM is being used in the grower and finisher phase. Only one study was identified in finisher pigs, where up to 8% of BSF meal was replacing up to 36% of SBM, with improved FCR and ADG at intermediate BSF levels (Yu et al., 2019). More studies in grower and finisher pigs are required to identify if a greater replacement of SBM for BSF results in similar performance, as this single study might suggest it may not be the case. The notion that alternative feedstuffs each have their own constraints when being used as 'soya replacers' comes largely from studies where these alternatives are addressed in isolation. However, an alternative approach would be to set a limit in the diet formulation to remove or restrain SBM inclusion and provide all other alternatives. A best-cost formulation may also look to include sustainability metrics to create a 'low carbon' or 'home-grown' diet which has other benefits to sustainable livestock production. Including BSF and other alternatives would help to result in a greater reduction in SBM, with the emphasis on using upper limits of each alternative below their anti-nutritional thresholds but so as to not produce a diet with too many ingredients. Such approaches have been shown successful in replacing all SBM in fast growing broilers (Houdijk et al., 2024), which arguably have a digestive system similar to that of pigs. Thus, realising that this can be achieved in broilers, this opens the notion to combine BSF with other protein sources such as pulses and oilseed meals from non-soya origin (e.g. rapeseed meal, sunflower meal). Indeed, within a background of a combination of SBM and rapeseed meal as main protein sources, a complete replacement of SBM with home-grown pulses for growing and finishing pigs has been observed under both experimental conditions (Smith et al., 2013; White et al., 2015) and commercial conditions (Houdijk et al., 2013).

Like SBM, FM has also been under pressure from a sustainability perspective, given that fish stocks are a finite resource (concerns of overfishing). FM also has environmental implications associated with its transport and processing (especially drying), and such factors have made it increasingly expensive and supply volatile. As such, we usually only find FM in the most specialist rations for young pigs (creep feeds, weaner feeds) as most of the FM available in the market is used in pet nutrition and aquaculture diets. The historic use of FM in young pig rations was not only driven by its high biological value, but also its associated health benefits, including gut health (see below) arising from a favourable fatty acid and amino acid profile. Therefore, replacing FM with BSF meal could not only provide a cost-effective alternative as well as one with improved sustainability credentials and supply benefits.

4.2 Gut health

Several studies have reported on the gut health benefits of BSF and extracted products such as insect oil, chitin, and antimicrobial peptides (AMPs) (Gasco et al., 2018; Biasato et al., 2019; Boontiam et al., 2022). While further work is needed to understand whether there is greater value in feeding these as extracted components or as the whole insect it should be noted that regulatory hurdles may also play a role in this decision.

As mentioned, chitin is the main carbohydrate incorporated in the insect exoskeleton and can impact protein and mineral bioavailability (Finke, 2007; Jonas-Levi and Martinez, 2017; Henriques et al., 2020). While pigs do not sufficiently produce chitinases to easily digest the chitin, it has been shown that the gut microbiota can secrete chitinase and gain value in this component (Šimůnek et al., 2001). The extent of this is likely related to the age of the pig and gut health before feeding a chitin-based compound.

Antimicrobial peptides are known to be produced by insects (Harlystiarini et al., 2019) as part of their immune response. These proteins inhibit the growth of harmful pathogens in the insect (Lu et al., 2014), but they also confer benefits to livestock feeding and may be a factor in the improvements in health seen in the papers mentioned above, most notably (Boontiam et al., 2022). *In vitro* BSF extracts have been shown to be strongly antibacterial against species of *Salmonella* and *E. coli* (Harlystiarini et al., 2019); both pathogens are important in pig production and may provide added value when feeding whole or extracted BSF products.

The composition of BSF oil is similar to that of coconut oil (Dayrit, 2015), predominantly containing lauric acid, with up to 52% recently having been reported (Ewald et al., 2020). Lauric acid is known to have antimicrobial properties (Spranghers et al., 2018). The content of lauric acid in BSF fat is affected by diet type and suitability but it is positively correlated to larvae weight and is hypothesised to be synthesised by the larvae (Spranghers et al., 2017) as it is found in high concentrations even when fed at very low levels in the insect substrate. Coconut oil and lauric acid are known to be beneficial in pig diets historically, with coconut oil being included in creep and nursery diets for antimicrobial benefits, but care must be taken to manage the antioxidant capacity of the diet when including highly saturated fatty acids.

Feeding of BSF has been shown to modify the gut microbiota (Håkenåsen et al., 2021; Boontiam et al., 2022; Liu et al., 2023), with increases seen in *Ruminococcaceae*, *Faecalibacterium*, *Butyricoccus*, and *Lactobacillus* spp. These changes may be due to the components mentioned above and changes to the macronutrients available in the diet. Further work is needed to consistently use BSF to improve gut health, but it is clear to see that gut health benefits can be gained from their feeding, and this should be harnessed to fully realise the value of BSF in pig diets.

4.3 Reducing carbon footprint of diets

It has been suggested that insect-based diets can reduce the carbon emissions of livestock feeding when replacing FM or SBM (van Huis and Oonincx, 2017; Ffoulkes et al., 2021). However, insect feedstocks and climate control in insect rearing must be considered (Oonincx, 2021) and the processing of insects post-harvest when compared to mass-produced feed protein such as SBM. In terms of locality and transport emissions, BSF can provide a highly localised protein source and may even be able to be produced on the site where the protein is to be used. This form of circular economy can make great use of locally available co-products, which would still fit with regulatory requirements within the EU. Insects such as BSF are classed as farmed livestock within the EU and as such must be reared and fed in accordance with relevant regulations (Alagappan et al., 2022). However, insects can be succesfully reared on vegetable surplus from supermarkets or pre-consumer sources. This would increase the nutrient suitability of these substrates and reduce unwanted variation in nutrient provision for livestock feeding (van Heugten et al., 2022). It may also be that insects play a role alongside anaerobic digesters, utilising waste heat and being used as a primary bio-converter in this process. We are increasingly seeing sustainable livestock farming look to utilise all on-farm primary and secondary resources to reduce waste, improve carbon capture and enhance overall resource efficiency; in this context, insects such as BSF can play a vital piece in this complicated puzzle (van Huis and Oonincx, 2017).

4.4 Animal welfare following feeding of BSF

A wealth of data supports the notion that insects support good animal welfare, especially regarding pig, poultry, and aquaculture feeding. This is mainly because most of the diet of juvenile monogastrics and fish would be supported by insects, either terrestrial or aquatic. Therefore, insect feeding supports key welfare criteria in livestock farming, providing towards freedoms 1 (freedom from thirst, hunger, and malnutrition) and 5 (freedom to express normal behaviour) (Mellor, 2016). The highest value regarding animal welfare from feeding insects may be attributed to live or whole insect feeding as they have then been attributed to providing stimulation in the form of manipulable materials in pig and poultry rearing (Star et al., 2020; Ipema et al., 2021a,b; 2022). However, wider use may be found in the use of insects in pelleted feeds, which would likely be in the form of a protein powder or extracted oil. This use may also benefit animal welfare, especially if the gut and animal health benefits reported above can be fully realised.

5 Challenges

5.1 Feed safety

A few studies have examined aspects of the safety of feeding BSF to pigs. Potential feed safety concerns such as microbial contamination and heavy metal accumulation need to be investigated and addressed (reviewed by van der Fels-Klerx et al. (2018)). An advantage is that, unlike other insect species, BSF are not considered disease vectors as the adults do not consume decayed organic material nor lay their eggs on organic material (van Huis, 2013). However, as with other animal-derived proteins, there is a risk of prion diseases, although there is (currently) no evidence that insects carry prions (DiGiacomo and Leury, 2019). BSF larvae have been found to accumulate some heavy metals such as cadmium from their diet but not others such as chromium, arsenic, nickel, and mercury (Charlton et al., 2015; Cai et al., 2018). Zinc concentration decreases in BSF larvae as its concentration increases in the rearing substrate (Diener et al., 2015). There are differing results regarding lead accumulation with some studies reporting higher concentrations in the BSF larvae than in the rearing substrate (Gao et al., 2017; Purschke et al., 2017; Cai et al., 2018) while others report the opposite (Diener et al., 2015). On a positive note regarding feed safety, BSF larvae have the ability to remove mycotoxins from contaminated feed without subsequent accumulation (Cai et al., 2018; van der Fels-Klerx et al., 2018). Additionally, BSF larvae reared on pesticide-spiked substrates did not show detectable levels of the pesticides in their tissues (Purschke et al., 2017). Thus, although when acting as bio-converters, some degree of bioaccumulation is always expected, the evidence that this is greatly impacting BSF quality is not strong. Therefore, existing quality assurance schemes in animal feed production that guarantee traditional feed safety are likely also applicable for the safe use of BSF.

5.2 Feed suitability

The knowledge of the nutritional value of BSF and any potential anti-nutritional factors it may contain will greatly determine its feed suitability. Some possible impacts of fibrous chitin have already been mentioned, including its prebiotic potential at relatively low levels and its fibre-like constraints at relatively high levels. The microbiome can make use of chitin as a source of energy by action of their chitinases. This opens the suggestions to develop chitinase as a possible feed additive, in the same way how, e.g. commercial phytases, xylanases, and proteases have been developed. This might assist the production of oligosaccharide-type structures from chitin to promote prebiotic properties and overall fermentability, and this contribution to host energy supply and physicochemical modification of gut content. Such an approach might also address the possible issue of allergen risk or allergy-promoting molecules present within the insects. Chitin may play a role in allergenicity as it is recognised by the immune system and can activate various immune cells (Komi et al., 2018). Despite chitin's role in immune and potentially allergic responses, the mechanisms by which it does so are still not fully understood (van Huis and Oonincx, 2017; van der Fels-Klerx et al., 2018). Another aspect of feed suitability is palatability. However, if palatability was a serious constraint, this would have been observed as reduction in feed intake in the performance studies undertaken. Since the latter has not been reported, it might be concluded that the palatability of BSF is unlikely a constraint.

6 Applications

6.1 Inclusion into pelleted feed

As has already been discussed earlier in the chapter, oil extracted BSF protein is a more suitable comparison to conventional protein sources (SBM and FM) due to the low level of oil in these materials. Pelleted monogastric feeds generally contain no more than 3.5% crude fat as higher levels can negatively affect pelleting and can reduce pellet quality as well as increasing the risk of oxidation and potentially leading to the early breakdown of feed vitamins.

Commercially, BSF is available and increasingly as a fat extracted protein over the whole unprocessed product. In terms of legislation, the EU classifies whole unprocessed terrestrial invertebrates and live-fed terrestrial invertebrates as currently permitted for feeding to all livestock. However, oil extraction would categorise the insect protein as processed animal protein (PAP) and is then regulated under different criteria. It is therefore in some cases more accessible to feed BSF as live and, or unprocessed rather than through the PAP regulations. For widespread use of BSF in pig diets it is likely that further processing will be needed to reduce limiting nutrients such as chitin and enable easy inclusion into pelleted feeds which makes up the majority of pig feed volume.

6.2 Live fed on farm

As noted above, this is generally an easier application due to the avoidance of the PAP regulations which came into force to reduce the risk of bovine spongiform encephalitis (BSE). Live feeding of BSF on farms is one of the main ways BSF are currently fed to livestock in the EU and UK, such as laying hens in the aim of reducing carbon footprint per egg. However, it is noted that current schemes are not fully effective due to the high costs of heating insect-rearing units and the limitations in legislation surrounding the feeding of insects. For the full value of BSF feeding to be realised it is likely that legislation surrounding permitted insect feeding substrates needs to be reviewed and relaxed to enable a wider array of lower value organic streams to be utilised.

7 Conclusion

The current body of research demonstrates that BSF has excellent potential as an insect species to be used for feeding commercial livestock, especially pigs. BSF is able to partially, and in some cases fully, replace SBM and FM without negative consequences for growth performance, gut health, or meat quality. BSF also has the ability to have less of an ecological footprint compared to the current plant protein sources but is hampered by current legislative frameworks. Ultimately, more research is needed at a commercial scale to ensure effective inclusion in commercial pig diets as a safe, cost-effective, and sustainable protein.

8 Acknowledgements

SRUC receives support from the Scottish Government (RESAS), including through an initiative around insect products for food production.

9 References

Alagappan, S., Rowland, D., Barwell, R., Mantilla, S. M. O., Mikkelsen, D., James, P., Yarger, O., and Hoffman, L. C. 2022. Legislative landscape of black soldier fly (Hermetia illucens) as feed. *J Insects Food Feed* 8:343–355.

Barragan-Fonseca, K. B., Dicke, M. and van Loon, J. J. A. 2017. Nutritional value of the black soldier fly (Hermetia illucens L.) and its suitability as animal feed: a review. *J Insects Food Feed* 3:105–120.

Biasato, I., Renna, M., Gai, F., Dabbou, S., Meneguz, M., Perona, G., Martinez, S., Lajusticia, A. C. B., Bergagna, S., Sardi, L., Capucchio, M. T., Bressan, E., Dama, A., Schiavone, A. and Gasco, L. 2019. Partially defatted black soldier fly larva meal inclusion in piglet diets: effects on the growth performance, nutrient digestibility, blood profile, gut morphology and histological features. *J Anim Sci Biotechnol* 10.

Boland, M. J., Rae, A. N., Vereijken, J. M., Meuwissen, M. P. M., Fischer, A. R. H., van Boekel, M. A. J. S., Rutherfurd, S. M., Gruppen, H., Moughan, P. J. and Hendriks, W. H. 2013. The future supply of animal-derived protein for human consumption. *Trends Food Sci Technol* 29:62–73.

Boontiam, W., Phaengphairee, P., Hong, J. and Kim, Y. Y. 2022. Full-fatted Hermetia illucens larva as a protein alternative: effects on weaning pig growth performance, gut health, and antioxidant status under poor sanitary conditions. *J Appl Anim Res* 50:732–739.

Cai, M., Hu, R., Zhang, K., Ma, S., Zheng, L., Yu, Z. and Zhang, J. 2018. Resistance of black soldier fly (Diptera: Stratiomyidae) larvae to combined heavy metals and potential application in municipal sewage sludge treatment. *Environ Sci Pollut Res* 25:1559–1567.

Charlton, A. J., Dickinson, M., Wakefield, M. E., Fitches, E., Kenis, M., Han, R., Zhu, F., Kone, N., Grant, N., Devic, E., Bruggeman, G., Prior, R. and Smith, R. 2015. Exploring the chemical safety of fly larvae as a source of protein for animal feed. *J Insects Food Feed* 1:7–16.

Chia, S. Y., Tanga, C. M., Osuga, I. M., Alaru, A. O., Mwangi, D. M., Githinji, M., Dubois, T., Ekesi, S., van Loon, J. J. A. and Dicke, M. 2021. Black soldier fly larval meal in feed enhances growth performance, carcass yield and meat quality of finishing pigs. *J Insects Food Feed* 7:433–447.

Chia, S. Y., Tanga, C. M., Osuga, I. M., Alaru, A. O., Mwangi, D. M., Githinji, M., Subramanian, S., Fiaboe, K. K. M., Ekesi, S., van Loon, J. J. A. and Dicke, M. 2019. Effect of dietary replacement of fishmeal by insect meal on growth performance, blood profiles and economics of growing pigs in Kenya. *Animals* 9. https://doi.org/10.3390/ani9100705

Crosbie, M., Zhu, C., Karrow, N. A. and Huber, L. A. 2021. The effects of partially replacing animal protein sources with full fat black soldier fly larvae meal (Hermetia illucens) in nursery diets on growth performance, gut morphology, and immune response of pigs. *Transl Anim Sci* 5. https://doi.org/10.1093/tas/txab057

Dayrit, F. M. 2015. The properties of lauric acid and their significance in coconut oil. *J Am Oil Chem Soc* 92:1–15.

Diener, S., Zurbrügg, C. and Tockner, K. 2015. Bioaccumulation of heavy metals in the black soldier fly, Hermetia illucens and effects on its life cycle. *J Insects Food Feed* 1:261–270.

Dierick, N. A., Decuypere, J. A., Molly, K., Van Beek, E. and Vanderbeke, E. 2002. The combined use of triacylglycerols containing medium-chain fatty acids (MCFAs) and exogenous lipolytic enzymes as an alternative for nutritional antibiotics in piglet nutrition. *Livest Prod Sci* 75:129–142.

DiGiacomo, K. and Leury, B. J. 2019. Review: insect meal: a future source of protein feed for pigs? *Animal* 13:3022–3030.

Driemeyer, H. 2016. Evaluation of black soldier fly (Hermetia illucens) larvae as an alternative protein source in pig creep diets in relation to production, blood and manure microbiology parameters. Available at https://scholar.sun.ac.za

Ewald, N., Vidakovic, A., Langeland, M., Kiessling, A., Sampels, S. and Lalander, C. 2020. Fatty acid composition of black soldier fly larvae (Hermetia illucens) – Possibilities and limitations for modification through diet. *Waste Manag* 102:40–47. https://doi.org/10.1016/j.wasman.2019.10.014

Ffoulkes, C., Illman, H., O'connor, R., Lemon, F., Behrendt, K., Wynn, S., Wright, P., Godber, O., Ramsden, M., Adams, J., Metcalfe, P., Walker, L.,. Gittins, J., Wickland, K., Nanua, S. and Sharples, B. 2021. Development of a roadmap to scale up insect protein production in the UK for use in animal feed. https://www.wwf.org.uk/sites/default/files/2021-06/the_future_of_feed_technical_report.pdf

Finke, M. D. 2007. Estimate of chitin in raw whole insects. *Zoo Biol* 26:105–115.

Finke, M. D. 2013. Complete nutrient content of four species of feeder insects. *Zoo Biol* 32:27–36.

Gao, Q., Wang, X., Wang, W., Lei, C., and Zhu, F. 2017. Influences of chromium and cadmium on the development of black soldier fly larvae. *Environ Sci Pollut Res* 24:8637–8644.

Gasco, L., Finke, M. and van Huis, A. 2018. Can diets containing insects promote animal health? *J Insects Food Feed* 4:1–4.

Go, Y. B., Lee, J. H., Lee, B. K., Oh, H. J., Kim, Y. J., An, J. W., Chang, S. Y., Song, D. C., Cho, H. A., Park, H. R., Chun, J. Y. and Cho, J. H. 2022. Effect of insect protein and protease on growth performance, blood profiles, fecal microflora and gas emission in growing pig. *J Anim Sci Technol* 64:1036–1076.

Gupta, M., Hart, P., Krebsbach, S. G., Da Gama, L., Baungaard, C., Trewern, J., Halevy, S., Walsh, L., Blandon, A., Weir, C., Ryan, A., Wakefield, S., Tesco, D. E., Webb, L., Nordmann, H. D., Weis, B., Umeasiegbu, K., Ffoulkes, C., Illman, H., Behrendt, K., Godber, O., Ramsden, M., Adams, J., Metcalfe, P., Walker, L., Gittins, J., Wynn, S., O'connor, R., Lemon, F., Wickland, K., Nanua, S., Sharples, B., Wright, P., Keller, E., Perkins, R. and Salter, D. 2021. The future of feed: a WWF roadmap to accelerating insect protein in UK feeds. www.wwf.org.uk/press-release/insects-animal-feed-report. Text © 2021 WWF-UK. All rights reserved.

Håkenåsen, I. M., Grepperud, G. H., Hansen, J. Ø., Øverland, M., Ånestad, R. M. and Mydland, L. T. 2021. Full-fat insect meal in pelleted diets for weaned piglets: effects on growth performance, nutrient digestibility, gastrointestinal function, and microbiota. *Anim Feed Sci Technol* 281. https://doi.org/10.1016/j.anifeedsci.2021.115086

Harlystiarini, H., Mutia, R., Wibawan, I. W. T. and Astuti, D. A. 2019. In vitro antibacterial activity of black soldier fly (Hermetia illucens) larva extracts against Gram-negative bacteria. *Buletin Peternakan* 43:125–129.

Hawkey, K., Brameld, J., Parr, T., Salter, A. and Hall, H. 2021. *Suitability of insects for animal feeding.Pages 26–38 in Insects as animal feed: novel ingredients for use in pet, aquaculture and livestock diets*. CABI, Wallingford, UK.

Hawkey, K. and Hall, H. 2023. Insects as animal feed. Pages 372–373 in *The Encyclopedia of Animal Nutrition*. Phillips, C., Ed. 2nd ed. CABI, GB.

Henchion, M., Hayes, M., Mullen, A. M., Fenelon, M. and Tiwari, B. 2017. Future protein supply and demand: strategies and factors influencing a sustainable equilibrium. *Foods* 6:1–21.

Henriques, B. S., Garcia, E. S., Azambuja, P. and Genta, F. A. 2020. Determination of chitin content in insects: an alternate method based on calcofluor staining. *Front Physiol* 11:1–10.

Heuel, M., Sandrock, C., Leiber, F., Mathys, A., Gold, M., Zurbrüegg, C., Gangnat, I. D. M., Kreuzer, M. and Terranova, M. 2022. Black soldier fly larvae meal and fat as a replacement for soybeans in organic broiler diets: effects on performance, body N retention, carcase and meat quality. *Br Poult Sci* 63:650–661.

Houdijk, J. G. M., Marchal, L., Bello, A., Gibbs, K. and Dersjant-Li, Y. 2024. Complete replacement of soya products with alternative ingredients for fast growing broilers. *British Poultry* Abstracts. in press.

Houdijk, J. G. M., Smith, L. A., Tarsitano, D., Tolkamp, B. J., Topp, C. E. F., Masey O'Neill, H. V., White, G., Wiseman, J., Kightley, S. and Kyriazakis, I. 2013. Peas and faba beans as home grown alternatives for soya bean meal in grower and finisher pig diets. Pages 145–175 *in Recent Advances in Animal Nutrition*. Garnsworthy, P.C., Wiseman, J., Eds. Nottingham University Press, Nottingham, UK.

Ipema, A. F., Bokkers, E. A. M., Gerrits, W. J. J., Kemp, B. and Bolhuis, J. E. 2021a. Providing live black soldier fly larvae (Hermetia illucens) improves welfare while maintaining performance of piglets post-weaning. *Sci Rep* 11. https://doi.org/10.1038/s41598-021-86765-3

Ipema, A. F., Gerrits, W. J. J., Bokkers, E. A. M., Kemp, B. and Bolhuis, J. E. 2021b. Live black soldier fly larvae (Hermetia illucens) provisioning is a promising environmental enrichment for pigs as indicated by feed- and enrichment-preference tests. *Appl Anim Behav Sci* 244.

Ipema, A. F., Gerrits, W. J. J., Bokkers, E. A. M., van Marwijk, M. A., Laurenssen, B. F. A., Kemp, B. and Bolhuis, J. E. 2022. Assessing the effectiveness of providing live black soldier fly larvae (Hermetia illucens) to ease the weaning transition of piglets. *Front Vet Sci* 9.

Janssen, R. H., Vincken, J.-P., van den Broek, L. A. M., Fogliano, V. and Lakemond, C. M. M. 2017. Nitrogen-to-protein conversion factors for three edible insects: Tenebrio molitor, Alphitobius diaperinus, and Hermetia illucens. *J Agric Food Chem* 65:2275–2278.

Jonas-Levi, A. and Martinez, J. J. I. 2017. The high level of protein content reported in insects for food and feed is overestimated. *J Food Compos Anal* 62:184–188. http://dx.doi.org/10.1016/j.jfca.2017.06.004

Kim, S. W., Less, J. F., Wang, L., Yan, T., Kiron, V., Kaushik, S. J. and Lei, X. G. 2019. Meeting global feed protein demand: challenge, opportunity, and strategy. *Annu Rev Anim Biosci* 7:221–243.

Komi, D. E. A., Sharma, L. and Dela Cruz, C. S. 2018. Chitin and its effects on inflammatory and immune responses. *Clin Rev Allergy Immunol* 54:213–223.

Liu, S., Wang, J., Li, L., Duan, Y., Zhang, X., Wang, T., Zang, J., Piao, X., Ma, Y. and Li, D. 2023. Endogenous chitinase might lead to differences in growth performance and intestinal health of piglets fed different levels of black soldier fly larva meal. *Anim Nutr* 14:411–424. Available at https://linkinghub.elsevier.com/retrieve/pii/S2405654523000665

Lu, A., Zhang, Q., Zhang, J., Yang, B., Wu, K., Xie, W., Luan, Y. X. and Ling, E. 2014. Insect prophenoloxidase: the view beyond immunity. *Front Physiol* 5:1–15.

Mellor, D. J. 2016. Updating animalwelfare thinking: moving beyond the "five freedoms" towards "A lifeworth living." *Animals* 6. https://doi.org/10.3390/ani6030021

Newton, G. L., Booram, C. V., Barker, R. W. and Hale, O. M. 1977. Dried Hermetia illucens larvae meal as a supplement for swine. *J Anim Sci* 44:395–400. Available at https://academic.oup.com/jas/article/44/3/395-400/4698064

OECD/FAO. 2023. *OECD-FAO Agricultural Outlook 2023–2032*. OECD Publishing, Paris, FR.

Oonincx, D. G. A. B. 2021. *Environmental impact of insect rearing.Pages 53–59 in Insects as animal feed: novel ingredients for use in pet, aquaculture and livestock diets*. CABI, UK.

Purschke, B., Scheibelberger, R., Axmann, S., Adler, A. and Jäger, H. 2017. Impact of substrate contamination with mycotoxins, heavy metals and pesticides on the growth performance and composition of black soldier fly larvae (Hermetia illucens) for use in the feed and food value chain. *Food Addit Contam Part A Chem Anal Control Expo Risk Assess* 34:1410–1420. https://doi.org/10.1080/19440049.2017.1299946

Rathore, A. S. and Gupta, R. D. 2015. Chitinases from bacteria to human: properties, applications, and future perspectives. *Enzyme Res* 2015. https://doi.org/10.1155/2015/791907

Šimůnek, J., Hodrová, B., Bartoňová, C. and Kopečný, J. 2001. Chitinolytic bacteria of the mammal digestive tract. *Folia Microbiol (Praha)* 46:76–78.

Smith, L. A., Houdijk, J. G. M., Homer, D. and Kyriazakis, I. 2013. Effects of dietary inclusion of pea and faba bean as a replacement for soybean meal on grower and finisher pig performance and carcass quality. *J Anim Sci* 91:3733–3741.

Spranghers, T., Michiels, J., Vrancx, J., Ovyn, A., Eeckhout, M., De Clercq, P. and De Smet, S. 2018. Gut antimicrobial effects and nutritional value of black soldier fly (Hermetia illucens L.) prepupae for weaned piglets. *Anim Feed Sci Technol* 235:33–42.

Spranghers, T., Ottoboni, M., Klootwijk, C., Ovyn, A., Deboosere, S., De Meulenaer, B., Michiels, J., Eeckhout, M., De Clercq, P. and De Smet, S. 2017. Nutritional composition of black soldier fly (Hermetia illucens) prepupae reared on different organic waste substrates. *J Sci Food Agric* 97:2594–2600.

Star, L., Arsiwalla, T., Molist, F., Leushuis, R., Dalim, M. and Paul, A. 2020. Gradual provision of live black soldier fly (Hermetia illucens) larvae to older laying hens: effect on production performance, egg quality, feather condition and behavior. *Animals* 10. https://doi.org/10.3390/ani10020216

van der Fels-Klerx, H. J., Camenzuli, L., Belluco, S., Meijer, N. and Ricci, A. 2018. Food safety issues related to uses of insects for feeds and foods. *Compr Rev Food Sci Food Saf* 17:1172–1183.

van Heugten, E., Martinez, G., McComb, A. and Koutsos, E. A. 2022. Improvements in performance of nursery pigs provided with supplemental oil derived from black soldier fly (Hermetia illucens) larvae. *Animals* 12. https://doi.org/10.3390/ani12233251

van Huis, A. 2013. Potential of insects as food and feed in assuring food security. *Annu Rev Entomol* 58:563–583.

van Huis, A. and Oonincx, D. G. A. B. 2017. The environmental sustainability of insects as food and feed: a review. *Agron Sustain Dev* 37. https://doi.org/10.1007/s13593-017-0452-8

Veldkamp, T. and Vernooij, A. G. 2021. Use of insect products in pig diets. *J Insects Food Feed* 7:781–793.

White, G. A., Smith, L. A., Houdijk, J. G. M., Homer, D., Kyriazakis, I. and Wiseman, J. 2015. Replacement of soya bean meal with peas and faba beans in growing/finishing pig diets: Effect on performance, carcass composition and nutrient excretion. *Anim Feed Sci Technol* 209:202–210.

Widowski, T. M., Curtis, S. E. and Graves, C. N. 1989. The neutrophil:lymphocyte ratio in pigs fed cortisol. *Can J Anim Sci* 69:501–504.

Yu, M., Li, Z., Chen, W., Rong, T., Wang, G., Li, J. and Ma, X. 2019a. Use of Hermetia illucens larvae as a dietary protein source: effects on growth performance, carcass traits, and meat quality in finishing pigs. *Meat Sci* 158:107837. https://doi.org/10.1016/j.meatsci.2019.05.008

Yu, M., Li, Z., Chen, W., Rong, T., Wang, G., Li, J. and Ma, X. 2019b. Hermetia illucens larvae as a potential dietary protein source altered the microbiota and modulated mucosal immune status in the colon of finishing pigs. *J Anim Sci Biotechnol* 10. https://doi.org/10.1186/s40104-019-0358-1

Yu, M., Li, Z., Chen, W., Rong, T., Wang, G., Li, J. and Ma, X. 2020a. Evaluation of full-fat Hermetia illucens larvae meal as a fishmeal replacement for weanling piglets: Effects on the growth performance, apparent nutrient digestibility, blood parameters and gut morphology. *Anim Feed Sci Technol* 264. https://doi.org/10.1016/j.anifeedsci.2020.114431

Yu, M., Li, Z., Chen, W., Rong, T., Wang, G., Li, J. and Ma, X. 2020b. Hermetia illucens larvae as a fishmeal replacement alters intestinal specific bacterial populations and immune homeostasis in weanling piglets. *J Anim Sci* 98:1–13.

Zhu, M., Liu, M., Yuan, B., Jin, X., Zhang, X., Xie, G., Wang, Z., Lv, Y., Wang, W. and Huang, Y. 2022. Growth performance and meat quality of growing pigs fed with black soldier fly (Hermetia illucens) larvae as alternative protein source. *Processes* 10. https://doi.org/10.3390/pr10081498

Chapter 5

Use of protein from black soldier flies in fish feed

Laura Gasco, University of Torino, Italy; Vlastimil Stejskal and Hung Quang Tran, University of South Bohemia in Ceske Budejovice, Czech Republic; Michele Costantini, Michele Zoli and Jacopo Bacenetti, University of Milano, Italy; and Francesco Gai, National Research Council, Italy

1 Introduction

The global population is increasing rapidly, and with it comes a growing demand for food, particularly protein sources. The aquaculture industry, which provides fish for human consumption, is expanding to meet this demand. The FAO defines aquaculture as the 'farming of aquatic organisms, including fish, molluscs, crustaceans, and aquatic plants'. Among the aquaculture aims, the one performed for the human consumption purposes is of utmost importance in providing food, income, and guarantying food security for several communities around the world (FAO, 2022). Since the year 1980, the sector increased at a pace of over 7.5% per year (FAO, 2022). In the period 2010–2020, the production of aquatic animals went from 59.9 to 87.5 million tons, and following the increasing demand of the growing population, the sector is in continuous expansion. Aquaculture is seen as a powerful means of meeting the world's growing demand for protein (Boyd et al., 2022).

http://dx.doi.org/10.19103/AS.2024.0139.06

The aquaculture expansion is not without impacts and constraints. Among them, the feed production is of major concern (Tacon et al., 2022; Wilfart et al., 2023). Aquafeed is a specialized type of feed designed to provide the essential nutrients required for the growth and development of fish and other aquatic organisms in aquaculture. The composition of aquafeed varies depending on the species being raised, their stage (age) of growth, and the type of aquaculture system being used (intensive vs. extensive) (Glencross et al., 2007; Turchini et al., 2019). Typically, aquafeed is composed of a combination of protein, lipids, carbohydrates, vitamins, and minerals. Protein is the most important nutrient in aquafeed: it provides essential amino acids (EAA), and it is essential for muscle growth and tissue repair (Kangsen et al., 2022). Overall, the composition of aquafeed is carefully designed to meet the specific nutritional requirements of the species considered, while also taking into account the environmental impact of feed production (Naylor et al., 2021a,b).

Carnivorous finfish species commonly raised in aquaculture, such as salmon, trout, sea bream, and sea bass, are preferred compared with omnivorous species, highly requested by consumers, and highly profitable from an economic point of view (Petereit et al., 2022). However, they present several issues associated with their production that need to be addressed. One major issue with carnivorous fish species is their high feed requirements. In particular, these species require large amounts of protein in their diets. In the past, aquafeed production made extensive use of fishmeal (FM), which was produced from ground-up fish. However, to reduce the reliance on FM and the pressure on wild fish stocks, alternative protein sources have increasingly been used (Hua et al., 2019; Glencross et al., 2020). Fish by-products represent an interesting alternative (Gasco et al., 2020a; Malcorps et al., 2021), but at present, their potential is not fully exploited (Šimat et al., 2019).

Nowadays aquaculture feeds are mainly composed of plants and animal by-products.

Plant-based proteins include legumes such as soybean meal (SBM), rapeseed meal, pea protein concentrate, and corn and wheat gluten. The transition toward proteins of vegetable origin, initially considered as the best solution to solve the environmental problems and high costs due to the use of FM, is currently facing several challenges and concerns (Fry et al., 2016; Pahlow et al., 2015). Indeed, many of the inputs and externalities linked with the agriculture (deforestation, water consumption, pesticides and fertilizers, areal footprint, and pollution) are now a matter of concern for aquaculture (Colombo et al., 2023). Moreover, plant-based proteins present a lower amount of crude protein with a less-optimal EAA profile when compared with marine proteins. They may also contain carbohydrates not always well tolerated by carnivorous species. Additionally, these plant proteins can contain antinutritional factors (i.e. saponins, lectins, phytic acid, glucosinolates, phytic acid, and protease inhibitors) that can cause fish health issues (Colombo, 2020; Colombo et al., 2023).

Processed animal proteins (PAPs) derived from by-products of the pork and poultry food production chains are valuable protein sources for aquafeed. Meat and bone meals from ruminant origin (bovine and ovine) are also interesting protein sources, but restrictions on their use apply in some countries (such as in all European countries) due to the bovine spongiform emergency (Woodgate et al., 2022). The origin and the large availability qualify PAPs as sustainable ingredients for aquafeed (Woodgate et al., 2022). Moreover, they are less expensive than FM, have a higher crude protein content, and better balanced EAA profile compared with plant-based proteins.

Emergent PAPs that have been gaining attraction in recent years are represented by insect-derived meals. It has been shown how some insects can efficiently bioconvert low-value and nonotherwise valorized plant biomasses into ingredients of high nutritional value. These nutrient sources are suitable for animal feed purposes (Gasco et al., 2023a,b; van Huis and Gasco, 2023) and are considered a sustainable and environmentally friendly alternative to conventional protein sources (Maiolo et al., 2020). Indeed, being produced using organic waste, they help reduce the amount of waste that ends up in landfills, therefore mitigating their environmental impact and potentially contributing to a reduction of greenhouse gas emissions (Bosch et al., 2019; Smetana et al., 2019; Gasco et al., 2020b). They are also a perfect example of a circular economy able to meet several sustainable development goals (Gasco et al., 2023a).

The use of insects in fish feed is not a new concept, but it has gained increased attention in recent years (van Huis and Gasco, 2023). Since 2017, the European Union approved the use of insect-derived proteins in aquaculture feed, and nowadays, PAPs derived from eight insect species are allowed (EC, 2017, 2021). Therefore, several companies have started producing insect-based feed for fish and other animals (IPIFF, 2019a,b). Among the eight insect-derived PAPs allowed, the black soldier fly (BSF) (*Hermetia illucens*) seems to be one of the most promising (van Huis, 2022). BSF larvae are voracious eaters and can consume a wide range of organic matter. At the end of their growth, and before becoming prepupae or pupae, larvae are processed into meals containing about 45% crude protein and 35–40% lipids on a dry matter basis. A further defatting process leads to the production of a raw material containing up to 60% of crude protein and about 8–14% of residual lipid (Ravi et al., 2019, 2020). Additionally, BSF meals have a balanced amino acid profile, providing nearly all of the EAA required by aquaculture species for growth and maintenance.

PAPs derived from insects have shown to sustain aquaculture species growth (Gougbedji et al., 2022; Liland et al., 2021; Hua, 2021; Tran et al., 2022b). Some insect proteins could maintain the growth performance of fed organisms compared with insect-free diets at plausible inclusion levels, for example, *Gryllus bimaculatus* (threshold, 50.3%), *Oxya fuscovittata* (19.2%), *Zonocerus variegatus* (15.1%), and *Cirina butyrospermi* (32.4%) (Tran et al., 2022b). A meta-regression analysis revealed that dietary BSF of as high as 13.1% could enhance growth performance, as illustrated by the specific growth rate (SGR) of aquatic species relative to control groups (Tran et al., 2022b).

Using insects in aquaculture not only provides valuable protein to aquaculture species but also has positive impacts on health due to the presence of bioactive compounds (antimicrobial peptides, chitin, and lauric acid). These compounds have demonstrated the ability to exert antimicrobial or prebiotic effects or to boost the animal immune system (Gasco et al., 2021; Veldkamp et al., 2022). These bioactive compounds can help face diseases and parasite issues, which normally require the use of antibiotics and other treatments to control them. It is well known how the large use of drugs in animal nutrition has led to the development of antibiotic-resistant bacteria, with negative impacts on human health and the environment (Ahmad and Khan, 2019; Radhouani et al., 2014; WHO, 2018).

A large proportion of aquafeed composition is currently derived from human-edible food (Froehlich et al., 2018a,b). This food–feed competition poses challenges to the existing food system, impacting the environment and resource utilization costs (Sandström et al., 2022). The insect could be a part of the solution while providing a good source of protein and reducing the use of food-competition feedstuff in aquafeed (van Riel et al., 2023).

Several studies have investigated the environmental impact of the production of BSF meal from a life cycle perspective. Particular focus was given to climate change, but other impact categories, such as acidification and eutrophication potential and land and water use, were also considered. Life cycle assessment (LCA) studies generally confirm BSF meal's great potential, but better performance than conventional protein sources cannot be taken for granted, as it is highly dependent on a number of production factors, such as substrate supply and energy consumption for larval growth and processing. For impacts on mitigation to take place, both (i) the use of by-products or waste from other supply chains as substrates and (ii) optimization of energy consumption and process efficiency play a fundamental role. On the other hand, there is limited research on the environmental consequences of using BSF meal as aquafeed.

This chapter reviews the potential use of BSF-derived proteins in aquaculture. In particular, the review focuses on the effect of BSF larva meal on fish performances and health as well as on economic and environmental parameters.

2 Performances and digestibility of BSF and feed contained different BSF

Due to its high protein content, benefits for the circular economy, and local availability (recent and future), BSF meal is a prominent animal-based protein source being investigated in aquafeed formulations to replace FM in recent years (Tran et al., 2022b). It can be the next alternative source for aquafeed production that is both generally environmentally sustainable and nutritionally desirable (Alagappan et al., 2022). The feasibility of replacing FM completely or partially with BSF is verified by widely available literature. However, fish showed higher heterogeneity in the results on available studies, which might be explained by a variety of reasons, including variation in the nutritional quality, palatability, and digestibility of BSF or variation regarding the type of FM used. Therefore, a meta-analysis approach was used to summarize the effectiveness and success of BSF inclusion in aquafeeds as a FM substitute due to the well-documented variability in success of BSF inclusions (Tran et al., 2022b). Moreover, protein from BSF might be tailored toward better utilization by aquatic animal species through culture substrate, which may enhance beneficial compounds, such as omega-3 fatty acids, carotenoids, and selenium (Ferrari et al., 2022; Hoc et al., 2021; Leni et al., 2023). Consequently, the enriched BSF can be utilized to enhance fillet quality and offer some health advantages in terms of enhanced immune function and improved resistance against pathogens.

2.1 Effect of black soldier fly on production performance

Similar to other animal-based proteins, the BSF-based meals can vary in compositional quality. This is mostly due to potential variations in raw material composition, quality, and processing demands, including temperature, time, and pressure. Such variance leads to variations in digestibility, greater ash content, and deficits in some important amino acids. However, improvements in processing technologies in BSF production facilities have overcome these difficulties through computerized process control, which is crucial in defining the quality of the finished product. BSF products have, therefore, been evaluated to replace FM or as complementary feed ingredient in various dietary inclusion levels for a variety of finfish and shellfish species, with differing degrees of effectiveness. Therefore, BSF meal in different forms was tested in the main groups of farmed fish as categorized by Tacon and Metian (2008) or Kok et al. (2020).

In rainbow trout (*Oncorhynchus mykiss*) diets, researchers evaluated BSF inclusion levels up to 45% (Randazzo et al., 2021) in high-FM diet formulations. The first attempt to test BSF in trout diets was made by St-Hilaire et al. (2007) in large trout (final body weight over 200–300 g), with successful inclusion levels up to 15% with no adverse effect on SGR and feed conversion rate (FCR).

Most of trout studies were performed on small juvenile fish with initial body weight from 5 g to 50 g and a 2.5-month trial duration. For small juvenile fish, BSF inclusion levels from 12% to 36% seem to be suitable, as demonstrated in studies with fish of initial body weight less than 20 g (Melenchón et al., 2022; Ratti et al., 2023; Sealey et al., 2011). Several authors tried to evaluate different levels of BSF inclusion for the on-growing part of the trout production cycle ending with marketable-size fish (Dumas et al., 2018; Randazzo et al., 2021; Renna et al., 2017). While Randazzo et al. (2021) observed slight improvement in SGR with an increased inclusion level of up to 45%, and Renna et al. (2017) were able to keep SGR on the same level as the control group up to 40%, Dumas et al. (2018) observed a decrease of SGR from BSF-based diets with inclusion higher than 13%. Mid-term (4 months) exposures of trout to diets containing BSF were performed by Italian authors (Biasato et al., 2022; Caimi et al., 2021) with successful inclusion levels up to 15% and 32%, respectively. Those diets can reduce the demand for FM in practical diets by 50–100%, representing the huge potential for fed aquaculture with less dependence on marine resources. On the other hand, BSF-based effective formulations require supplementation with crystalline amino acids (methionine and lysine) and minerals.

In low-FM diets, which contain plant-derived proteins, the potential for employing BSF-based products has only recently begun to be investigated at modest inclusion levels (3–15%) showing a new potential direction of trout feed formulations (Caimi et al., 2021).

There is still a lack of evaluation of BSF-based trout diets for the whole production cycle from on-growing (10 g) to marketable fish (250–400 g depending on the market) for growth, possible health issues, and economic feasibility. Moreover, different ratios among major alternative protein sources, and particularly the ratio between poultry by-product meal, plant-based protein sources, and BSF products, could be a crucial aspect and needs to be optimized in further trials to allow optimal performance to be attained through diets deprived of FM.

Atlantic salmon (*Salmo salar*) has been recognized as one of the top users of FM and fish oil within the fed aquaculture sector, and huge efforts have been made in the past two decades to utilize FM alternatives (especially plant-based protein sources) in this industry. Nowadays, insect meal, including BSF meal, is most widely tested in salmon aquafeeds due to a short life cycle, higher FCR, and efficiency in bioconversion and recovery of nutrients from a range of organic materials. The history of testing BSF-based diets in salmon is dated since 2016, when Lock et al. (2016) performed their first feeding trial testing inclusion levels up to 25%. They found out that BSF inclusion levels of 10% do not change SGR and even improve FCR in post-smolts. Later, Kousoulaki et al. (2022) confirmed that even 25% BSF inclusion is possible in post-smolts without significantly affecting SGR and FCR. For smolt fish, BSF inclusion levels up to 60% seem to be suitable, as demonstrated in studies with fish of initial body weight less than 50 g (Belghit et al., 2018). BSF inclusion of up to 20% was successfully tested in salmon fry as it maintains SGR and improves FCR (Fisher et al., 2020). In marketable salmon, BSF inclusion level of 9.8% maintains SGR and FCR unchanged compared with the control FM-based diet. With this formulation, the authors have decreased FM need by 67% without growth impairment, which is another big step to decrease resilience of the salmon industry on marine sources (Belghit et al., 2019). Two studies tested 100% replacement of FM with BSF in salmon feed, and based on their results, 100% FM replacement is impossible with BSF products (Belghit et al., 2019; Lock et al., 2016). Similar to trout diets, long-term analyses of BSF-based diets all over the entire production cycle, from on-growing to marketable fish (3000–4000 g depending on the market), are still lacking. These analyses should consider growth, potential health problems, and profitability.

Farmed marine fish belongs to one of the most significant consumers of FM and fish oil in the aquaculture industry. Among them, seabass and seabream are important farmed fish from a total volume production point of view, especially in the Mediterranean. Diets with BSF have been tested in farmed marine fish since 2012, when Kroeckel et al. (2012) performed the first study with turbot (*Psetta maxima*), observing a strong negative effect of BSF inclusion on SGR and FCR. Literature related to the use of BSF in the diet of marine fish is extensive (n = 27), and the authors have tested inclusion levels up to 60% (Vargas-Abúndez et al., 2019) for ornamental clown fish. In marketable-size fish (final body weight of 396 g), where inclusion of BSF has the greatest impact on saving marine resources, it is possible to replace up to 50% of FM with BSF inclusion at a level of 15.7%. The most complex study covered growth intervals from 48 g to 340 g in gilthead seabream (*Sparus aurata*) (Pulido-Rodriguez et al., 2021). It proved that BSF inclusion level of up to 32.4% is possible in FM-free diets with no adverse effect on SGR or FCR. They were able to reduce the FM need for effective formulation by 100% and, at the same time, amount of fish oil needed by 27.9%. Similarly, Di Rosa et al. (2023) succeeded with 50% replacement in marketable gilthead seabream, resulting in a decrease of FM needs by 50% for effective diets. For European seabass, *Dicentrarchus labrax,* as the second most important cultured marine fish, 30–50% FM replacement seems to be feasible, resulting in saving of FM up to 50% (Abdel-Tawwab et al., 2020; Mastoraki et al., 2020; Magalhães et al., 2017). Similar results allowing 48% FM replacement without impairing SGR and FCR were found for Japanese seabass *Lateolabrax japonicus* (Wang et al., 2019). For barramundi (*Lates calcarifer*), it appears feasible to replace up to 30% of FM by the inclusion of 21.9% BSF in classic FM-based diets (Hender et al., 2021). However, 100% FM-free diets are possible by complementary use of BSF and poultry by-products in barramundi (Chaklader et al., 2019). Groupers seem to tolerate inclusion levels up to 30%, resulting in a decrease in FM need by 30% for this premium fish (Mohamad-Zulkifli et al., 2019).

Since 2020, when Stejskal et al. (2020) and Vongvichith et al. (2020) conducted their initial investigation with Eurasian perch (*Perca fluviatilis*) and climbing perch (*Anabas testudineus*), diets containing BSF have been investigated in farmed freshwater carnivorous fish. Moreover, BSF meals were successfully tested in other carnivorous freshwater species like largemouth bass *Micropterus salmonids* (Fischer et al., 2022; Wang et al., 2022), pikeperch *Sander lucioperca* (Tran et al., 2021), and snakehead *Channa gachua* (Hariati et al., 2022). Those studies proved that BSF inclusion could be performed up to 9.5% (17% FM replacement), 18% (50% FM replacement), 37.5% (75% FM replacement), and 40% (42% FM replacement) for largemouth bass, pikeperch, climbing perch, and Eurasian perch, respectively (Mapanao et al., 2023; Stejskal et al., 2020; Tran et al., 2021; Wang et al., 2022). Based on the abovementioned studies, savings for FM in effective diets can be quantified as 25% for largemouth bass, 50% for pikeperch, 75% for climbing perch, and 42% for Eurasian perch. Total replacement of FM with BSF is not suitable for this category of fish.

BSF meal was tested in other freshwater species that are not considered pure carnivorous and have diverse nutritional physiology. Those fish include valuable fish like Siberian sturgeon (*Acipenser baerii*) (Caimi et al., 2020b; Rawski et al., 2020), rice field eel (*Monopterus albus*) (Hu et al., 2020), and Japanese eel (*Anguilla japonica*) (Kuo et al., 2022). For sturgeon, Caimi et al. (2020b) reported a maximal inclusion level of 18.5%, while Rawski et al. (2020) observed even slightly elevated SGR and improved FCR when including 30% of BSF. This formulation saved up to 61% of FM in effective sturgeon feed formula. A trend of improving SGR and FCR was observed in Japanese eel when including up to 18.4% of BSF, while growth performance was maintained similarly to the control diet with a higher inclusion level of 45.8%. A similar trend was observed by Hu et al. (2020). Saving FM at 19% and 75% without compromising growth performance is possible for rice field eel and Japanese eel, respectively (Hu et al., 2020; Kuo et al., 2022).

Muin et al. (2024) made the first attempt to test BSF-based diets in tilapia juveniles (initial body weight >3 g) with significantly improved SGR and FCR when including 15% of BSF (corresponded to 50% FM replacement). They could even fully replace FM in juvenile tilapia diets by including 30%, corresponding to 100% FM replacement, and no negative impact on SGR and FCR when compared with the control FM-based diet. It has been proven that 100% FM replacement is possible in fry (Limbu et al., 2022) and fingerlings (Agbohessou et al., 2021; Tippayadara et al., 2021) without impairment of SGR or FCR. In adult fish, 16.8–19.5% inclusion resulted in an unchanged (Nairuti et al., 2021) or superior growth rate (Wachira et al., 2021). Unfortunately, no study covering the whole production interval is currently available. Available studies also do not present a final common marketable size, and only two studies are available for tilapias with a final body weight over 120 g (Nairuti et al., 2021; Wachira et al., 2021). On the other hand, the study of Wachira et al. (2021) lasted 140 days and covered growth interval from 35 g to 176.5 g.

According to Shepherd et al. (2017), shrimp aquaculture is from a global perspective industry that utilizes a large portion of FM and oil, respectively. In 2017, Cummins et al. (2017) performed the first study with BSF-based diets in Pacific whiteleg shrimp (*Litopenaeus vannamei*), proving that 100% replacement of FM with BSF is closely related to the decrease of SGR and increased FCR. Practical BSF-based feed formulations can potentially replace up to 60% FM and decrease the need for FM by 45–60% (Cummins et al., 2017; Wang et al., 2021). Wang et al. (2021) were also able to decrease fish oil need by 53% while still being effective. Despite the fact that no study is available for larger shrimp (20–40 g) studies, Cummins et al. (2017) and Wang et al. (2021) represent significant contributions to the sustainable development of the aquafeed shrimp industry.

Since 2017, there has been a significant increase in information addressing the nutritional value of BSF meals for use in carp feeds, including the first study performed by Li et al. (2017).

Omnivorous fish, like cyprinids, are less sensitive to dietary reductions in FM and fish oil than carnivorous species. Therefore, with FM/fish oil-free BSF-based diets, growth rate and feed conversion ratio were not adversely impaired. For common carp, *Cyprinus carpio,* 100% FM replacement has been proven to be possible with no adverse effects on SGR (Li et al., 2017; Xu et al., 2020), while Gebremichael et al. (2021) reported a slight improvement in SGR and FCR in carp-fed dietary BSF at an inclusion level of 12% (100% FM replacement). Those results are in line with findings in another cyprinid goldfish, *Carassius auratus*, where 100% replacement and BSF inclusion ranged from 21.0% to 33.5%, resulting in unchanged SGR and FCR (Kamalii et al., 2022), while inclusion levels of up to 21% improved growth performance indices (Kamalii et al., 2022; Khieokhajonkhet et al., 2022). A similar trend was observed in tench *Tinca tinca,* where inclusion levels of 34.8% (45% FM replacement) improved SGR and FCR for tench fry, while lower inclusion levels (10.9%) were favorable for tench fingerlings (Fabrikov et al., 2020; Hidalgo et al., 2022). Unfortunately, no long-term studies (max trial duration 100 days) or studies ending with marketable-size fish are available for cyprinids.

Available documentation of the nutritional value of BSF considered as candidates for use in catfish feeds has become substantial since 2017 when Hu et al. (2017) performed their first trial testing FM replacement up to 30% using BSF in less-cultured yellow catfish (*Pelteobagrus fulvidraco*). Their findings show that BSF may replace 20% of the FM without significantly affecting SGR and FCR. In the same species, Xiao et al. (2018) were able to improve SGR and FCR by gradually increasing BSF levels up to 22.3% for market-size fish. Growth parameters were unaffected even at an inclusion level of 34.3%, and worsening of growth parameters occurred when BSF meal was introduced at a level of 46.2%, representing 80% FM replacement. Similar results were achieved in striped catfish, *Pangasianodon hypophthalmus,* where 60% FM replacement with BSF (inclusion level of 17.4%) resulted in unchanged SGR and FCR (Sudha et al., 2022). In African catfish, *Clarias gariepinus,* effective inclusion levels of BSF ranged from 7.5% for fry (Adeoye et al., 2020) to 20.6% for fingerlings (Maranga et al., 2022). No data for marketable fish are available for both African catfish and striped catfish. The inclusion of BSF as an alternative to FM alternative allows the saving of FM at a level of 60% for yellow catfish and striped catfish and 66% for African catfish in effective diets (Maranga et al., 2022; Sudha et al., 2022; Xiao et al., 2018).

3 Fish health effects

Besides being considered one of the most promising sources of alternative proteins, insects are also sources of compounds able to exert some effects on gut health parameters and gut microbiota (Foysal and Gupta, 2022), fish oxidative status, and animal's immune system, hence improving health status and decreasing susceptibility to diseases (Gasco et al., 2018b).

A summary of the studies, carried out in different fish species of interest for aquaculture, on the effects of FM replacement by BSF meals in fish gut morphology, gut microbiota, oxidative status, immune function, and resistance against common pathogens, is reported in the following subparagraphs.

The mucosal villi and the muscular layer are the main indicators of gut health and functional status (Elia et al., 2018). Generally, a healthy gut is characterized by elongated villi, thick mucosa, and muscular layers (Józefiak et al., 2019). As far as nutritional trials with different inclusion levels of BSF meals are concerned, several studies evaluated these gut health indicators. In particular, in rainbow trout (Renna et al., 2017) and in juvenile Siberian sturgeon (Caimi et al., 2020a), unaffected villus height was observed when compared with the control fish, while in juvenile barramundi (*L. calcarifer*) fed poultry by-product-based diets supplemented with BSF meal, an increased microvilli height was reported by Chaklader et al. (2019). Concerning mucosa and muscular layers, in juvenile Siberian sturgeon fed BSF meals, a higher muscular layer thickness and a lower mucosa thickness have been reported (Józefiak et al., 2019). A shortening of the villi in the anterior gut of rainbow trout fed diets with 26.4% of BSF meal was observed by Dumas et al. (2018), while Cardinaletti et al. (2019) utilizing diets containing 10.5% and 21% of BSF meal in the same fish recorded a significant reduction in the intestinal fold length. On the other hand, these studies did not find any correlation between negative alterations in the gut morphology and growth performance with the latter resulted unaffected in the majority of cases (Cardinaletti et al., 2019; Józefiak et al., 2019), therefore, the authors suggest that insect meal utilization has no remarkable effects on the fish intestinal absorption surface.

Another interesting approach related to the gut health studies is the evaluation of the expression of the pro-inflammatory marker genes associated with gut inflammation indicators such as submucosa cellularity. In this sense, in a study carried out in Atlantic salmon fed a diet containing 14.75% of BSF meal, a higher degree of submucosa cellularity not accompanied by a differential expression of the pro-inflammatory marker genes was reported by Li et al. (2020). In the same species, Weththasinghe et al. (2021) reported that the dietary inclusion of BSF meal and BSF paste at 25% and 6.7%, respectively, significantly improved intestinal health without any inflammation but produced slight effects on protein levels and skin mucus. In another salmonid, rainbow trout, the results of the IL-1β, IL-10, TGF-β, COX-2, and TCR-β gene expression analysis revealed that experimental poultry by-products and BSF prepupae meal-based diets did not induce any inflammation (Gaudioso et al., 2021). Moreover, BSF meal-supplemented diets improved cytokine expression, mucin secretion in the gut and skin of juvenile barramundi, *L. calcarifer* compared with control (FM) diet-supplemented groups (Hender et al., 2021).

When assessing the intestinal mucin composition, unaffected mucin staining intensity was reported in rainbow trout (Elia et al., 2018) and juvenile Siberian sturgeon (Caimi et al., 2020a) fed BSF-based diets in comparison with the control group. Elia et al. (2018) also observed a predominance of neutral mucins over the acidic subtype in both the BSF- and the control-fed fish. Despite the information about the gut mucin composition in rainbow trout being very scarce, the neutral mucins are known to promote the nutrient absorption and transportation (especially carbohydrates and fatty acids) (Caimi et al., 2020a). A greater number of acidic mucins was also identified in the juvenile barramundi fed BSF-supplemented diets, thus protecting fish by binding and preventing the adherence of pathogenic bacteria in the intestinal epithelium (Chaklader et al., 2019).

The effects of BSF on overall health and immunity of aquatic animals could be evaluated through the use of an excellent indicator, the modulation of the intestinal microbiota, defined as a complex, dynamic biological system composed of commensal, symbiotic, and pathogenic microbial communities. As the results of the developments of recent sequencing techniques, the physiological gut microbiota in fish has been characterized and Proteobacteria, Firmicutes, and Bacteroidetes were identified as the dominant bacterial phyla. Foysal and Gupta (2022) summarized the results of a recent systematic meta-analysis concerning different studies in rainbow trout. The authors sum that Proteobacteria were the dominant bacteria in the fish gut fed FM while Actinobacteria and Firmicutes the most abundant in the gut of BSF and fish fed BSF diets. The higher abundance of *Actinomyces*, *Bacillus*, *Lactobacillus*, and *Enterococcus* at the genus level contributed to the augmentation of Actinobacteria and Firmicutes in the fish gut.

Similar findings, in terms of increase in *Clostridium leptum* subgroup and *Clostridium coccoides* – Eubacterium rectale cluster (butyrate-producing bacteria), Bacillus, Carnobacterium, Enterococcus, and Lactobacillus (lactic acid bacteria), were also identified in juvenile Acipenser baeri fed 15% of BSF meal (Józefiak et al., 2019). Moreover, higher Enterobacteriaceae and *Aeromonas* (potential pathogenic bacteria) count were reported by the same authors in BSF-fed fish compared with the control group.

The processing of the BSF meal also impacted the gut microbiota diversity. In particular, defatted BSF was linked to a specific group of bacteria, which might be due to the lack of cholesterol in the defatted BSF attributed to the level of cholesterol in the gut environment. Gut microbiota of the BSF larvae could also be widely affected by the rearing substrate with a direct consequence on the intestinal microbiota of fish (Osimani et al., 2019). In particular, an abundance of Enterobacteriaceae increase and Clostridia decrease has been reported in zebrafish-fed diets containing meal from BSF larvae reared on vegetable substrate when compared with the coffee by-product (Osimani et al., 2019).

Overall, the available data regarding the effects of insect meal utilization (especially the BSF) on fish gut microbiota suggest that a beneficial bacteria (lactic acid and butyrate producers) selection could be modulated by the insect meal to the detriment of potential pathogens (Proteobacteria).

The nutritive value of alternative protein sources for FM could also be evaluated in terms of changes in antioxidant capacity. Insect meal effects on fish oxidative status have been only partially addressed when compared with plant feedstuffs.

The mechanisms of action of insect meals on fish oxidative status are related to an increase of the antioxidant potential or a decrease of the oxidative damage. As far as the increase of the antioxidant potential is concerned, Caimi et al. (2020a) found an increased superoxide dismutase and glutathione reductase activities in the liver and kidney of Siberian sturgeon fed a 37.5% dietary BSF meal inclusion level, while an enhanced serum catalase activity was found in Jian carp and African catfish fed diets including BSF meal at 11.5% and 17%, respectively (Li et al., 2017; Fawole et al., 2020).

Looking at the other insect meal effect, the capability of diminishing oxidative damage, is related to the decreasing of lipid-peroxidation-specific markers, such as malond aldehyde (MDA). Wang et al. (2019) reported this effect in the serum of Japanese sea bass (*L. japonicus*) fed up to 19% BSF meal.

The antioxidant properties of insect meals can be related to their chitin content and/or with dietary composition, especially with the lipid fraction. Thus, *in vitro* and *in vivo* studies showed that chitin and its derivatives, such as chitosan, were able to increase antioxidant enzyme activities and decrease lipid peroxidation levels (Ngo and Kim, 2014). In contrast, dietary chitin content was related with the lower GPx activity recorded in both the liver and kidney of Siberian sturgeon and rainbow trout fed BSF meal diets (Elia et al., 2018; Caimi et al., 2020a). Such an effect was associated with the capability of chitin to bind dietary inorganic (selenite and selenite) and organic (selenomethionine and selenocysteine) selenium (Se) forms (Elia et al., 2018; Caimi et al., 2020a), thus making Se physiologically unavailable. Since the catalytic center of SeGPx contains selenocysteine, its unavailability due to chitin bond may impair SeGPx activity. Besides chitin content, dietary lipid levels and composition can also interfere with antioxidant enzymes activities and peroxidation processes. In Siberian sturgeon fed isolipidic HI-based diets, the higher levels of total fatty acids (FA), saturated FA (SFA) and monounsaturated FAs, and the lower PUFA/SFA ratio recorded in the 50% FM replacement level compared with the control, could in part explain the higher antioxidant enzyme activities of this group (Caimi et al., 2020a). As the susceptibility of fish tissues to suffer lipid peroxidation is highly dependent on the degree of unsaturation of dietary lipids, the lower MDA level recorded in African catfish fed with BSF meal-based diets was probably related with the lower PUFAs of these diets compared with the FM control (Fawole et al., 2020).

In conclusion, the majority of the studies reported an absence of effect or even a positive influence of insect meals on fish oxidative status. As already pointed by Martínez-Alvarez et al. (2005), insect meal effects on fish antioxidant defenses seem to be dependent on feeding behavior and on nutritional factors. Nutritional factors such as dietary chitin and lipid contents and lipid composition can also deeply interfere with insect meals' antioxidant capacity (Caimi et al., 2020a; Fawole et al., 2020).

Another interesting aspect of the use of the insect meals, evaluated in fish nutritional studies, is their effects on animal's immune system and the related possibility to decrease susceptibility to diseases (Gasco et al., 2018b). However, until now, only few studies (Xiao et al., 2018; Chaklader et al., 2019; Li et al., 2019; Stenberg et al., 2019; Fawole et al., 2020) have addressed the effects of FM replacement by insect meals in fish immune function and resistance against common pathogens.

In order to assess fish health status, different hematological parameters are considered as important tools. Among them, white blood cells' (WBC) and red blood cells' (RBC) total counts, differential leukocyte counts, hemoglobin (Hb), hematocrit (Ht), and erythrocyte indices are usually taken into account. The few studies available showed an absence or even a positive effect of dietary insect meals on fish hematological parameters (Ogunji et al., 2008; Taufek et al., 2016; Fawole et al., 2020). Thus, inclusion of 6%, 11.5%, and 17% of BSF meal in diets for African catfish (*C. gariepinus*) did not affect Ht, Hb, WBC, RBC, MCV, MCH, MCHC, and differential leukocyte counts, meaning that BSF meal did not compromise O_2 transportation to vital organs, body capability of defense against invading pathogens, and fish health status (Fawole et al., 2020). Although with no effect on Ht, a higher Hb concentration was recorded in Nile tilapia (*Oreochromis niloticus*) fed with 25–100% dietary maggot meal (*Musca domestica*) (Ogunji et al., 2008). A 35% dietary inclusion of cricket (*G. bimaculatus*) meal had positive effects on African catfish Ht, Hb, MCH, and MCHC, thus indicating a higher health status in this group compared with fish fed the FM-based diet (Taufek et al., 2016).

Survival rate after pathogens' challenge is a common approach to evaluate fish disease resistance. Although scarcely assessed, the majority of the studies pointed that dietary insect meals have the capability of improving fish disease resistance against common pathogens (Ming et al., 2013; Ido et al., 2015, 2019; Su et al., 2017; Song et al., 2018; Chaklader et al., 2019; Li et al., 2019; Stenberg et al., 2019). Thus, barramundi fed 45% poultry by-product diet supplemented with 10% BSF meal showed higher survival rate against *Vibrio harveyi* than fish fed the FM-based diet (Chaklader et al., 2019). However, in Atlantic salmon, a reference diet and a diet with 60% of BSF meal inclusion led to similar plasma antibody titers after vaccination against infectious pancreatic necrosis (IPNV) (Li et al., 2019). Also, in Atlantic salmon, dietary inclusion of 10% and 15% BSF meal, prior to isolation of head kidney cells, had no effect on the transcription of inflammatory-related genes during a viral (polyinosinic acid:polycytidylic acid, poly I:C) or a bacterial (lipopolysaccharide (LPS)) stimulation (Stenberg et al., 2019). A 2.5% inclusion of maggot meal in diets for black carp (*Mylopharyngodon piceus*) and a 5% inclusion in diets for red sea bream (*Pargus major*), were effective in increasing fish resistance against *Aeromonas hydrophila* and *Edwardsiella tarda*, respectively (Ming et al., 2013; Ido et al., 2015). Survival rate of pearl gentian grouper (*Epinephelus lanceolatus male* × *Epinephelus fuscoguttatus female*) against *V. harveyi*, red sea bream against *E. tarda*, and yellow catfish (*P. fulvidraco*) against *Edwardsiella ictaluri*, was improved in fish fed with 7.5%, 10%, and 27% of TM meal, respectively (Su et al., 2017; Song et al., 2018; Ido et al., 2019).

Lysozyme and the complement system are markers of the nonspecific immune response being used as main indicators of innate immunity in fish. Lysozyme is a bactericidal enzyme that lyses the constituents of the peptidoglycan layer of gram-positive bacterial cell walls. Besides bacteriolysis, lysozyme also promotes phagocytosis by directly activating phagocyte cells or indirectly due to its opsoninic nature (Saurabh and Sahoo, 2008). Complement is a system of serum proteins responsible for several immune functions such as elimination of invading pathogens and altered cells, phagocyte recruitment, promotion of inflammatory responses, and modulation of adaptive immune responses (Nakao et al., 2011). In barramundi, a 45% poultry by-product diet supplemented with 10% BSF led to higher serum lysozyme and bactericidal activities, and head kidney relative expression of complement components 3 and 4, in fish challenged with *V. harveyi* compared with the same group challenged before (Chaklader et al., 2019). A 2.5% inclusion of maggot meal in diets for black carp was enough to improve serum lysozyme and complement components 3 and 4 (Ming et al., 2013). Besides the mentioned positive effects on fish non-specific immune response, insect meals can also enhance components of the adaptive immune system, and other immune elements such as pro-inflammatory cytokines. Thus, in yellow catfish, 9% and 18% TM meal increased the liver and spleen transcription levels of the immunoglobulin M (IgM), the predominant antibody type in fish, whereas IgM plasma levels were enhanced at 18% and 27% TM meal inclusion levels (Su et al., 2017). Additionally, an upregulation of the gene expression of the major histocompatibility complex (MHC), which plays a key role in transporting peptide antigens to T cells, was observed in the spleen and kidney of yellow catfish fed with 18% TM meal before and after an *E. ictaluri* challenge, respectively (Su et al., 2017). In contrast, feeding barramundi with a control and a 10% BSF meal-supplemented diets had no effect on head kidney MHC class II gene expression, both before and after challenge groups with *V. harveyi* (Chaklader et al., 2019). The gene expression of the pro-inflammatory cytokine IL-1β, a commonly used immune-regulatory gene in fish, was upregulated in the liver and spleen of fish fed 9% and 18% TM, respectively, before and after bacterial challenge (Su et al., 2017). In Atlantic salmon, a 60% BSF meal dietary inclusion level, increased both proximal and distal intestine transcription levels of cd3γδ and foxp3, meaning that BSF meal seemed capable to stimulate regulatory T-cell activity in these intestinal segments (Li et al., 2019).

The enhanced immune function and improved resistance against pathogens reported in fish fed with dietary insect meals might be related with insect's immune-activating polysaccharides such as chitin, silkrose, or dipterose (Esteban et al., 2000, 2001; Cuesta et al., 2003; Gopalakannan and Arul, 2006; Harikrishnan et al., 2012; Ohta et al., 2014, 2016; Shanthi et al., 2014), and/or with the presence of antimicrobial peptides (Hou et al., 2007; Fu et al., 2009; Elhag et al., 2017). Thus, crustacean chitin provided by intraperitoneal injection, incubated with head kidney leukocytes, or included in the diet, shown to have an immunostimulatory effect on both innate humoral and cellular immune response of gilthead sea bream (Esteban et al., 2000, 2001; Cuesta et al., 2003). Dietary chitin and a chitin derivative, chitosan, enhanced the innate immune system of grouper (*Epinephelus bruneus*), common carp (*C. carpio*), and *Cirrhina mrigala*, thus conferring disease resistance against

Vibrio alginolyticus, *A. hydrophila,* and *Aphanomyces invadans*, respectively (Gopalakannan and Arul, 2006; Harikrishnan et al., 2012; Shanthi Mari et al., 2014). The mechanisms by which chitin modulates the immune system are related with its ability to act as pathogen-associated molecular patterns (PAMPs) by binding to pattern recognition receptors (PRRs), thus inducing cytokine and chemokine production (Lee et al., 2008). Besides chitin, other immune-activating polysaccharides, such as dipterose isolated from melon fly pupae (*Bactrocera cucurbitae*) and silkrose isolated from silkmoth pupae (*Antheraea yamamai*), were able to activate mammalians' innate immune response against various pathogenic microorganisms and viral infections (Ohta et al., 2014, 2016). Insects also possess antimicrobial peptides that can exhibit activity against microbial diseases. Elhag et al. (2017) isolated from BSF seven new gene fragments of three types of antimicrobial peptides, with inhibitory activity against gram-positive and gram-negative bacteria and fungus. Antifungal and antibacterial activity substances were also obtained from the hemolymph and extracts of the housefly (Hou et al., 2007; Fu et al., 2009).

Overall, insect meals seem capable of positively modulating fish immune response and disease resistance, thus being promising feedstuffs to reduce antibiotic use in fish farming. However, the specific role of insect's immune-activating polysaccharides and/or antimicrobial peptides in promoting fish health needs to be further investigated.

In addition, complete replacement of FM with BSF has been investigated on Nile tilapia juveniles for a period of 59 days. At the end of the feeding trial experiment, BSF meal supplementation did not affect the growth and innate immunity of *O. niloticus* (Agbohessou et al., 2021).

Fishmeal was replaced with different levels (0%, 13%, 25%, 37%, 48%, 68%, 85%, and 100%) of BSF larvae for 65 days and the authors showed that the fish group supplemented with 25% of BSF larvae meal significantly enhanced the growth performance and immunocompetence of *P. fulvidraco* (Xiao et al., 2018).

4 Economic fish-in:fish-out ratio for black soldier fly meal in aquafeeds

All methods of producing food, including aquaculture, can be evaluated in terms of how it affects the environment and how many resources are used in the process. According to Tacon (2020), the entire quantity of intensive aquaculture feed consumed in 2017 was 51.2 million tons, and the quantity needed in 2025 is predicted to be around 73.2 million tons. It also suggests that in order to meet the global growth in aquaculture, an additional 22 million tons will need to be produced each year. Furthermore, it is projected that fish harvested from marine sources for the production of FM will reach the ecological limit of population sustainability by 2037 (Cottrell et al., 2020), making further use of this resource unsustainable over the long term. The economic fish-in/fish-out (eFIFO) ratio, which takes into consideration the principle of economic allocation frequently employed in LCAs, represents the quantity of FM and fish oil necessary for producing one weight equivalent of farmed fish back to wild fish weight equivalents (typically a kilogram or ton). eFIFO is now used to quantify marine resources used in the farming of farmed fish. Nine taxonomic groups that contributed to the eFIFO data were categorized. To determine if BSF-based aquafeed can satisfy worldwide demand FM, fish oil shifting effects on BSF-based diets were constructed as three-dimensional surface plots (Fig. 1). The potential for insect meals to transform the aquaculture sector from a net consumer to a net producer of fish is demonstrated by the possibility that substantial or complete replacement of FM with insect meals might lower eFIFO to 1.0 (Tran et al., 2022a). It appears possible in all taxa to reduce the amount of FM and fish oil in aquafeeds while increasing insect meal to fulfill the anticipated eFIFO by 2025. To achieve optimal performance while feeding diets free of FM, it may also be crucial to optimize the various ratios among the main alternative protein sources, particularly the ratio between poultry by-product meal and BSF meal in the diet (Chaklader et al., 2019; Chaklader et al., 2022; Randazzo et al., 2021).

5 Food-feed competition aspects of using black soldier fly in aquaculture species

The aquafeed industry has moved toward less reliance on marine resources for FM and fish oil, the essential protein and lipid sources, respectively, by incorporating terrestrial materials, such as plant and animal by-products. This shift, on the other hand, impairs production performance due to nutritional constraints and exerts significant food–feed competition, meaning that several feed ingredients, such as soybean protein and wheat, would otherwise be destined for humans (Crona et al., 2023; Naylor et al., 2021a; Sandström et al., 2022). Regarding the former issue, there is a common consensus that dietary BSF (*H. illucens*) at low-to-medium levels as a replacement for FM could sustain the growth performance of aquaculture species (Tran et al., 2022a). Recently, the food–feed competition aspects in the aquaculture sector have been thoroughly investigated, revealing that the complementarity of novel protein sources – insect meal, micro- and macroalgae, and yeast – and by-products could reduce edible food-fed animal usage (van Riel et al., 2023).

Our synthesis of 108 published articles, spanning seven important aquaculture species worldwide, demonstrated that dietary BSF could significantly reduce the proportion of food-competition ingredients in aquafeed, meaning that less than 50% of total materials in aquafeed are derived

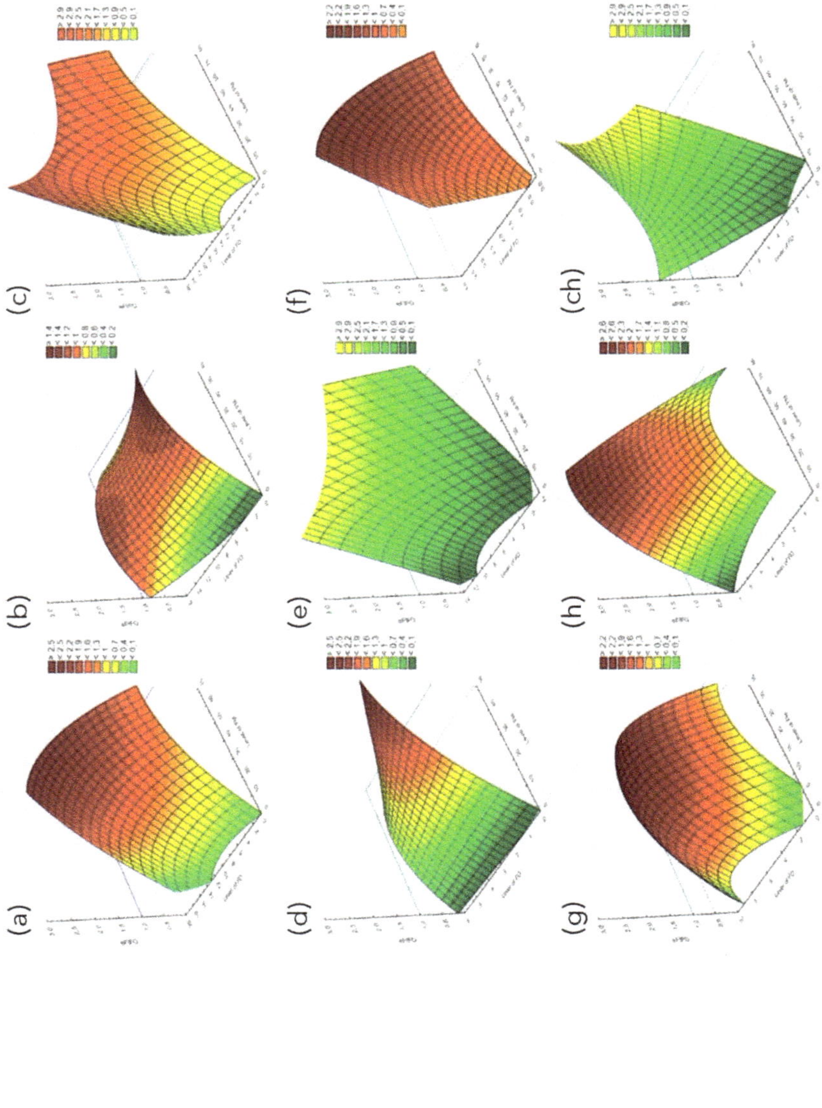

Figure 1 Effect of fish meal and fish oil level on eFIFO in BSF-based diets for (a) trout (n = 19), (b) salmon (n = 7), (b) marine fish (n = 26), (d) tilapia (n = 10), (e) cyprinids (n =10), (f) catfish (n = 7), (g) shrimps (n = 4), (h) freshwater carnivorous (n = 8), and (ch) other freshwater fish (n = 6). Values under the blue line represent net producers. Dashed pink line represents data for common diet of each farmed fish group. Data were retrieved

from nonfood resources, as illustrated by negative linear regression model ($P < 0.05$, $R^2 > 0.5$) (Fig. 2). When examining the use of human-food materials in aquafeed, we discovered that among seven aquaculture species: marine fish, salmon, and trout consumed a significant amount (more than 50% of total aquafeed formula), whereas feed for carp, catfish, and tilapia contained a large proportion of nonfood materials. It has been determined that shrimp feeds have maintained a relatively balanced ratio of edible and inedible materials (Fig. 3).

The animal and aquaculture feed industry continues to recruit a variety of materials that humans would otherwise consume (Cheng et al., 2022; Costello et al., 2020; Naylor et al., 2021b). This would contribute to the food system's constraints, particularly with the growing population and planetary resource boundary in the coming decades. Many secondary products generated from the agriculture sector have been identified as potential for animal feed input (Sandström et al., 2022), which would divert food to humans in the future (Cheng et al., 2022; Crona et al., 2023). In practice, approximately 70% of feed components for carp are nonfood materials, such as rapeseed meal, SBM, wheat bran, and distiller grains, while wheat and maize, the edible food, continue being incorporated in this omnivorous feed at nearly 30%. On the contrary, salmon production largely relies (approximately 70%) on food-competition feedstuff, such as FM, fish oil, and oilseeds. A similar phenomenon is observed in trout aquaculture (Sandström et al., 2022). Our compiled data demonstrated that BSF could be part of an inedible-food-for-feed solution, particularly the possibility of reducing the use of FM or forage fish from marine resources destined for desired consumers. The previous study demonstrated that dietary insect meals, including BSF, could significantly reduce economic fish-in:fish-out ratio and, consequently, FM in diets for aquaculture species while maintaining the growth performance of fed organisms (Tran et al., 2021). Aside from providing protein for aquafeed, BSF oil could also be a potential lipid source for some aquaculture species (Dumas et al., 2018; Li et al., 2016; Xu et al., 2021b), thereby sparing the use of oilseed products.

BSF has been used in conjunction with other plants and animal by-products to replace human-food FM in the diets of a number of aquaculture species and gained recognition (Chaklader et al., 2019; Estévez et al., 2022; Randazzo et al., 2021). Furthermore, the use of secondary products from insect meal farming, such as frass (Banavar et al., 2022; Yildirim-Aksoy et al., 2020a,b, 2022) and pulp (Peng et al., 2021, 2022; Xu et al., 2020, 2021a) could also provide a solution to valorize materials to produce high-quality protein from seafood while less dependence of food materials.

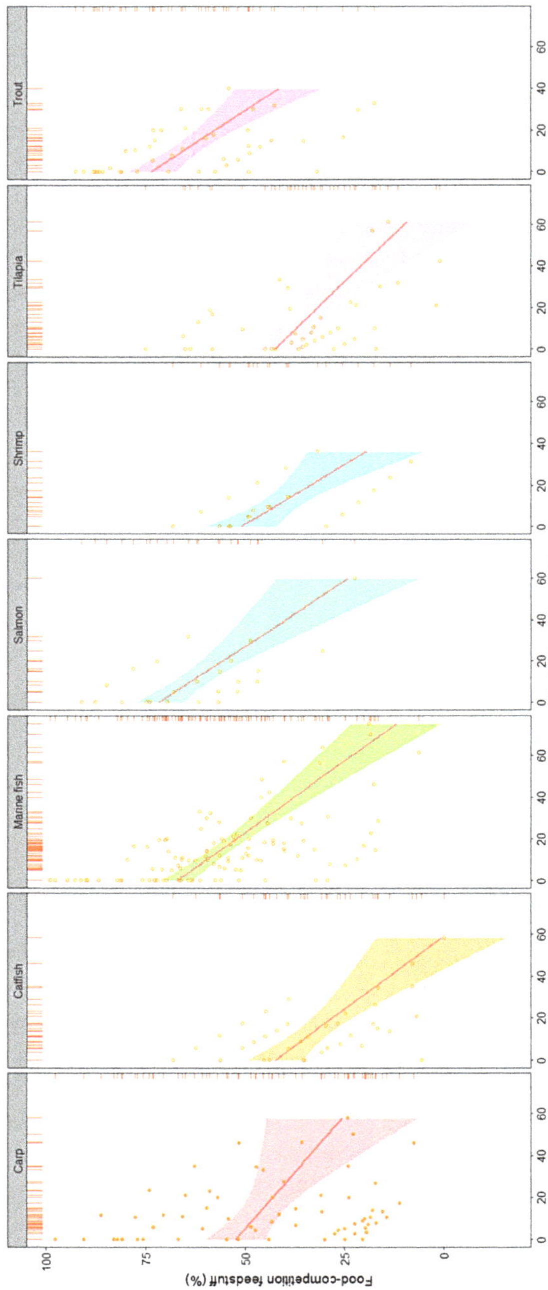

Figure 2 Food–feed competition aspects of using black soldier fly (*H. illucens*) as alternative protein sources in diets for important aquaculture species. Solid lines and shaded areas represent fitted models with the mean and 95% confidence interval, respectively. Dots are observed data points. Rugs represent the distribution of data in the corresponding axis.

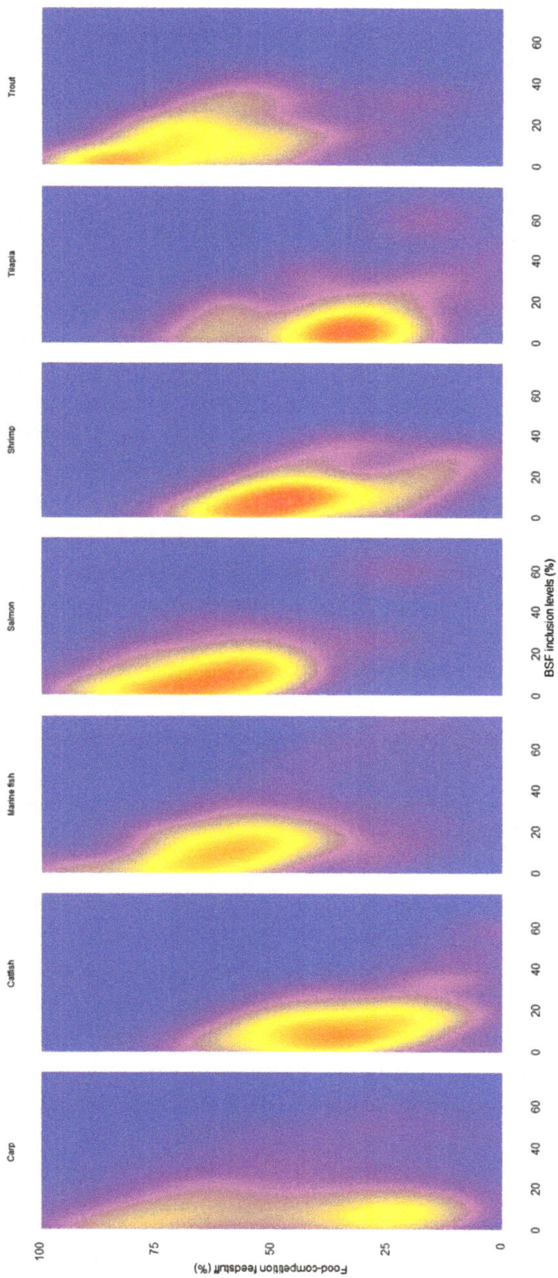

Figure 3 Density map elicits the use of food-competition feedstuff in aquafeed for different aquatic animal species.

6 Waste output

Based on a compiled dataset of 28 published articles investigating BSF as a replacement for FM in aquatic animal diets, we stimulated solid waste output, including total solid waste (TSW), solid nitrogenous waste (SNW), and solid phosphorous waste (SPW), according to previous description (Tran et al., 2021, 2022c):

$$\text{TSW} = (\text{feed consumed} \times (100 - \text{ADC DM (\%)}) = \text{unconsumed feed}$$

SNW or SPW = [N or P consumed × (1 − ADC of N or P)] + (N or P of waste feed), where DM is dry matter, N, P are the content of nitrogen and phosphorus in diets, respectively, and ADC is the apparent digestibility coefficient. The calculated solid waste output values were converted to response ratio, which represents the ratio of measured indices in experimental and control groups.

The correlation analysis indicated that dietary BSF in aquaculture species significantly correlated with solid waste and nitrogen waste as illustrated by the generalized additive model – GAM ($P = 0.003$, adjusted $R^2 = 0.219$, deviance explained = 27.6%) and ($P < 0.001$, adjusted $R^2 = 0.46$, deviance explained = 47.5%), respectively. At the same time, there was no statistical correlation between phosphorous waste generated by organisms fed dietary BSF ($P = 0.51$, $R^2 = 0.03$) (Fig. 4). The data visualization reveals that inclusion levels as high as 20% dietary BSF could result in a slight increase in nitrogen discharge compared to BSF-free diets. Beyond this threshold, nutrient load accelerates. Phosphorous solid waste appeared consistent while feeding organisms with dietary BSF, as indicated by regression equation with relatively low intercept (0.96) and slope (0.00099).

Few studies have been conducted on the fecal output from aquaculture species fed dietary BSF. Weththasinghe et al. (2021) reported that incorporating BSF at the expense of FM in Atlantic salmon (*S. salar*) diets significantly decreased phosphorus excrement while positively correlating with nitrogen load. Supporting this finding, Shaw et al. (2022) demonstrated that dietary BSF fed to Nile tilapia (*O. niloticus*) significantly increased total inorganic nitrogen as well as decreased soluble reactive phosphorus. A similar finding indicated that dietary insect meal (*G. bimaculatus*) resulted in decreased phosphorus excretion, but unlike prior study, the dietary treatment significantly reduced fecal nitrogen excrements in striped snakehead (*Channa striata*) (Prachom et al., 2023).

The GAM for TSW indicates that fecal waste outputs would attain a maximum as high as 35% BSF inclusion before significantly decreasing. This efficacy requires further investigation due to the limited observed data for BSF inclusion levels greater than 35%. Supporting this, Tran et al. (2022c)

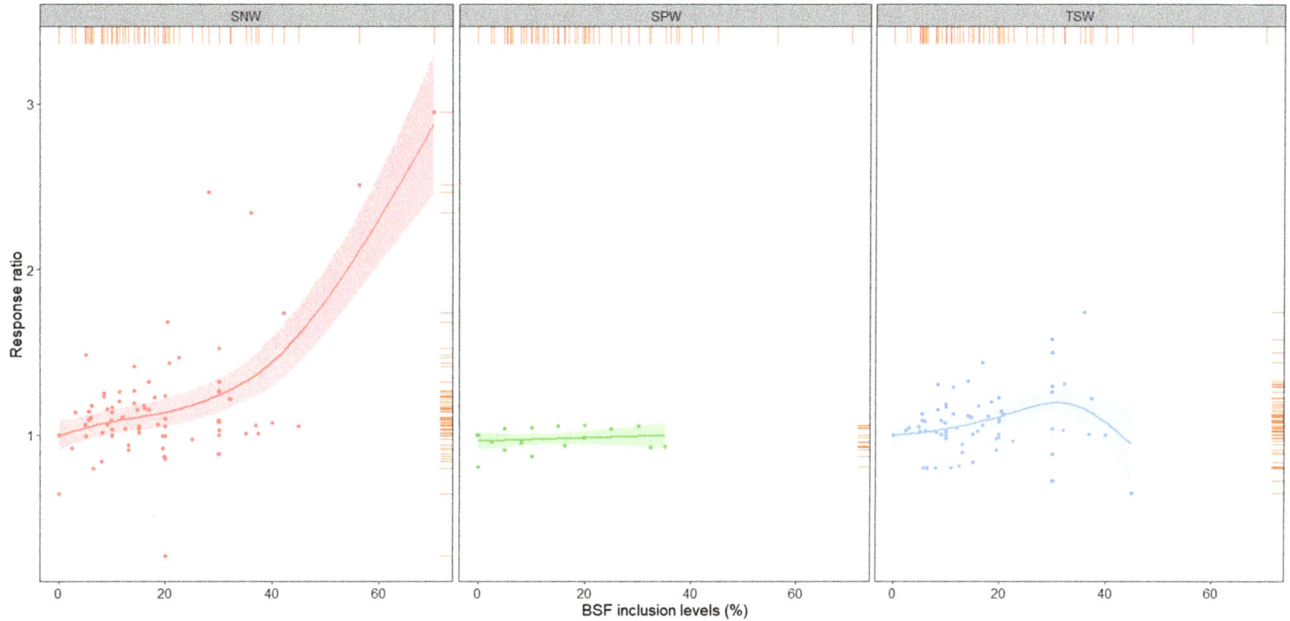

Figure 4 Solid waste output generated from aquatic animals fed dietary BSF (*H. illucens*). Solid lines and shaded areas represent fitted models with the mean and 95% confidence interval, respectively. Dots are observed data points. Rugs represent the distribution of data in the corresponding axis. Data were retrieved from Tran et al. (2021) and recent studies (Biasato et al., 2022; Bordignon et al., 2022; Hariati et al., 2022; Kishawy et al., 2022; Moutinho et al., 2022; Stejskal et al., 2023; Takakuwa et al., 2022).

documented that substantial replacement FM by insect meal (*Tenebrio molitor*) reduced the total waste output of European perch (*P. fluviatilis*).

Incorporating BSF into aquafeed formulation could impose an environmental burden associated with nitrogenous excretion and total fecal waste but can be mitigated by a negligible or minor decrease in phosphorus. The former issue could be effectively managed by integrating insect-fed fish in an aquaponic system, which has been recently initiated (Bordignon et al., 2022; Stathopoulou et al., 2022; Zarantoniello et al., 2021), or by supplement of exogenous enzymes, protease (Bolton et al., 2021; Gasco et al., 2016), and chitin-degraded probiotics (Rangel et al., 2022; Zhang et al., 2014).

7 Environmental effects

The growing interest in alternative protein sources is in part due to the growing environmental concerns of consumers and society at large. Supply chains in the agri-food sector are increasingly called upon first to trace the environmental impact of their production processes, and secondly to mitigate it as much as possible.

Generally, the high feed conversion efficiency of some insects, including BSF, makes them considerable as a potentially sustainable protein source. In fact, FCR is one of the main drivers of environmental sustainability of animal productions, as it directly links production to the resources needed to obtain it.

However, it is important to evaluate the inputs (material and energy consumed) and outputs (products, by-products, and emissions) at each step of the supply chain, in LCA approach. The dried BSF meal production process can essentially be divided into inputs such as material supply, larval growth, harvesting and drying (Salomone et al., 2015). Substrate supply impacts depend mainly on their source. The greatest efficiency in terms of larvae yield and protein content is usually obtained by using agricultural products as growing substrates (e.g. rye or barley flour, soybean meal, alfalfa meal, etc.), but when fed with these, insect rearing is still associated with significant impacts 'embedded' by substrate production. At the same time, several studies have verified excellent larval growth attitudes also on various types of low-value agri-food by-products and other organic waste (Bava et al., 2019; Siddiqui et al., 2022). It follows that the latter constitutes, in terms of sustainability and circularity, the most promising substrates (Smetana et al., 2016), whose production can be considered environmental burden-free. Transportation of substrates is also crucial. The best practice would be to locate production sites in areas close to sources of excess organic matter, whether the by-product of the food industry (e.g. milling plants) or organic waste, whether of urban, agricultural, or industrial origin. At the same time, it must be considered that using substrates with high-humidity content, such as food waste, can make transport more impactful. Furthermore, some substrates may require pretreatment (e.g. sorting and pulping) before entering the process, thus increasing overall energy requirements. Afterwards, the larval growth requires heat and humidity and maintaining optimal growing conditions can also be an environmental sticking point, again due to energy requirements. Proximity should therefore be sought for waste heat sources from other supply chains, e.g. biomass energy conversion plants, finding agreements for reuse. Bioconversion, depending on the conditions, can also lead to nonnegligible methane emissions from the substrates, which should also be accounted and reported to its influence on climate change. Collection and separation of larvae from the digested substrate usually take place manually, therefore without direct impact on the environment. Washing and drying, on the other hand, involve water and energy consumption.

In a case study from Salomone et al. (2017), BSF meal resulted in a higher GWP and energy use compared with soybean meal (mainly due to the transport of substrates and the drying process). At the same time, these results can be highly variable depending on the energy source, as evidenced by a sensitivity analysis. Conversely, significant benefits related to land use were observed. Boakye-Yiadom et al. (2022) reviewed GHG emissions from BSF larvae bioconversion activities based on different gas sensing techniques while highlighting their influencing factors and analyzed LCA studies on BSF larvae products. In 12 LCA studies focused on BSF larvae production, the impact for climate change ranged from 1.36 kg to 15.1 kg CO_2eq/kg of protein, reflecting again the possible variability based on the different production factors described above.

Finally, it should be reported that the environmental concerns related to the insect production stage are also due to the current low technology readiness level (TRL) of the process (Smetana et al., 2016). It is well known that emerging technologies (low TRL) involve greater impacts per product unit than processes that are optimized and adaptable to large-scale production (Arvidsson et al., 2018), and this only promises ample possibilities for improvement in the energy and environmental efficiency of the production of insect meals.

Moving to the impact of aquaculture instead, feed production is by far the main hotspot to energy demand, climate change, land competition, and use of biotic resources, especially in intensive systems producing high trophic-level species such as salmon, trout, or sea bass (Aubin, 2013). In more detail, many studies on different fish species report that feed contributes from 70% of the impact of climate change (e.g. for sea bass and sea bream, Abdou et al., 2017) to over 90% in trout (d'Orbcastel et al., 2009) and salmon farms (Pelletier et al., 2009). This implies that impacts are closely correlated with the FCR, as reported by Bohnes et al. (2019). FCR is influenced by multiple factors, such as the technology used, fish species, and site mortality (Pelletier et al., 2009).

In general, management possibilities exist to decrease FCR. However, these findings undoubtedly also draw attention to the composition of the feed. The impact of aquafeed comes mainly from the production of raw materials, and in particular FM and fish oil (FMFO) ingredients, due to the fuel-intensive use of fishing fleets and the low efficiency of processing plants (Bohnes et al., 2019). Also, soybean meal has a high impact on climate change since it is mostly imported from South America and associated with deforestation. Insect meal, e.g. from BSF, has been proposed as a valid conventional protein substitute to improve aquaculture environmental performance (Goglio et al., 2022). Insects are a valuable protein source for the growing need for aquafeed, able, at the same time, to develop an efficient organic by-product management, and reintroducing in the food chain organic matter whose traditional management may involve environmental issues (Papargyropoulou et al., 2014). Indeed, according to Mertenat et al. (2019), biotreatment of organic waste from BSF has the potential to reduce GHG emissions compared with composting by up to 55%.

Smarason et al. (2017) quantified the environmental impacts of 1 kg of three different feed types: (i) standard feed with high inclusion levels of marine ingredients; (ii) experimental feed with high inclusion levels of agricultural ingredients; and (iii) a hypothetical BSF larvae-based feed. The results showed that the latter has the best environmental performance even if, from an energetic point of view, it is the most energy intensive due to heating and drying. Modahl and Brekke (2022) calculated the environmental impact of insect meal with six different growth substrates. The results were also shown with respect to GHG for the most important salmon protein feed ingredients in Norway. They conclude that insect meals based on vegetable substrates with low economic value have the potential to compete with the protein' sources commonly used so far and could even cover 10–15% of today's imported proteins in the European Union.

Although there are several LCA studies aimed at assessing the environmental impact of alternative protein feeds, only a few refer to the whole fish production life cycle with alternative feeds. Goglio et al. (2022) compared the environmental performance of salmon production fed with alternative and conventional feeds. In the alternative diet tested, all soybean products, and most FMFO (reduced to 5.9% and 4.5% inclusion, respectively) were replaced by BSF meal (17.7% inclusion) and spray-dried heterotrophic microalgae (*Schizochytrium limacinum*, 6.75%). However, as fish production performance was negatively affected by the alternative diet (conventional diet FCR: 0.89, alternative diet FCR: 0.96), the impact of climate change for kg of salmon was 35.8% higher in the alternative diet scenario. Notably, in this impact category, the production of the substrate, consisting mainly of rye middling, had a relevant contribution (46.4% of the total). Le Feon et al. (2019) also reported a similar trend by comparing four trout production scenarios, two of which replacing

15% and 30% of fish meal with mealworm meal. The impact of kg of trout produced was the highest in the scenario with 30% inclusion in five of the seven categories analyzed, due to the high impact of meal production. However, in the biotic resource use and water consumption categories, mealworm meal reduced the impact.

8 Conclusion

In conclusion, the main impact drivers related to the insect meal production phase are substrate production and supply and energy consumption. Depending on the substrate, production technology, and observation of best practices, the impact of insect meal may be less than that of conventional protein sources. Regarding substrates, the evidence discussed above leads to a push for using matrices with low economic value, i.e. by-products and waste from other supply chains, even if the regulations on the subject can often be stringent. Currently, research is aimed at demonstrating the qualitative and sanitary profile of insects reared on such substrates, as well as at validating their environmental profile, in order to push for updates to current regulations. Finally, from a life cycle perspective, further studies analyzing the entire life cycle of fish are needed, as also production and growth performances are key aspects in guiding development toward more sustainable production.

9 References

Abdel-Tawwab, M., Khalil, R. H., Metwally, A. A., Shakweer, M. S., Khallaf, M. A. and Abdel-Latif, H. M. R. 2020. Effects of black soldier fly (*Hermetia illucens* L.) larvae meal on growth performance, organs-somatic indices, body composition, and hemato-biochemical variables of European sea bass, *Dicentrarchus labrax*. *Aquaculture* 522, 735136. https://doi.org/10.1016/j.aquaculture.2020.735136.

Abdou, K., Aubin, J., Romdhane, M. S., Le Loc'h, F. and Lasram, F. B. R. 2017. Environmental assessment of seabass (*Dicentrarchus labrax*) and seabream (*Sparus aurata*) farming from a life cycle perspective: a case study of a Tunisian aquaculture farm. *Aquaculture* 471, 204–212.

Adeoye, A. A., Akegbejo-Samsons, Y., Fawole, F. J. and Davies, S. J. 2020. Preliminary assessment of black soldier fly (*Hermetia illucens*) larval meal in the diet of African catfish (*Clarias gariepinus*): impact on growth, body index, and hematological parameters. *Journal of the World Aquaculture Society* 51(4), 1024–1033. https://doi.org/10.1111/jwas.12691.

Agbohessou, P. S., Mandiki, S. N. M., Gougbédji, A., Megido, R. C., Hossain, Md. S., De Jaeger, P., Larondelle, Y., Francis, F., Lalèyè, P. A. and Kestemont, P. 2021. Total replacement of fish meal by enriched-fatty acid *Hermetia illucens* meal did not substantially affect growth parameters or innate immune status and improved whole body biochemical quality of *Nile tilapia* juveniles. *Aquaculture Nutrition* 27(3), 880–896. https://doi.org/10.1111/anu.13232.

Ahmad, M. and Khan, A. U. 2019. Global economic impact of antibiotic resistance: a review. *Journal of Global Antimicrobial Resistance* 19, 313–316. https://doi.org/10.1016/j.jgar.2019.05.024.

Alagappan, S., Rowland, D., Barwell, R., Mantilla, S. M. O., Mikkelsen, D., James, P., Yarger, O. and Hoffman, L. C. 2022. Legislative landscape of black soldier fly (*Hermetia illucens*) as feed. *Journal of Insects as Food and Feed* 8(4), 343–355. https://doi.org/10.3920/JIFF2021.0111.

Arvidsson, R., Tillman, A., Sandén, B. A., Janssen, M., Nordelöf, A., Kushnir, D. and Molander, S. 2018. Environmental assessment of emerging technologies: recommendations for prospective LCA. *Journal of Industrial Ecology* 22(6), 1286–1294.

Aubin, J. 2013. Life cycle assessment as applied to environmental choices regarding farmed or wild-caught fish. *CABI Reviews*. CABI International. https://doi.org/10.1079/PAVSNNR20138011.

Banavar, A., Amirkolaei, S. K., Duscher, L., Khairunisa, B. H., Mukhopadhyay, B., Schwarz, M., Urick, S. and Ovissipour, R. 2022. Nutritional evaluation of black soldier fly frass as an ingredient in florida pompano (*Trachinotus carolinus* L.) Diets. *Animals* 12(18), 2407.

Bava, L., Jucker, C., Gislon, G., Lupi, D., Savoldelli, S., Zucali, M. and Colombini, S. 2019. Rearing of *Hermetia illucens* on different organic by-products: influence on growth, waste reduction, and environmental impact. *Animals* 9(6), 289.

Belghit, I., Liland, N. S., Waagbø, R., Biancarosa, I., Pelusio, N., Li, Y., Krogdahl, Å. and Lock, E.-J. 2018. Potential of insect-based diets for Atlantic salmon (*Salmo salar*). *Aquaculture* 491, 72–81. https://doi.org/10.1016/j.aquaculture.2018.03.016.

Belghit, I., Liland, N. S., Gjesdal, P., Biancarosa, I., Menchetti, E., Li, Y., Waagbø, R., Krogdahl, Å. and Lock, E.-J. 2019. Black soldier fly larvae meal can replace fish meal in diets of sea-water phase Atlantic salmon (*Salmo salar*). *Aquaculture* 503, 609–619. https://doi.org/10.1016/j.aquaculture.2018.12.032.

Biasato, I., Chemello, G., Odcon, S. B., Ferrocino, I., Corvaglia, M. R., Caimi, C., Resconi, A., Paul, A., van Spankeren, M., Capucchio, M. T., Colombino, E., Cocolin, L., Gai, F., Schiavone, A. and Gasco, L. 2022. *Hermetia illucens* meal inclusion in low-fishmeal diets for rainbow trout (*Oncorhynchus mykiss*): effects on the growth performance, nutrient digestibility coefficients, selected gut health traits, and health status indices. *Animal Feed Science and Technology* 290, 115341. https://doi.org/10.1016/j.anifeedsci.2022.115341.

Boakye-Yiadom, K. A., Ilari, A. and Duca, D. 2022. Greenhouse gas emissions and life cycle assessment on the black soldier fly (*Hermetia illucens* L.). *Sustainability* 14(16), 10456.

Bohnes, F. A., Hauschild, M. Z., Schlundt, J. and Laurent, A. 2019. Life cycle assessments of aquaculture systems: a critical review of reported findings with recommendations for policy and system development. *Reviews in Aquaculture* 11(4), 1061–1079.

Bolton, C. M., Muller, N., Hyland, J., Johnson, M. P., Valente, C. S., Davies, S. J. and Wan, A. H. L. 2021. Black soldier fly larval meal with exogenous protease in diets for rainbow trout (*Oncorhynchus mykiss*) production meeting consumer quality. *Journal of Agriculture and Food Research* 6, 100232.

Bordignon, F., Gasco, L., Birolo, M., Trocino, A., Caimi, C., Ballarin, C., Bortoletti, M., Nicoletto, C., Maucieri, C. and Xiccato, G. 2022. Performance and fillet traits of rainbow trout (*Oncorhynchus mykiss*) fed different levels of *Hermetia illucens* meal in a low-tech aquaponic system. *Aquaculture* 546, 737279.

Bosch, G., van Zanten, H. H. E., Zamprogna, A., Veenenbos, M., Meijer, N. P., van der Fels-Klerx, H. J. and van Loon, J. J. A. 2019. Conversion of organic resources by black soldier fly larvae: legislation, efficiency and environmental impact. *Journal of Cleaner Production* 222, 355–363.

Boyd, C. E., McNevin, A. A. and Davis, R. P. 2022. The contribution of fisheries and aquaculture to the global protein supply. *Food Security* 14(3), 805–827. https://doi.org/10.1007/s12571-021-01246-9.

Caimi, C., Gasco, L., Biasato, I., Malfatto, V., Varello, K., Prearo, M., Pastorino, P., Bona, M. C., Francese, D. R., Schiavone, A., Elia, A. C., Dörr, A. J. M. and Gai, F. 2020a. Could dietary black soldier fly meal inclusion affect the liver and intestinal histological traits and the oxidative stress biomarkers of Siberian sturgeon (*Acipenser baerii*) juveniles? *Animals* 10(1), 155.

Caimi, C., Renna, M., Lussiana, C., Bonaldo, A., Gariglio, M., Meneguz, M., Dabbou, S., Schiavone, A., Gai, F., Elia, A. C., Prearo, M. and Gasco, L. 2020b. First insights on black soldier fly (*Hermetia illucens* L.) larvae meal dietary administration in Siberian sturgeon (*Acipenser baerii* Brandt) juveniles. *Aquaculture* 515, 734539. https://doi.org/10.1016/j.aquaculture.2019.734539.

Caimi, C., Biasato, I., Chemello, G., Oddon, S. B., Lussiana, C., Malfatto, V. M., Capucchio, M. T., Colombino, E., Schiavone, A., Gai, F., Trocino, A., Brugiapaglia, A., Renna, M. and Gasco, L. 2021. Dietary inclusion of a partially defatted black soldier fly (*Hermetia illucens*) larva meal in low fishmeal-based diets for rainbow trout (*Oncorhynchus mykiss*). *Journal of Animal Science and Biotechnology* 12(1), 50. https://doi.org/10.1186/s40104-021-00575-1.

Cardinaletti, G., Randazzo, B., Messina, M., Zarantoniello, M., Giorgini, E., Zimbelli, A., Bruni, L., Parisi, G., Olivotto, I. and Tulli, F. 2019. Effects of graded dietary inclusion level of full-fat *Hermetia illucens* prepupae meal in practical diets for rainbow trout (*Oncorhynchus mykiss*). *Animals* 9(5), 1–15.

Chaklader, M. R., Siddik, M. A. B., Fotedar, R. and Howieson, J. 2019. Insect larvae, *Hermetia illucens* in poultry by-product meal for barramundi, *Lates calcarifer* modulates histomorphology, immunity and resistance to *Vibrio harveyi*. *Scientific Reports* 9(1), 16703. https://doi.org/10.1038/s41598-019-53018-3.

Chaklader, M. R., Chung, W. H., Howieson, J. and Fotedar, R. 2022. A combination of *Hermetia illucens* reared on fish waste and poultry by-product meal improves sensory and physicochemical quality of farmed *Barramundi filets*. *Frontiers in Nutrition* 8, 1190.

Cheng, L., Zhang, X., Reis, S., Ren, C., Xu, J. and Gu, B. 2022. A 12% switch from monogastric to ruminant livestock production can reduce emissions and boost crop production for 525 million people. *Nature Food* 3(12), 1040–1051.

Colombo, S. M., Roy, K., Mraz, J., Wan, A. H. L., Davies, S. J., Tibbetts, S. M., Øverland, M., Francis, D. S., Rocker, M. M., Gasco, L., Spencer, E., Metian, M., Trushenski, J. T. and Turchini, G. M. 2023. Towards achieving circularity and sustainability in feeds for farmed blue foods. *Reviews in Aquaculture* 15(3), 1115–1141. https://doi.org/10.1111/raq.12766.

Colombo, S. M. 2020. Chapter 2Physiological considerations in shifting carnivorous fishes to plant-based diets. In: Benfey, T. J., Farrell, A. P. and Brauner, C. J. (Eds). *Fish Physiology* (vol. 38). Academic Press, 53–82.

Costello, C., Cao, L., Gelcich, S., Cisneros-Mata, M. Á., Free, C. M., Froehlich, H. E., Golden, C. D., Ishimura, G., Maier, J., Macadam-Somer, I., Mangin, T., Melnychuk, M. C., Miyahara, M., de Moor, C. L., Naylor, R., Nøstbakken, L., Ojea, E., O'Reilly, E., Parma, A. M., Plantinga, A. J., Thilsted, S. H. and Lubchenco, J. 2020. The future of food from the sea. *Nature* 588, 95–100.

Crona, B. I., Wassénius, E., Jonell, M., Koehn, J. Z., Short, R., Tigchelaar, M., Daw, T. M., Golden, C. D., Gephart, J. A., Allison, E. H., Bush, S. R., Cao, L., Cheung, W. W. L., DeClerck, F., Fanzo, J., Gelcich, S., Kishore, A., Halpern, B. S., Hicks, C. C., Leape, J. P., Little, D. C., Micheli, F., Naylor, R. L., Phillips, M., Selig, E. R., Springmann, M., Sumaila, U. R., Troell, M., Thilsted, S. H. and Wabnitz, C. C. C. 2023. Four ways blue foods can help achieve food system ambitions across nations. *Nature* 616(7955), 104–112.

Cummins, V. C., Rawles, S. D., Thompson, K. R., Velasquez, A., Kobayashi, Y., Hager, J. and Webster, C. D. 2017. Evaluation of black soldier fly (*Hermetia illucens*) larvae meal as partial or total replacement of marine fish meal in practical diets for Pacific white shrimp (*Litopenaeus vannamei*). *Aquaculture* 473, 337–344. https://doi.org/10.1016/j.aquaculture.2017.02.022.

Di Rosa, A. R., Caccamo, L., Pansera, L., Oteri, M., Chiofalo, B. and Maricchiolo, G. 2023. Influence of *Hermetia illucens* larvae meal dietary inclusion on growth performance, gut histological traits and stress parameters in *Sparus aurata*. *Animals* 13(3), 339. https://doi.org/10.3390/ani13030339.

d'Orbcastel, E. R., Blancheton, J. P. and Aubin, J. 2009. Towards environmentally sustainable aquaculture: comparison between two trout farming systems using Life Cycle Assessment. *Aquacultural Engineering* 40(3), 113–119.

Dumas, A., Raggi, T., Barkhouse, J., Lewis, E. and Weltzien, E. 2018. The oil fraction and partially defatted meal of black soldier fly larvae (*Hermetia illucens*) affect differently growth performance, feed efficiency, nutrient deposition, blood glucose and lipid digestibility of rainbow trout (*Oncorhynchus mykiss*). *Aquaculture* 492, 24–34. https://doi.org/10.1016/j.aquaculture.2018.03.038.

EC 2017. Commission Regulation (EU) 2017/893 of 24 May 2017 amending Annexes I and IV to Regulation (EC) No. 999/2001 of the European Parliament and of the Council and Annexes X, XIV and XV to Commission Regulation (EU) No 142/2011 as regards the provisions on processed animal protein. *Official Journal* L138, 92–116.

EC 2021. Commission Regulation (EU) 2021/1925 of 5 November 2021 certain Annexes to Regulation (EU) No. 142/2011 as regards the requirements for placing on the market of certain insect products and the adaptation of a containment method. *Official Journal* L393, 4–8.

Elia, A. C., Capucchio, M. T., Caldaroni, B., Magara, G., Dörr, A. J. M., Biasato, I., Biasibetti, E., Righetti, M., Pastorino, P., Prearo, M., Gai, F., Schiavone, A. and Gasco, L. 2018. Influence of *Hermetia illucens* meal dietary inclusion on the histological traits, gut mucin composition and the oxidative stress biomarkers in rainbow trout (*Oncorhynchus mykiss*). *Aquaculture* 496, 50–57.

Estévez, A., Blanco, B., Fernández, L., Ferreira, M. and Soula, M. 2022. Effects of alternative and sustainable ingredients, insect meal, microalgae and protein and lipid from tuna cooking water, on meagre (*Argyrosomus regius*) growth, food conversion and muscle and liver composition. *Aquaculture* 548, 737549.

Fabrikov, D., Sánchez-Muros, M. J., Barroso, F. G., Tomás-Almenar, C., Melenchón, F., Hidalgo, M. C., Morales, A. E., Rodriguez-Rodriguez, M. and Montes-Lopez, J. 2020. Comparative study of growth performance and amino acid catabolism in *Hermetia illucens*, *Tinca tinca* and *Sparus aurata* and the catabolic changes in response to insect meal inclusion in the diet. *Aquaculture* 529, 735731. https://doi.org/10.1016/j.aquaculture.2020.735731.

FAO 2022. *The State of World Fisheries and Aquaculture 2022: Towards Blue Transformation*. Rome: FAO. https://doi.org/10.4060/cc0461en.

Fawole, F. J., Adeoye, A. A., Tiamiyu, L. O., Ajala, K. I., Obadara, S. O. and Ganiyu, I. O. 2020. Substituting fishmeal with *Hermetia illucens* in the diets of African catfish (*Clarias gariepinus*): effects on growth, nutrient utilization, haematophysiological response, and oxidative stress biomarker. *Aquaculture* 518, 734849.

Ferrari, L., Sele, V., Silva, M., Bonilauri, P., De Filippo, F., Selmin, F., Ørnsrud, R., Pinotti, L. and Ottoboni, M. 2022. Biofortification of selenium in black soldier fly (*Hermetia illucens*) prepupae reared on seaweed or selenium enriched substrates. *Journal of Insects as Food and Feed* 8(8), 887–899. https://doi.org/10.3920/JIFF2021.0153.

Fischer, H., Romano, N., Renukdas, N., Kumar, V. and Sinha, A. K. 2022. Comparing black soldier fly (*Hermetia illucens*) larvae versus prepupae in the diets of largemouth bass, *Micropterus salmoides*: effects on their growth, biochemical composition, histopathology, and gene expression. *Aquaculture* 546, 737323. https://doi.org/10.1016/j.aquaculture.2021.737323.

Fisher, H. J., Collins, S. A., Hanson, C., Mason, B., Colombo, S. M. and Anderson, D. M. 2020. Black soldier fly larvae meal as a protein source in low fish meal diets for *Atlantic salmon* (*Salmo salar*). *Aquaculture* 521, 734978. https://doi.org/10.1016/j.aquaculture.2020.734978.

Foysal, M. J. and Gupta, S. K. 2022. A systematic meta-analysis reveals enrichment of Actinobacteria and Firmicutes in the fish gut in response to black soldier fly (*Hermetica illucens*) meal-based diets. *Aquaculture* 549, 737760.

Froehlich, H. E., Jacobsen, N. S., Essington, T. E., Clavelle, T. and Halpern, B. S. 2018a. Avoiding the ecological limits of forage fish for fed aquaculture. *Nature Sustainability* 1(6), 298–303.

Froehlich, H. E., Runge, C. A., Gentry, R. R., Gaines, S. D. and Halpern, B. S. 2018b. Comparative terrestrial feed and land use of an aquaculture-dominant world. *Proceedings of the National Academy of Sciences of the United States of America* 115(20), 5295–5300.

Fry, J. P., Love, D. C., MacDonald, G. K., West, P. C., Engstrom, P. M., Nachman, K. E. and Lawrence, R. S. 2016. Environmental health impacts of feeding crops to farmed fish. *Environment International* 91, 201–214.

Gasco, L., Henry, M., Piccolo, G., Marono, S., Gai, F., Renna, M., Lussiana, C., Antonopoulou, E., Mola, P. and Chatzifotis, S. 2016. *Tenebrio molitor* meal in diets for European sea bass (*Dicentrarchus labrax* L.) juveniles: growth performance, whole body composition and in vivo apparent digestibility. *Animal Feed Science and Technology* 220, 34–45.

Gasco, L., Acuti, G., Bani, P., Dalle Zotte, A., Danieli, P. P., De Angelis, A., Fortina, R., Marino, R., Parisi, G., Piccolo, G., Pinotti, L., Prandini, A., Schiavone, A., Terova, G., Tulli, F. and Roncarati, A. 2020a. Insect and fish by-products as sustainable alternatives to conventional animal proteins in animal nutrition. *Italian Journal of Animal Science* 19(1), 360–372.

Gasco, L., Biancarosa, I. and Liland, N. S. 2020b. From waste to feed: a review of recent knowledge on insects as producers of protein and fat for animal feeds. *Current Opinion in Green and Sustainable Chemistry* 23, 67–79.

Gasco, L., Józefiak, A. and Henry, M. 2021. Beyond the protein concept: health aspects of using edible insects on animals. *Journal of Insects as Food and Feed* 7(5), 715–741. https://doi.org/10.3920/JIFF2020.0077.

Gasco, L., Biasato, I., Enes, P. and Gai, F. 2023a. Chapter 16 - Potential and challenges for the use of insects as feed for aquaculture. In: Morales-Ramos, J. A., Rojas, M. G. and Shapiro-Ilan, D. I. (Eds). *Mass Production of Beneficial Organisms* (2nd edn.). Academic Press, 465–492. https://doi.org/10.1016/B978-0-12-822106-8.00009-9.

Gasco, L., Renna, M., Bellezza Oddon, S., Rezaei Far, A., Naser El Deen, S. and Veldkamp, T. 2023b. Insect meals in a circular economy and applications in monogastric diets. *Animal Frontiers*, in press.

Gaudioso, G., Marzorati, G., Faccenda, F., Weil, T., Lunelli, F., Cardinaletti, G., Marino, G., Olivotto, I., Parisi, G., Tibaldi, E., Tuohy, K. M. and Fava, F. 2021. Processed animal proteins from insect and poultry by-products in a fish meal-free diet for rainbow trout: impact on intestinal microbiota and inflammatory markers. *International Journal of Molecular Sciences* 22(11), 5454.

Gebremichael, A., Hancz, C. and Kucska, B. 2021. Effect of total or partial replacing of fishmeal with black solider fly (*Hermetia illucens*) meal on growth performance and body condition indices of common carp (*Cyprinus carpio*). *Bioflux* 14(4), 2280–2286.

Glencross, B. D., Booth, M. and Allan, G. L. 2007. A feed is only as good as its ingredients – a review of ingredient evaluation strategies for aquaculture feeds. *Aquaculture Nutrition* 13(1), 17–34.

Glencross, B. D., Baily, J., Berntssen, M. H. G., Hardy, R., MacKenzie, S. and Tocher, D. R. 2020. Risk assessment of the use of alternative animal and plant raw material resources in aquaculture feeds. *Reviews in Aquaculture* 12(2), 703–758. https://doi.org/10.1111/raq.12347.

Goglio, P., Van Den Burg, S., Kousoulaki, K., Skirtun, M., Espmark, Å. M., Kettunen, A. H. and Abbink, W. 2022. The environmental impact of partial substitution of fish-based feed with algae-and insect-based feed in salmon farming. *Sustainability* 14(19), 12650.

Gougbedji, A., Detilleux, J., Lalèyè, P. A., Francis, F. and Caparros Megido, R. 2022. Can insect meal replace fishmeal? A meta-analysis of the effects of black soldier fly on fish growth performances and nutritional values. *Animals* 12(13), 1700. https://doi.org/10.3390/ani12131700.

Hariati, A. M., Yuniarti, A., Arifin, N. B., Fakhri, M. and Wiadnya, D. G. R. 2022. The effect of maggot formulated feed on essential amino acids composition, digestibility and growth of dwarf snakehead, *Channa gachua* fry. *Sains Malaysiana* 51(10), 3143–3151.

Hender, A., Siddik, M. A. B., Howieson, J. and Fotedar, R. 2021. Black soldier fly, *Hermetia illucens* as an alternative to fishmeal protein and fish oil: impact on growth, immune response, mucosal barrier status, and flesh quality of juvenile barramundi, *Lates calcarifer* (Bloch, 1790). *Biology* 10(6), 505. https://doi.org/10.3390/biology10060505.

Hidalgo, M. C., Morales, A. E., Pula, H. J., Tomás-Almenar, C., Sánchez-Muros, M. J., Melenchón, F., Fabrikov, D. and Cardenete, G. 2022. Oxidative metabolism of gut and innate immune status in skin and blood of tench (*Tinca tinca*) fed with different insect meals (*Hermetia illucens* and *Tenebrio molitor*). *Aquaculture* 558, 738384. https://doi.org/10.1016/j.aquaculture.2022.738384.

Hoc, B., Francis, F., Carpentier, J., Mostade, L., Blecker, C., Purcaro, G. and Caparros Megido, R. 2021. ω3-enrichment of *Hermetia illucens* (L. 1758) prepupae from

oilseed byproducts. *Journal of the Saudi Society of Agricultural Sciences* 20(3), 155–163. https://doi.org/10.1016/j.jssas.2021.01.001.
Hu, J., Wang, G., Huang, Y., Sun, Y., He, F., Zhao, H. and Li, N. 2017. Effects of substitution of fish meal with black soldier fly (*Hermetia illucens*) larvae meal, in yellow catfish (*Pelteobagrus fulvidraco*) diets. *Israeli Journal of Aquaculture-Bamidgeh* 69.
Hu, Y., Huang, Y., Tang, T., Zhong, L., Chu, W., Dai, Z., Chen, K. and Hu, Y. 2020. Effect of partial black soldier fly (*Hermetia illucens* L.) larvae meal replacement of fish meal in practical diets on the growth, digestive enzyme and related gene expression for rice field eel (*Monopterus albus*). *Aquaculture Reports* 17, 100345. https://doi.org/10.1016/j.aqrep.2020.100345.
Hua, K., Cobcroft, J. M., Cole, A., Condon, K., Jerry, D. R., Mangott, A., Praeger, C., Vucko, M. J., Zeng, C., Zenger, K. and Strugnell, J. M. 2019. The future of aquatic protein: implications for protein sources in aquaculture diets. *One Earth* 1(3), 316–329. https://doi.org/10.1016/j.oneear.2019.10.018.
Hua, K. 2021. A meta-analysis of the effects of replacing fish meals with insect meals on growth performance of fish. *Aquaculture* 530, 735732.
IPIFF 2019a. The European insect sector today: challenges, opportunities and regulatory landscape. Available at: https://ipiff.org/wp-content/uploads/2019/12/2019IPIFF_VisionPaper_updated.pdf.
IPIFF 2019b. Building bridges between the insect production chain, research and policymakers. Available at: https://ipiff.org/brochures/.
Kamalii, A., Antony, C., Ahilan, B., Uma, A. and Prabu, E. 2022. Dietary protein replacement of fish meal with black soldier fly larvae meal: effects on growth, whole-body composition, digestive enzyme activity, muscle-growth-related gene expression and haemato-biochemical responses of juvenile goldfish, *Carassius auratus*. *Turkish Journal of Fisheries and Aquatic Sciences* 23(2). https://doi.org/10.4194/TRJFAS21837.
Kangsen, M., Xue, M., He, G., Xie, S. Q. and Kaushik, S. J. 2022. Chapter 4 - Protein and amino acids. In: Hardy, R. W. and Kaushik, S. J. (Eds). *Fish Nutrition* (4th edn.). Academic Press, 181–302. https://doi.org/10.1016/B978-0-12-819587-1.00012-4.
Khieokhajonkhet, A., Uanlam, P., Ruttarattanamongkol, K., Aeksiri, N., Tatsapong, P. and Kaneko, G. 2022. Replacement of fish meal by black soldier fly larvae meal in diet for goldfish *Carassius auratus*: growth performance, hematology, histology, total carotenoids, and coloration. *Aquaculture* 561, 738618. https://doi.org/10.1016/j.aquaculture.2022.738618.
Kishawy, A. T. Y., Mohammed, H. A., Zaglool, A. W., Attia, M. S., Hassan, F. A. M., Roushdy, E. M., Ismail, T. A. and Ibrahim, D. 2022. Partial defatted black solider larvae meal as a promising strategy to replace fish meal protein in diet for *Nile tilapia* (*Oreochromis niloticus*): performance, expression of protein and fat transporters, and cytokines related genes and economic efficiency. *Aquaculture* 555, 738195.
Kok, B., Malcorps, W., Tlusty, M. F., Eltholth, M. M., Auchterlonie, N. A., Little, D. C., Harmsen, R., Newton, R. W. and Davies, S. J. 2020. Fish as feed: using economic allocation to quantify the Fish In: Fish Out ratio of major fed aquaculture species. *Aquaculture* 528, 735474. https://doi.org/10.1016/j.aquaculture.2020.735474.
Kousoulaki, K., Sveen, L., Norén, F. and Espmark, Å. 2022. Atlantic salmon (*Salmo salar*) performance fed low trophic ingredients in a fish meal and fish oil free diet. *Frontiers in Physiology* 13, 884740. https://doi.org/10.3389/fphys.2022.884740.
Kroeckel, S., Harjes, A.-G. E., Roth, I., Katz, H., Wuertz, S., Susenbeth, A. and Schulz, C. 2012. When a turbot catches a fly: evaluation of a pre-pupae meal of the black

soldier fly (*Hermetia illucens*) as fish meal substitute – growth performance and chitin degradation in juvenile turbot (*Psetta maxima*). *Aquaculture* 364–365, 345–352. https://doi.org/10.1016/j.aquaculture.2012.08.041.

Kuo, I. P., Liu, C. S., Yang, S. D., Liang, S. H., Hu, Y. F. and Nan, F. H. 2022. Effects of replacing fishmeal with defatted black soldier fly (*Hermetia illucens* Linnaeus) larvae meal in Japanese eel (*Anguilla japonica*) diet on growth performance, fillet texture, serum biochemical parameters, and intestinal histomorphology. *Aquaculture Nutrition*, 1866142. https://doi.org/10.1155/2022/1866142.

Le Feon, S., Thévenot, A., Maillard, F., Macombe, C., Forteau, L. and Aubin, J. 2019. Life cycle assessment of fish fed with insect meal: case study of mealworm inclusion in trout feed, in France. *Aquaculture* 500, 82–91. https://doi.org/10.1016/j.aquaculture.2018.06.051.

Józefiak, A., Nogales-Mérida, S., Rawski, M., Kierończyk, B. and Mazurkiewicz, J. 2019. Effects of insect diets on the gastrointestinal tract health and growth performance of Siberian sturgeon (*Acipenser baerii* Brandt, 1869). *BMC Veterinary Research* 15(1), 348.

Leni, G., Maistrello, L., Pinotti, G., Sforza, S. and Caligiani, A. 2023. Production of carotenoid-rich *Hermetia illucens* larvae using specific agri-food by-products. *Journal of Insects as Food and Feed* 9(2), 171–181. https://doi.org/10.3920/JIFF2022.0016.

Li, S., Ji, H., Zhang, B., Tian, J., Zhou, J. and Yu, H. 2016. Influence of black soldier fly (*Hermetia illucens*) larvae oil on growth performance, body composition, tissue fatty acid composition and lipid deposition in juvenile Jian carp (*Cyprinus carpio* var. *Jian*). *Aquaculture* 465, 43–52.

Li, S., Ji, H., Zhang, B., Zhou, J. and Yu, H. 2017. Defatted black soldier fly (*Hermetia illucens*) larvae meal in diets for juvenile Jian carp (*Cyprinus carpio* var. *Jian*): growth performance, antioxidant enzyme activities, digestive enzyme activities, intestine and hepatopancreas histological structure. *Aquaculture* 477, 62–70. https://doi.org/10.1016/j.aquaculture.2017.04.015.

Li, Y., Kortner, T. M., Chikwati, E. M., Munang'andu, H. M., Lock, E. J. and Krogdahl, Å. 2019. Gut health and vaccination response in pre-smolt Atlantic salmon (*Salmo salar*) fed black soldier fly (*Hermetia illucens*) larvae meal. *Fish and Shellfish Immunology* 86, 1106–1113.

Li, Y., Kortner, T. M., Chikwati, E. M., Belghit, I., Lock, E.-J. and Krogdahl, Å. 2020. Total replacement of fish meal with black soldier fly (*Hermetia illucens*) larvae meal does not compromise the gut health of Atlantic salmon (*Salmo salar*). *Aquaculture* 520, 734967.

Liland, N. S., Araujo, P., Xu, X. X., Lock, E.-J., Radhakrishnan, G., Prabhu, A. J. P. and Belghit, I. A meta-analysis on the nutritional value of insects in aquafeeds. *Journal of Insects as Food and Feed* 7(5), 743–759. https://doi.org/10.3920/JIFF2020.0147.

Limbu, S. M., Shoko, A. P., Ulotu, E. E., Luvanga, S. A., Munyi, F. M., John, J. O. and Opiyo, M. A. 2022. Black soldier fly (*Hermetia illucens* L.) larvae meal improves growth performance, feed efficiency and economic returns of *Nile tilapia* (*Oreochromis niloticus* L.) fry. *Aquaculture, Fish and Fisheries* 2(3), 167–178. https://doi.org/10.1002/aff2.48.

Lock, E. R., Arsiwalla, T. and Waagbø, R. 2016. Insect larvae meal as an alternative source of nutrients in the diet of *Atlantic salmon* (*Salmo salar*) postsmolt. *Aquaculture Nutrition* 22(6), 1202–1213. https://doi.org/10.1111/anu.12343.

Maiolo, S., Parisi, G., Biondi, N., Lunelli, F., Tibaldi, E. and Pastres, R. 2020. Fishmeal partial substitution within aquafeed formulations: life cycle assessment of four alternative protein sources. *International Journal of Life Cycle Assessment* 25(8), 1455–1471. https://doi.org/10.1007/s11367-020-01759-z.

Malcorps, W., Newton, R. W., Sprague, M., Glencross, B. D. and Little, D. C. 2021. Nutritional characterisation of European aquaculture processing by-products to facilitate strategic utilisation. *Frontiers in Sustainable Food Systems* 5. https://doi.org/10.3389/fsufs.2021.720595.

Mapanao, R., Jiwyam, W., Nithikulworawong, N. and Weeplian, T. 2023. Effects of black soldier fly (*Hermetia illucens*) larvae as a fish meal replacement on growth performance, feed utilisation, morphological characters and carcass composition of Thai climbing perch (*Anabas testudineus*). *Journal of Applied Aquaculture* 35(1), 1–15. https://doi.org/10.1080/10454438.2021.1923609.

Maranga, B., Kagali, R., Mbogo, K., Orina, P., Munguti, J. and Ogello, E. 2022. Growth performance of African catfish (*Clarias gariepinus*) fed on diets containing black soldier fly (*Hermetia illucens*) larvae under aquaponic system. *Aquaculture Studies* 23(5). https://doi.org/10.4194/AQUAST910.

Martínez-Alvarez, R. M., Morales, A. E. and Sanz, A. 2005. Antioxidant defenses in fish: biotic and abiotic factors. *Reviews in Fish Biology and Fisheries* 15(1–2), 75–88.

Mastoraki, M., Mollá Ferrándiz, P., Vardali, S. C., Kontodimas, D. C., Kotzamanis, Y. P., Gasco, L., Chatzifotis, S. and Antonopoulou, E. 2020. A comparative study on the effect of fish meal substitution with three different insect meals on growth, body composition and metabolism of European sea bass (*Dicentrarchus labrax* L.). *Aquaculture* 528, 735511. https://doi.org/10.1016/j.aquaculture.2020.735511.

Melenchón, F., de Mercado, E., Pula, H. J., Cardenete, G., Barroso, F. G., Fabrikov, D., Lourenço, H. M., Pessoa, M. F., Lagos, L., Weththasinghe, P., Cortés, M. and Tomás-Almenar, C. 2022. Fishmeal dietary replacement up to 50%: a comparative study of two insect meals for rainbow trout (*Oncorhynchus mykiss*). *Animals* 12(2), 179. https://doi.org/10.3390/ani12020179.

Mertenat, A., Diener, S. and Zurbrügg, C. 2019. Black soldier fly biowaste treatment–assessment of global warming potential. *Waste Management* 84, 173–181.

Modahl, I. S. and Brekke, A. 2022. Environmental performance of insect protein: a case of LCA results for fish feed produced in Norway. *SN Applied Sciences* 4(6), 183.

Mohamad-Zulkifli, N. F. N., Yong, A. S. K., Kawamura, G., Lim, L.-S., Senoo, S., Devic, E., Mustafa, S. and Shapawi, R. 2019. Apparent digestibility coefficient of black soldier fly (*Hermetia illucens*) larvae in formulated diets for hybrid grouper (*Epinephelus fuscoguttatus* ♀ x *Epinephelus lanceolatus* ♂). *Aquaculture, Aquarium, Conservation & Legislation* 12(2), 513–522.

Moutinho, S., Oliva-Teles, A., Martínez-Llorens, S., Monroig, Ó. and Peres, H. 2022. Total fishmeal replacement by defatted *Hermetia illucens* larvae meal in diets for gilthead seabream (*Sparus aurata*) juveniles. *Journal of Insects as Food and Feed* 8(12), 1455–1468.

Muin, H. and Taufek, N. M. 2024. Evaluation of growth performance, feed efficiency and nutrient digestibility of red hybrid tilapia fed dietary inclusion of black soldier fly larvae (*Hermetia illucens*). *Aquaculture and Fisheries* 9(1), 46–51. https://doi.org/10.1016/j.aaf.2022.09.006.

Nairuti, R. N., Munguti, J. M., Waidbacher, H. and Zollitsch, W. 2021. Growth performance and survival rates of *Nile tilapia* (*Oreochromis niloticus* L.) reared on diets containing black soldier fly (*Hermetia illucens* L.) larvae meal. *Die Bodenkultur* 72(1), 9–19. https://doi.org/10.2478/boku-2021-0002.

Naylor, R. L., Hardy, R. W., Buschmann, A. H., Bush, S. R., Cao, L., Klinger, D. H., Little, D. C., Lubchenco, J., Shumway, S. E. and Troell, M. 2021a. A 20-year retrospective review of global aquaculture. *Nature* 591(7851), 551–563.

Naylor, R. L., Kishore, A., Sumaila, U. R., Issifu, I., Hunter, B. P., Belton, B., Bush, S. R., Cao, L., Gelcich, S., Gephart, J. A., Golden, C. D., Jonell, M., Koehn, J. Z., Little, D. C., Thilsted, S. H., Tigchelaar, M. and Crona, B. 2021b. Blue food demand across geographic and temporal scales. *Nature Communications* 12(1), 5413.

Ngo, D. H. and Kim, S. K. 2014. Antioxidant effects of chitin, chitosan and their derivatives. *Advances in Food and Nutrition Research* 73, 15–31.

Ogunji, J. O., Kloas, W., Wirth, M., Neumann, N. and Pietsch, C. 2008. Effect of housefly maggot meal (magmeal) diets on the performance, concentration of plasma glucose, cortisol and blood characteristics of *Oreochromis niloticus* fingerlings. *Journal of Animal Physiology and Animal Nutrition* 92(4), 511–518.

Ogunji, J. O., Nimptsch, J., Wiegand, C., Schulz, C. and Rennert, B. 2011. Effect of housefly maggot meal (magmeal) diets on catalase, and glutathione S-transferase in the liver and gills of carp *Cyprinus carpio* fingerling. *International Aquatic Research* 3, 11–20.

Pahlow, M., van Oel, P. R., Mekonnen, M. M. and Hoekstra, A. Y. 2015. Increasing pressure on freshwater resources due to terrestrial feed ingredients for aquaculture production. *Science of the Total Environment* 536, 847–857.

Papargyropoulou, E., Lozano, R., Steinberger, J. K., Wright, N. and Ujang, Zb. 2014. The food waste hierarchy as a framework for the management of food surplus and food waste. *Journal of Cleaner Production* 76, 106–115. https://doi.org/10.1016/j.jclepro.2014.04.020.

Pelletier, N., Tyedmers, P., Sonesson, U., Scholz, A., Ziegler, F., Flysjo, A., Kruse, S., Cancino, B. and Silverman, H. 2009. Not all salmon are created equal: life cycle assessment (LCA) of global salmon farming systems. *Environmental Science and Technology* 43(23), 8730–8736. https://doi.org/10.1021/es9010114.

Peng, K., Mo, W., Xiao, H., Wang, G. and Huang, Y. 2021. Effects of black soldier fly pulp on growth performance, histomorphology and lipid metabolism gene expression of *Micropterus salmoides*. *Aquaculture Reports* 20, 100737.

Peng, K., Mo, W., Xiao, H., Hu, J., Zhu, X., Huang, Y. and Wang, G. 2022. Dietary black soldier fly pulp affects growth, antioxidant and immune capacity of *Micropterus salmoides*. *Journal of Insects as Food and Feed* 8(11), 1197–1203.

Petereit, J., Hoerterer, C. and Krause, G. 2022. Country-specific food culture and scientific knowledge transfer events – do they influence the purchasing behaviour of seafood products? *Aquaculture* 560, 738590. https://doi.org/10.1016/j.aquaculture.2022.738590.

Prachom, N., Yuangsoi, B., Pumnuan, J., Ashour, M., Davies, S. J. and El-Haroun, E. 2023. Effects of substituting the two-spotted cricket (*Gryllus bimaculatus*) meal for fish meal on growth performances and digestibility of striped snakehead (*Channa striata*) juveniles. *Life* 13(2), 594.

Pulido-Rodriguez, L. F., Cardinaletti, G., Secci, G., Randazzo, B., Bruni, L., Cerri, R., Olivotto, I., Tibaldi, E. and Parisi, G. 2021. Appetite regulation, growth performances and fish quality are modulated by alternative dietary protein ingredients in gilthead sea bream (*Sparus aurata*) culture. *Animals* 11(7), 1919. https://doi.org/10.3390/ani11071919.

Radhouani, H., Silva, N., Poeta, P., Torres, C., Correia, S. and Igrejas, G. 2014. Potential impact of antimicrobial resistance in wildlife, environment and human health. *Frontiers in Microbiology* 5. Available at: https://www.frontiersin.org/articles/10.3389/fmicb.2014.00023.

Randazzo, B., Zarantoniello, M., Cardinaletti, G., Cerri, R., Giorgini, E., Belloni, A., Contò, M., Tibaldi, E. and Olivotto, I. 2021. *Hermetia illucens* and poultry by-product meals as alternatives to plant protein sources in gilthead seabream (*Sparus aurata*) diet: a multidisciplinary study on fish gut status. *Animals* 11(3), 677. https://doi.org/10.3390/ani11030677.

Rangel, F., Santos, R. A., Monteiro, M., Lavrador, A. S., Gasco, L., Gai, F., Oliva-Teles, A., Enes, P. and Serra, C. R. 2022. Isolation of chitinolytic bacteria from European sea bass gut microbiota fed diets with distinct insect meals. *Biology* 11(7), 964.

Ratti, S., Zarantoniello, M., Chemello, G., Giammarino, M., Palermo, F. A., Cocci, P., Mosconi, G., Tignani, M. V., Pascon, G., Cardinaletti, G., Pacetti, D., Nartea, A., Parisi, G., Riolo, P., Belloni, A. and Olivotto, I. 2023. Spirulina-enriched substrate to rear black soldier fly (*Hermetia illucens*) prepupae as alternative aquafeed ingredient for rainbow trout (*Oncorhynchus mykiss*) diets: possible effects on zootechnical performances, gut and liver health status, and fillet quality. *Animals* 13(1), 173. https://doi.org/10.3390/ani13010173.

Ravi, H. K., Vian, M. A., Tao, Y., Degrou, A., Costil, J., Trespeuch, C. and Chemat, F. 2019. Alternative solvents for lipid extraction and their effect on protein quality in black soldier fly (*Hermetia illucens*) larvae. *Journal of Cleaner Production* 238, 1–13.

Ravi, H. K., Degrou, A., Costil, J., Trespeuch, C., Chemat, F. and Vian, M. A. 2020. Larvae mediated valorization of industrial, agriculture and food wastes: biorefinery concept through bioconversion, processes, procedures, and products. *Processes* 8(7), 857.

Rawski, M., Mazurkiewicz, J., Kierończyk, B. and Józefiak, D. 2020. Black soldier fly full-fat larvae meal as an alternative to fish meal and fish oil in Siberian sturgeon nutrition: the effects on physical properties of the feed, animal growth performance, and feed acceptance and utilization. *Animals* 10(11), 2119. https://doi.org/10.3390/ani10112119.

Renna, M., Schiavone, A., Gai, F., Dabbou, S., Lussiana, C., Malfatto, V., Prearo, M., Capucchio, M. T., Biasato, I., Biasibetti, E., De Marco, M., Brugiapaglia, A., Zoccarato, I. and Gasco, L. 2017. Evaluation of the suitability of a partially defatted black soldier fly (*Hermetia illucens* L.) larvae meal as ingredient for rainbow trout (*Oncorhynchus mykiss* Walbaum) diets. *Journal of Animal Science and Biotechnology* 8, 57. https://doi.org/10.1186/s40104-017-0191-3.

Salomone, R., Saija, G., Mondello, G., Giannetto, A., Fasulo, S. and Savastano, D. 2017. Environmental impact of food waste bioconversion by insects: application of life cycle assessment to process using *Hermetia illucens*. *Journal of Cleaner Production* 140, 890–905.

Sandström, V., Chrysafi, A., Lamminen, M., Troell, M., Jalava, M., Piipponen, J., Siebert, S., van Hal, O., Virkki, V. and Kummu, M. 2022. Food system by-products upcycled in livestock and aquaculture feeds can increase global food supply. *Nature Food* 3(9), 729–740.

Sealey, W. M., Gaylord, T. G., Barrows, F. T., Tomberlin, J. K., McGuire, M. A., Ross, C. and St-Hilaire, S. 2011. Sensory analysis of rainbow trout, *Oncorhynchus mykiss*, fed enriched black soldier fly prepupae, *Hermetia illucens*. *Journal of the World Aquaculture Society* 42(1), 34–45. https://doi.org/10.1111/j.1749-7345.2010.00441.x.

Shaw, C., Knopf, K. and Kloas, W. 2022. Fish feeds in aquaponics and beyond: a novel concept to evaluate protein sources in diets for circular multitrophic food production systems. *Sustainability* 14(7), 4064.

Siddiqui, S. A., Ristow, B., Rahayu, T., Putra, N. S., Widya Yuwono, N., Nisa', K., Mategeko, B., Smetana, S., Saki, M., Nawaz, A. and Nagdalian, A. 2022. Black soldier fly larvae (BSFL) and their affinity for organic waste processing. *Waste Management* 140, 1–13, https://doi.org/10.1016/j.wasman.2021.12.044.

Šimat, V., Vlahović, J., Soldo, B., Skroza, D., Ljubenkov, I. and Generalić Mekinić, I. 2019. Production and refinement of omega-3 rich oils from processing by-products of farmed fish species. *Foods* 8(4), 125.

Smarason, B. O., Ogmundarson, O., Arnason, J., Bjornsdottir, R. and Daviosdottir, B. 2017. Life cycle assessment of Icelandic arctic char fed three different feed types. *Turkish Journal of Fisheries and Aquatic Sciences* 17(1), 79–90. http://doi.org/10.4194/1303-2712-v17_1_10.

Smetana, S., Palanisamy, M., Mathys, A. and Heinz, V. 2016. Sustainability of insect use for feed and food: life cycle assessment perspective. *Journal of Cleaner Production* 137, 741–751. https://doi.org/10.1016/j.jclepro.2016.07.148.

Smetana, S., Schmitt, E. and Mathys, A. 2019. Sustainable use of *Hermetia illucens* insect biomass for feed and food: attributional and consequential life cycle assessment. *Resources, Conservation and Recycling* 144, 285–296.

Stathopoulou, P., Asimaki, A., Berillis, P., Vlahos, N., Levizou, E., Katsoulas, N., Karapanagiotidis, I. T., Rumbos, C. I., Athanassiou, C. G. and Mente, E. 2022. Aqua-Ento-Ponics: effect of insect meal on the development of sea bass, *Dicentrarchus labrax*, in co-culture with lettuce. *Fishes* 7(6), 397.

Stejskal, V., Tran, H. Q., Prokesova, M., Gebauer, T., Giang, P. T., Gai, F. and Gasco, L. 2020. Partially defatted *Hermetia illucens* larva meal in diet of Eurasian perch (*Perca fluviatilis*) juveniles. *Animals* 10(10), 1876. https://doi.org/10.3390/ani10101876.

Stejskal, V., Tran, H. Q., Prokesová, M., Zare, M., Gebauer, T., Policar, T., Caimi, C., Gai, F. and Gasco, L. 2023. Defatted black soldier fly (*Hermetia illucens*) in pikeperch (*Sander lucioperca*) diets: effects on growth performance, nutrient digestibility, fillet quality, economic and environmental sustainability. *Animal Nutrition* 12, 7–19.

Stenberg, O. K., Holen, E., Piemontese, L., Liland, N. S., Lock, E. J., Espe, M. and Belghit, I. 2019. Effect of dietary replacement of fish meal with insect meal on in vitro bacterial and viral induced gene response in *Atlantic salmon* (*Salmo salar*) head kidney leukocytes. *Fish and Shellfish Immunology* 91, 223–232.

Sudha, C., Ahilan, B., Felix, N., Uma, A. and Prabu, E. 2022. Effects of dietary protein substitution of fishmeal with black soldier fly larval meal on growth and physiological responses of juvenile striped catfish, *Pangasianodon hypophthalmus*. *Aquaculture Research* 53(6), 2204–2217. https://doi.org/10.1111/are.15739.

Tacon, A. G. J. and Metian, M. 2008. Global overview on the use of fish meal and fish oil in industrially compounded aquafeeds: trends and future prospects. *Aquaculture* 285(1–4), 146–158. https://doi.org/10.1016/j.aquaculture.2008.08.015.

Tacon, A. G. J., Metian, M. and McNevin, A. A. 2022. Future feeds: suggested guidelines for sustainable development. *Reviews in Fisheries Science and Aquaculture* 30(2), 135–142.

Taufek, N. M., Aspani, F., Muin, H., Raji, A. A., Razak, S. A. and Alias, Z. 2016. The effect of dietary cricket meal (*Gryllus bimaculatus*) on growth performance, antioxidant enzyme activities, and haematological response of African catfish (*Clarias gariepinus*). *Fish Physiology and Biochemistry* 42(4), 1143–1155.

Takakuwa, F., Tanabe, R., Nomura, S., Inui, T., Yamada, S., Biswas, A. and Tanaka, H. 2022. Availability of black soldier fly meal as an alternative protein source to fish meal in red sea bream (*Pagrus major*, Temminck & Schlegel) fingerling diets. *Aquaculture Research* 53(1), 36–49.

Tippayadara, N., Dawood, M. A. O., Krutmuang, P., Hoseinifar, S. H., Doan, H. V. and Paolucci, M. 2021. Replacement of fish meal by black soldier fly (*Hermetia illucens*) larvae meal: effects on growth, haematology, and skin mucus immunity of *Nile tilapia, Oreochromis niloticus*. *Animals* 11(1), 193. https://doi.org/10.3390/ani11010193.

Tran, H. Q., Prokešová, M., Zare, M., Gebauer, T., Elia, A. C., Colombino, E., Ferrocino, I., Caimi, C., Gai, F., Gasco, L. and Stejskal, V. 2021. How does pikeperch *Sander lucioperca* respond to dietary insect meal *Hermetia illucens*? Investigation on gut microbiota, histomorphology, and antioxidant biomarkers. *Frontiers in Marine Science* 8, 680942. https://doi.org/10.3389/fmars.2021.680942.

Tran, H. Q., Van Doan, H. and Stejskal, V. 2022a. Environmental consequences of using insect meal as an ingredient in aquafeeds: a systematic view. *Reviews in Aquaculture* 14(1), 237–251. https://doi.org/10.1111/raq.12595.

Tran, H. Q., Nguyen, T. T., Prokešová, M., Gebauer, T., Doan, H. V. and Stejskal, V. 2022b. Systematic review and meta-analysis of production performance of aquaculture species fed dietary insect meals. *Reviews in Aquaculture* 14(3), 1637–1655.

Tran, H. Q., Prokešová, M., Zare, M., Matoušek, J., Ferrocino, I., Gasco, L. and Stejskal, V. 2022c. Production performance, nutrient digestibility, serum biochemistry, fillet composition, intestinal microbiota and environmental impacts of European perch (*Perca fluviatilis*) fed defatted mealworm (*Tenebrio molitor*). *Aquaculture* 547, 737499.

Turchini, G. M., Trushenski, J. T. and Glencross, B. D. 2019. Thoughts for the future of aquaculture nutrition: realigning perspectives to reflect contemporary issues related to judicious use of marine resources in aquafeeds. *North American Journal of Aquaculture* 81(1), 13–39.

van Huis, A. 2022. Edible insects: challenges and prospects. *Entomological Research* 52(4), 161–177. https://doi.org/10.1111/1748-5967.12582.

van Huis, A. and Gasco, L. 2023. Insects as feed for livestock production. *Science* 379(6628), 138–139. https://doi.org/10.1126/science.adc9165.

van Riel, A. J., Nederlof, M. A. J., Chary, K., Wiegertjes, G. F. and de Boer, I. J. M. 2023. Feed-food competition in global aquaculture: current trends and prospects. *Reviews in Aquaculture* 15(3), 1142–1158. https://doi.org/10.1111/raq.12804.

Vargas-Abúndez, A. J., Randazzo, B., Foddai, M., Sanchini, L., Truzzi, C., Giorgini, E., Gasco, L. and Olivotto, I. 2019. Insect meal-based diets for clownfish: biometric, histological, spectroscopic, biochemical and molecular implications. *Aquaculture* 498, 1–11. https://doi.org/10.1016/j.aquaculture.2018.08.018.

Veldkamp, T., Dong, L., Paul, A. and Govers, C. 2022. Bioactive properties of insect products for monogastric animals – a review. *Journal of Insects as Food and Feed* 8(9), 1027–1040. https://doi.org/10.3920/JIFF2021.0031.

Vongvichith, B., Morioka, S., Sugita, T., Phousavanh, N., Phetsanghanh, N., Chanthasone, P., Pommachan, P. and Nakamura, S. 2020. Evaluation of the efficacy of aquaculture feeds for the climbing perch *Anabas testudineus*: replacement of fishmeal by black soldier fly *Hermetia illucens* prepupae. *Fisheries Science* 86(1), 145–151. https://doi.org/10.1007/s12562-019-01381-5.

Wachira, M. N., Osuga, I. M., Munguti, J. M., Ambula, M. K., Subramanian, S. and Tanga, C. M. 2021. Efficiency and improved profitability of insect-based aquafeeds for farming nile tilapia fish (*Oreochromis niloticus* L.). *Animals* 11(9), 2599. https://doi.org/10.3390/ani11092599.

Wang, B. T., Wu, H. L., Dong, Y. W., Jiang, B., Li, W., Su, Y. L., Huang, Y. H. and Liu, C. 2022. Effects of fishmeal substitution with *Hermetia illucens* L. on the growth, metabolism and disease resistance of *Micropterus salmoides*. *Journal of Insects as Food and Feed* 8(11), 1343–1353. https://doi.org/10.3920/JIFF2021.0201.

Wang, G., Peng, K., Hu, J., Yi, C., Chen, X., Wu, H. and Huang, Y. 2019. Evaluation of defatted black soldier fly (*Hermetia illucens* L.) larvae meal as an alternative protein ingredient for juvenile Japanese seabass (*Lateolabrax japonicus*) diets. *Aquaculture* 507, 144–154. https://doi.org/10.1016/j.aquaculture.2019.04.023.

Wang, G., Peng, K., Hu, J., Mo, W., Wei, Z. and Huang, Y. 2021. Evaluation of defatted *Hermetia illucens* larvae meal for *Litopenaeus vannamei*: effects on growth performance, nutrition retention, antioxidant and immune response, digestive enzyme activity and hepatic morphology. *Aquaculture Nutrition* 27(4), 986–997. https://doi.org/10.1111/anu.13240.

Weththasinghe, P., Hansen, J. Ø., Nøkland, D., Lagos, L., Rawski, M. and Øverland, M. 2021. Full-fat black soldier fly larvae (*Hermetia illucens*) meal and paste in extruded diets for Atlantic salmon (*Salmo salar*): effect on physical pellet quality, nutrient digestibility, nutrient utilization and growth performances. *Aquaculture* 530, 735785.

Wilfart, A., Garcia-Launay, F., Terrier, F., Soudé, E., Aguirre, P. and Skiba-Cassy, S. 2023. A step towards sustainable aquaculture: multiobjective feed formulation reduces environmental impacts at feed and farm levels for rainbow trout. *Aquaculture* 562, 738826. https://doi.org/10.1016/j.aquaculture.2022.738826.

Woodgate, S. L., Wan, A. H. L., Hartnett, F., Wilkinson, R. G. and Davies, S. J. 2022. The utilisation of European processed animal proteins as safe, sustainable and circular ingredients for global aquafeeds. *Reviews in Aquaculture* 14(3), 1572–1596. https://doi.org/10.1111/raq.12663.

World Health Organization 2018. *Antimicrobial Resistance and Primary Health Care (No. WHO/HIS/SDS/2018.56)*. World Health Organization.

Xiao, X., Jin, P., Zheng, L., Cai, M., Yu, Z., Yu, J. and Zhang, J. 2018. Effects of black soldier fly (*Hermetia illucens*) larvae meal protein as a fishmeal replacement on the growth and immune index of yellow catfish (*Pelteobagrus fulvidraco*). *Aquaculture Research* 49(4), 1569–1577. https://doi.org/10.1111/are.13611.

Xu, X., Ji, H., Yu, H. and Zhou, J. 2020. Influence of dietary black soldier fly (*Hermetia illucens* Linnaeus) pulp on growth performance, antioxidant capacity and intestinal health of juvenile mirror carp (*Cyprinus carpio* var. *specularis*). *Aquaculture Nutrition* 26(2), 432–443. https://doi.org/10.1111/anu.13005.

Xu, F. M., Hou, S. W., Wang, G. X., Gong, J. Y., Zhou, L., Huang, Y. H., Huang, X. D. and Liu, L. 2021a. Effects of zymolytic black soldier fly (*Hermetia illucens*) pulp as dietary supplementation in largemouth bass (*Micropterus salmoides*). *Aquaculture Reports* 21, 100823.

Xu, X., Ji, H., Belghit, I., Liland, N. S., Wu, W. and Li, X. 2021b. Effects of black soldier fly oil rich in n-3 HUFA on growth performance, metabolism and health response of juvenile mirror carp (*Cyprinus carpio* var. *specularis*). *Aquaculture* 533, 736144.

Yildirim-Aksoy, M., Eljack, R. and Beck, B. H. 2020a. Nutritional value of frass from black soldier fly larvae, *Hermetia illucens*, in a channel catfish, *Ictalurus punctatus*, diet. *Aquaculture Nutrition* 26(3), 812–819.

Yildirim-Aksoy, M., Eljack, R., Schrimsher, C. and Beck, B. H. 2020b. Use of dietary frass from black soldier fly larvae, *Hermetia illucens*, in hybrid tilapia (Nile x Mozambique, *Oreocromis niloticus* x *O. mozambique*) diets improves growth and resistance to bacterial diseases. *Aquaculture Reports* 17, 100373.

Yildirim-Aksoy, M., Eljack, R., Beck, B. H. and Peatman, E. 2022. Nutritional evaluation of frass from black soldier fly larvae as potential feed ingredient for Pacific white shrimp, *Litopenaeus vannamei*. *Aquaculture Reports* 27, 101353.

Zarantoniello, M., Randazzo, B., Nozzi, V., Truzzi, C., Giorgini, E., Cardinaletti, G., Freddi, L., Ratti, S., Girolametti, F., Osimani, A., Notarstefano, V., Milanović, V., Riolo, P., Isidoro, N., Tulli, F., Gioacchini, G. and Olivotto, I. 2021. Physiological responses of Siberian sturgeon (*Acipenser baerii*) juveniles fed on full-fat insect-based diet in an aquaponic system. *Scientific Reports* 11(1), 1057.

Zhang, Y., Zhou, Z., Liu, Y., Cao, Y., He, S., Huo, F., Qin, C., Yao, B. and Ringø, E. 2014. High-yield production of a chitinase from *Aeromonas veronii* B565 as a potential feed supplement for warm-water aquaculture. *Applied Microbiology and Biotechnology* 98(4), 1651–1662.

Part 2

Production and application of yellow mealworms and other insects

Chapter 6

Advances in mass breeding/ rearing techniques for yellow mealworms as a protein source

David Deruytter, Carl Coudron and Jonas Claeys, Insect Research Centre, Belgium

1 Introduction

In recent years, the yellow mealworm has undergone a significant transformation, from a global pest species to a prominent insect species for feed and food purposes. However, farming mealworms is still a relatively new business venture, offering ample opportunities for optimization and innovation. The mealworm value chain is in its infancy, resulting mostly in start-ups needing to manage various aspects of their farm, including their mealworm colony: feeding, reproduction, and harvesting. Additionally, they must manage logistics such as feed supply, sales, and product transportation. Furthermore, it is common that live insects undergo some level of processing, such as blanching and freezing, before being sold.

However, the division of attention for managing a start-up farm, combined with the gradual organic growth of the company, can lead to suboptimal outcomes and a decrease in overall efficiency. To address these challenges and improve efficiency within the mealworm farming industry, it is crucial for start-ups to explore innovative approaches and collaborations. This may involve sharing best practices, seeking specialized expertise, or establishing partnerships along

http://dx.doi.org/10.19103/AS.2024.0139.07

the value chain. By doing so, start-ups can streamline their operations, reduce costs, and enhance the overall productivity of their mealworm farming ventures.

In this chapter, the focus will be on rearing techniques for the mealworm.

2 Mealworm life cycle

The yellow mealworm (*Tenebrio molitor*) has a holometabolic life cycle. It starts its life as an egg, oval in shape and with a weight of approximately 0.6 –0.7 mg (Ludwig and Fiore, 1960; Morales-Ramos et al., 2019). As with all life stages for insects, the temperature has an unmistakable impact on the hatch and developmental rate of the eggs, from as low as 5.5 days up to 14.5 days, depending on the temperature (Ludwig and Fiore, 1960). Although the time to hatch may change, a temperature range between 15°C and 30°C and relative humidity (RH) between 40% and 80% does not seem to affect the hatch rate (Deruytter et al., 2023). The larvae after hatching weigh approximately half of this (0.379 mg at 7 days). Initially, these larvae can almost triple in weight in their first week, with some individuals eventually reaching over 200 mg. Although they are often harvested sooner as the majority of the population will pupate at a lower average weight. In contrast to many other insect species, the number of molts in the larvae stage is not fixed (9–23; Ribeiro et al., 2018). Park et al. (2014) observed imaginal molting with mealworms after their 14th instar up to their 20th and this may be a useful characteristic for future artificial selection programs. When a larva is fully grown, it will, on average, pupate, and after 6–9 days, a fresh white beetle will emerge (Ribeiro et al., 2018). Over the course of a few days, this beetle of 12–20 mm will darken to the black beetle (hence the name darkling beetle), which is the final adult form. An adult beetle can mate after a few days, and the females will start to lay eggs soon thereafter, with up to 12 eggs/day (Berggreen et al., 2018; Deruytter et al., 2019). The entire life cycle, egg to egg, can easily be 3 months or more.

3 Rearing environment

Mealworms are sturdy animals that can survive and even reproduce for a long time in suboptimal conditions. Yet efficient rearing of mealworms is much more difficult and may be time-consuming and/or capital-intensive. Although there are several methods and variations used to rear mealworms by the industry or academia, the following information will provide a general view of the most common rearing method.

3.1 Environment

As with all ectothermic animals, the rearing climate is very important for the successful and efficient production of mealworms. From the climate parameters, the two most important are temperature and RH, although also light, CO_2, and other gases may be important.

Mealworms are able to tolerate a broad temperature range, although their ability to withstand extreme temperatures depends on various factors such as their developmental stage, duration of exposure, interaction with RH, acclimatization, and genetics. Studies have shown that temperatures above 40°C for 24 h are lethal for all stages of mealworms (Allen et al., 2012; Punzo and Mutchmor, 1978; Punzo and Mutchmor, 1980). Bjørge et al. (2018) observed that the optimal temperature for mealworm growth is 31°C, with a positive net growth rate observed between 12°C and 39°C.

It is recommended to set the environmental temperature several degrees lower than the optimal temperature, as mealworms produce heat as a by-product of metabolic activity, thereby increasing the temperature in the crate compared with the outside temperature. This heat production is directly related to the number of mealworms in a crate and their average weight. Deruytter et al. (2022) observed temperature differences as high as 12°C in high-density crates with fully grown mealworms, and indicated that the temperature difference is directly linked to the biomass in the crate.

Furthermore, there are indications that mealworms require a less strict environmental temperature than is often claimed. Coudron et al. (2021) conducted a study comparing the growth of mealworms under controlled climate conditions with that of mealworms grown in a greenhouse. The researchers found that mealworms raised in the greenhouse, which had a more variable climate compared with the climate rooms (SD 3.3°C versus 0.3°C), experienced only a small decrease in their development rate compared to those raised under controlled conditions.

As with temperature, mealworms are able to live and thrive in a broad range of relative humidities. In highly humid conditions (88% RH and up), mealworms are capable of absorbing atmospheric water, but there is limited data available on the lower limits of RH for mealworms (Machin, 1975; Hansen et al., 2004). Punzo and Mutchmor (1980) observed that eggs are prone to dehydration when exposed to prolonged periods of low RH (12%) at 25°C and 35°C, with high mortality after 11–14 days. Yet, between 40% and 80% RH, the influence on the hatch rate of the eggs is small (Deruytter et al., 2023). For both small and large mealworms (<30 mg or >100 mg), Punzo and Mutchmor observed that there was no significant change in survival at 25°C with changing RH between 12% and 98%. A high RH in a production environment may not be ideal as it promotes mite proliferation, and there is the risk of condensation on colder walls, leading to the growth of molds and other microorganisms. Therefore, it is recommended to set the RH at 60–70%, which means that mealworms will not be able to benefit from atmospheric water absorption, and water will have to be supplied with the feed (Bjørge et al., 2018).

Ventilation of the rearing room is essential to decrease the concentration of (harmful) emission gases such as CO_2 and to reduce the concentration of airborne microorganisms (e.g. fungal spores). These can be harmful to both the animals and the employees. In a highly active production setting, ventilation may be needed to remove the metabolic heat and evaporated moisture ensuring an optimal environment. Nevertheless, even in a perfectly healthy environment, active ventilation is recommended to avoid differences in the microclimate and temperature striation, ensuring a homogeneous and known climate throughout the entire room.

As the mealworms are negatively phototactic, the mealworm prefers to grow in the dark (Eberle et al., 2022). Hence, the larvae can be kept in complete darkness and thereby improving the growth and reducing the rearing costs.

4 Rearing mealworms

Mealworms have four distinct life stages (egg, larvae, pupae, and adult), and preferably, one does not have more than two stages intermingling at the same time. Each of these life stages has specific needs and, therefore, requires specific knowledge and equipment. Here, the technique is outlined from egg to egg. Yet, in a production facility, it may be beneficial to focus on only a part of the cycle. This decentralized approach is also commonly applied in other fields of animal husbandry, such as poultry and pigs, where there are specialized breeders and fatteners. This approach may be beneficial in terms of the required expertise and financial investment.

4.1 Egg production and handling

Egg production starts with adding adult beetles to a crate or other system. Female beetles are sexually receptive throughout their adult lives and need to mate frequently to keep their egg production rate at level (Drnevich et al., 2001). More (female) beetles increase the number of total eggs produced in a certain time, yet a higher density may reduce the number of eggs per beetle when cannibalism is not controlled (Deruytter et al., 2019). The latter can be contained by putting beetles on a sieve (mesh opening size approximately 2 × 2 mm). Females will push their ovipositor through the openings in the mesh. This puts the eggs out of reach of other beetles, resulting in a higher egg yield. Females can lay over 12 eggs/day, yet this decreases significantly as they age (Morales-Ramos et al., 2012; Berggreen et al., 2018). Furthermore, the amount of eggs per female and the changes with beetle age may depend on the population used in the farm (Adamaki-Sotiraki et al., 2022). One beetle/cm^2 results in approximately 2000 beetles in a 60 × 40 cm crate (outer dimensions, frequently used in industry). If a natural 1:1 sex ratio between males and females is used, this would result in a theoretical egg yield of 12 000 eggs/day or 84 000 eggs/week. In reality, in our facility (Insect Research Centre of Inagro), we reach yields up to 60 000 larvae/week in those conditions. In part, this discrepancy can be linked to a less strict follow-up of the feeding regime, the usage of a mix of beetles of different ages in production compared to a rigorous scientific experiment, and the feed that the beetles are provided with in both nutritional and physical characteristics. The oviposition period should not be longer than 7 days, as the first eggs may start to hatch, but shorter periods will result in a more even growth of the larvae as the age difference is minimal.

After the oviposition period, the eggs can be either collected directly from the substrate if the particle size is smaller than 0.5 mm (e.g. flour) via sieving or left in the substrate to hatch, depending on the specific set-up. The eggs (or egg/substrate mix) can be stored at a temperature between 15°C and 30°C and a RH between 40% and 80%.

4.2 Growing the larvae

The eggs typically hatch within 7–8 days in optimal conditions, giving rise to a very small white larvae. At this stage, they are fragile as their exoskeleton is yet to harden and should be handled with care, or preferably avoided altogether. These neonate larvae can be kept at high densities, exceeding 50 larvae/cm^2. Maintaining young larvae at a high density has certain advantages. High-density larvae can generate heat through their metabolic processes. This self-generated heat can help maintain a favorable temperature within the larval rearing environment, thus reducing the need for external heating sources. In addition, neonate larvae have a low feed consumption rate, which can cause problems as the wet feed is prone to desiccation or spoilage (fungus/bacteria growth) when left unconsumed. Increasing larval density will increase the overall larvae consumption capacity, thus reducing the initial risk of overfeeding with wet feed. The downside of a high initial stocking density is that, in time, this will need to be reduced due to the sheer increase in larvae biomass. Therefore, at Inagro's Insect Research Centre, the larvae are redistributed to a lower density at the age of 4 weeks.

In a production facility, the optimal density will depend on the feed and level of automation. In Deruytter et al. (2022), evidence is provided that the growth of the larvae depends on the number of larvae per cm^3, not cm^2, if wheat bran is used. About 1 larva/cm^3 was deemed optimal if the larvae are fed once with dry feed (e.g. in a manual operation), but higher concentrations were used, and therefore higher final yield could be achieved if the larvae were fed more than once. A feed depth of 10 cm did not alter the growth speed, and for a 60 × 40 cm crate (2000 cm^2 inner dimensions), this resulted in 20 000 larvae. For reproduction, when the intended use of the larvae is to become a new generation of beetles, a lower density is recommended (5000 larvae/crate), as this will result in a more even growth, larger larvae, and more healthy/larger pupae for the next generation.

Regardless of whether the larvae are for production or reproduction, they do need to be fed with a dry feed or a wet feed. The dry feed can consist of several agricultural by-products (Morales-Ramos et al., 2020; Rumbos et al., 2022), with wheat bran as a frequently used feed and scientific reference. Wet feed should be provided regularly throughout their growing period. Traditionally, chopped carrots are frequently used, but other nonfood organic materials can be used without any issue and may even improve their nutritional content (Kotsou et al., 2023). The amount of wet feed provided should be balanced with the mealworm biomass present in the crate to ensure that the feed does not go to waste through desiccation, gets moldy, or starts to rot. The latter two may have detrimental effects on the health of the larvae. To keep the wet feed stable for a longer period, it is possible to ferment the feed as the pH does not seem to affect the growth of the larvae (Coudron et al., 2022a).

In addition to the quantity, the location plays a key role in the growth of the larvae. The mobility of mealworms also increases as they mature, impacting the distribution of wet feed within the crate. In a study by Deruytter et al. (2020), it was observed that the distribution of wet feed plays a significant role in the growth and density distribution of the larvae. The researchers recommend that larvae, particularly those smaller than 10 mg, should always have a wet feed source within 5 cm of their location. This ensures that larvae of all sizes have equal opportunities for growth and are evenly distributed throughout the crate.

When mealworms are fully grown, which takes around 8–12 weeks, they can be harvested. An indication of their readiness for harvest is the presence of multiple pupae in the crate. Preferably, all dry feed should be consumed, leaving behind only frass, which consists of undigested feed, dead insects, and feces. The frass can easily be sieved from the other material using a 0.5 mm sieve. It is advisable not to delay the harvest when pupae are observed. Some applications require the mealworms to be starved to prevent the insect's gut load to enter the food chain. However, this does not or only minimally affect the microbial numbers (Wynants et al., 2017; Mancini et al., 2019) and may result in less nutritious larvae and more pupae. Harvesting the mealworms can be done using a sieve with a mesh size of 2–3 mm, allowing the frass to fall through while retaining the mealworms on top. After sieving, there may be some exuviae left as they are approximately the same size as the mealworms but much lighter. They can be easily blown off and collected as a source of chitin.

4.3 Pupation

The cannibalistic tendencies of mealworms can cause damage to fresh pupae. Certain measures can be taken to prevent this. One approach is to reduce the density of mealworms within the crate and maintain a high ratio of feed to mealworms. However, it is strongly recommended to regularly separate the pupae from the larvae. Especially, if there are differences in the growth of the larvae, it may take several weeks for the majority of mealworms to undergo pupation, and a small fraction may not pupate at all.

Pupae and mealworms are fairly similar in appearance, but pupae are slightly shorter and slightly broader. This characteristic can be exploited to sieve out the pupae using a bar/rectangular sieve with a mesh width of approximately 3.5 mm, and a length longer than 20 mm can be used for this purpose. The specific dimensions may vary depending on the specific strain and feed being used but can easily be checked via microscopy and width measurements. The sieve will retain the pupae while allowing the mealworms and substrate to pass through. Ensure that the sieve has no sharp edges and that the retention time on the sieve is minimal, with minimal vibrations to optimize the survival of the pupae. It is advisable to collect the pupae in a crate lined with wheat bran. The bran serves a dual purpose: it acts as a cushion to prevent pupal damage when sieving, and it serves as a source of feed for the emerging young beetles, preventing them from feeding on other pupae. After 11 days, all healthy pupae should have emerged, and the new beetles can be added to the breeding stock.

4.4 Beetle population

The adult beetles have a similar feeding habit as their larvae, which requires a dry feed and a wet feed source. The provision of the latter seems to have a direct impact on their longevity and daily reproductive output, making it a crucial factor. When taken care well, the adult beetles lay eggs daily and can maintain this pattern for a prolonged period (at least 2 months), but with a decline in egg laying and hatch rate in older females (Adamaki-Sotiraki et al., 2022). Therefore, it may be optimal to keep a young and active population. Again different methods to achieve this goal are possible, we will describe a system that results in a high egg yield and predictable outcome.

Each time a batch of beetles is transferred to a new oviposition crate, they are sorted and replenished. The first step is separating the active beetles from the dead or inactive ones. This is done by exploiting the grab reflex of active beetles by tipping them on a slow-moving angled conveyor belt made of a material they can grasp onto. This will allow the active beetles to be transported up while the dead/inactive beetles drop to the bottom of the belt, ensuring a healthy population. The second step is to remove some of the active but older population, up to a total of 25% (dead + inactive + others), so that a quarter of the population can be replaced by fresh beetles. In that way, at least 75% of the beetles are 4 weeks or younger (Fig. 1). Hereafter, the beetles can be weighed

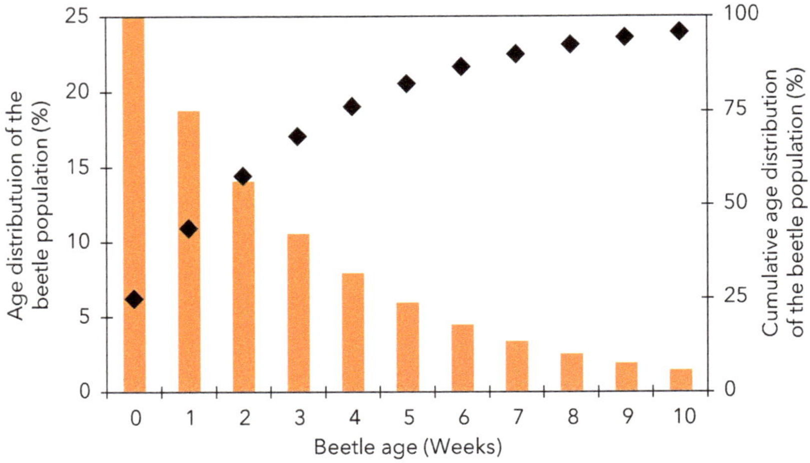

Figure 1 Final age structure of the beetle population when 25% new beetles are introduced each week. This theoretical model does not include a different mortality rate for older beetles.

and transferred back to the new crates. This homogenization of live and young beetles results in a high and consistent reproductive output across all crates in that batch.

5 Scale-up challenges

It is clear from the previous section that there are steps in the rearing of mealworms that pose a challenge when you want to scale up the production and those need to be carefully considered before building or designing a farm.

First of all, the environment may be more difficult to control in a high-density rearing facility. Although this does depend on the local climate, the (metabolic) heat production of the larvae may result in an overproduction of heat and may require cooling the rearing rooms rather than heating them. The total amount of carbon dioxide and water vapor will not be negligible at this point. This may increase the investment costs for a good HVAC system. Furthermore, the capacity of the HVAC system should be calculated correctly based on the future biomass as it is easy to underestimate based on a small-scale/hobby facility.

A second major challenge is feeding the mealworms with a wet feed (e.g. carrots). Due to the nature of the wet feed, neither a solid nor a liquid, it becomes inherently more difficult to transport. Each crate should receive this feed at least 3 times a week, and this should be well distributed throughout the crate. This means that each crate should be fed at least 18 times (start at week 3 and a total of 8 weeks of growth) excluding feeding in the different locations within a crate. Regardless, if one chooses to transport the crates to the feeding station or the other way around, it is a logistically complex and capital-intensive automation process is needed to reduce the manual labor cost.

Finally, in a large-scale facility, one may be inclined to build one very large climate room for the production of the larvae. Yet this may have serious drawbacks. Besides the impossibility to provide an age-appropriate optimal climate, it may be unfavorable for the control and spread of diseases and pest species throughout the facility. Separate rooms have the advantage that they can be sanitized thoroughly but may impose a large cost on both the building and accompanying automation.

6 Design of rearing facilities

Although the individual business choices, location, and existing infrastructure will determine the final outline of the rearing facility, there are some things each facility will have in common. Feed will arrive either from the local farm or further away and will be temporarily stored at the facility. In the next step, the feed is processed. This may be as simple as mixing additives in the feed but may be more complex depending on the specific operation. In the rearing rooms, the larvae are fed with the processed feed and those rooms may be subdivided by age. In many cases, the neonates (young larvae) are initially kept at a higher temperature than when they are fully grown larvae. At the end of the cycle, the larvae are harvested. The frass is collected as a biofertilizer, and, in general, the insects go further toward a processing line.

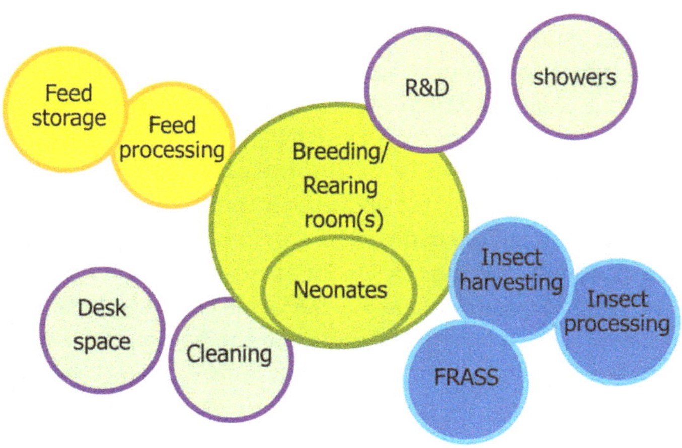

Within this scheme, there is a lot of room for flexibility and creativity, but there are some crucial points that need special attention to ensure a smooth operation:

1. **General building**: Although a gentle breeze on a hot summer day is liked by all, this is not something you want in the facility. The windows and doors should be shut at all times with a well-designed interior HVAC system to control the indoor climate. Each part of the building should be well thought of in terms of how it will be cleaned and maintained, including the floor, walls, and ceiling.
2. **Feed preparation and storage areas**: In many cases, the feed will need to be sanitized to ensure no unwanted pest species entering the facility. Thereafter, the feed will need to be stored in a proper way, cool and dry for the dry feed, and refrigerated or ensilaged for the wet feed.
3. **Rearing area**: To minimize the risk of contamination, it is recommended to position the rearing rooms at a considerable distance from the direct outside environment. It is advisable to have a minimum of two doors separating the outside environment from the rearing facility. One possible configuration could be door, desk, door, rearing facility but even better outside, door, desk, door, showers/changing rooms, door, and rearing facility.
4. **Sieving/harvesting area**: This area should be separated from the other rooms with a separate ventilation system as plenty of dust and accompanying hazards will be produced there while sieving the larvae. We strongly advise against sieving in any kind of general area, the fine dust produced, and airborne allergens do produce a potential health risk.

As mentioned before, it may not be necessary or even beneficial to perform each step of the insect value chain in a single company but rather have a decentralized industry. The equipment and knowledge needed for the reproduction, production, and processing of the insects are quite specific and they may be underused (reproduction) or very expensive in a small scale (processing).

Finally, innovative ways to rear mealworms or locate your mealworm farm may improve the economics of the endeavor by reducing the operational costs or maximize the generated income. One example is entomoponics (Coudron et al., 2022b), where mealworms are bred in (or in close proximity to) a greenhouse. This has the advantage that the larvae can be fed with greenhouse waste products, the heat from the greenhouse can be used to heat the larvae, and finally the carbon dioxide that the larvae produce can be used to stimulate the growth of the plants in the greenhouse.

7 Conclusion

Mealworm farming has great potential as emerging sector with the agro-industry. Although large strives forward have been made in recent years in our understanding of optimal rearing conditions and the needs of a mealworm farm, there are still some major knowledge gaps. Fortunately, besides the industry also, the research is booming with new information published on a near daily basis. The information provided above is part of that, but there are multiple ways to achieve success and rear these larvae.

8 Where to look for further information

Two interesting documents have been published that provide additional insights on the use of insects and mealworms:

- The FAO recommendation on edible insects: https://www.fao.org/3/i3253e/i3253e.pdf.
- The EFSA opinion on safety of dried mealworm larvae: Safety of dried yellow mealworm (*Tenebrio molitor* larva) as a novel food pursuant to Regulation (EU) 2015/2283 (wiley.com).

Regional and national insect associations are a great place to look for up-to-date information on legislation, fact sheets and guidelines. These include:

- IPIFF: Europe;
- NACIA: North America;
- AFFIA: Asia; and
- IPAA: Australia.

Finally, it is important to note that the amount of research on *Tenebrio molitor* is increasing, thus improving our knowledge on a near-daily basis. A free way to follow-up on the scientific publications is checking Google Scholar.

9 References

Adamaki-Sotiraki, C., Rumbos, C. I. and Athanassiou, C. G. (2022). Strain effect on the adult performance of the yellow mealworm, *Tenebrio molitor* L. *Journal of Insects as Food and Feed* 8(12), 1401–1410.

Berggreen, I. E., Offenberg, J., Calis, M. and Heckmann, L.-H. (2018). Impact of density, reproduction period and age on fecundity of the yellow mealworm *Tenebrio molitor* (Coleoptera: Tenebrionidae). *Journal of Insects as Food and Feed* 4(1), 43–50.

Bjørge, J. D., Overgaard, J., Malte, H., Gianotten, N. and Heckmann, L. H. (2018). Role of temperature on growth and metabolic rate in the tenebrionid beetles *Alphitobius diaperinus* and *Tenebrio molitor*. *Journal of Insect Physiology* 107, 89–96.

Coudron, C. L., Deruytter, D. and Claeys, J. (2022a). The influence of wet feed pH on the growth of *Tenebrio molitor* Larvae. *Sustainability* 14(13), 7841.

Coudron, C. L., Deruytter, D., Craeye, S. and Bleyaert, P. (2022b). Entomoponics: combining insect rearing and greenhouse vegetable production–a case study with

Tenebrio molitor and high-wire cucumber cultivation. *Journal of Insects as Food and Feed* 8(4), 427–437.

Deruytter, D., Coudron, C. L. and Teerlinck, S. (2019). Influence of crate size, oviposition time, number of adults and cannibalism on the reproduction of *Tenebrio molitor*. *Journal of Insects as Food and Feed* 5(4), 247–255.

Deruytter, D., Coudron, C. L. and Claeys, J. (2022). The effects of density on the growth and temperature production of *Tenebrio molitor* Larvae. *Sustainability* 14(10), 6234.

Deruytter, D., Coudron, C. L. and Claeys, J. (2023). Transporting *Tenebrio molitor* eggs: the effect of temperature, humidity and time on the hatch rate. *Sustainability* 15(7), 6231.

Drnevich, J. M., Papke, R. S., Rauser, C. L. and Rutowski, R. L. (2001). Material benefits from multiple mating in female mealworm beetles (*Tenebrio molitor* L.). *Journal of Insect Behavior* 14(2), 215–230.

Eberle, S., Schaden, L. M., Tintner, J., Stauffer, C. and Schebeck, M. (2022). Effect of temperature and photoperiod on development, survival, and growth rate of mealworms, *Tenebrio molitor*. *Insects* 13(4), 321.

Hansen, L. L., Ramløv, H. and Westh, P. (2004). Metabolic activity and water vapour absorption in the mealworm *Tenebrio molitor* L. (Coleoptera, Tenebrionidae): real-time measurements by two-channel microcalorimetry. *Journal of Experimental Biology* 207(3), 545–552.

Kotsou, K., Chatzimitakos, T., Athanasiadis, V., Bozinou, E., Adamaki-Sotiraki, C., Rumbos, C. I., Athanassiou, C. G. and Lalas, S. I. (2023). Waste orange peels as a feed additive for the enhancement of the nutritional value of *Tenebrio molitor*. *Foods* 12(4), 783.

Ludwig, D. and Fiore, C. (1960). Further studies on the relationship between parental age and the life cycle of the mealworm, *Tenebrio molitor*. *Annals of the Entomological Society of America* 53(5), 595–600.

Machin, J. (1975). Water balance in *Tenebrio molitor*, L. larvae; the effect of atmospheric water absorption. *Journal of Comparative Physiology* 101(2), 121–132.

Morales-Ramos, J. A., Rojas, M. G., Kay, S., Shapiro-Ilan, D. I. and Tedders, W. L. (2012). Impact of adult weight, density, and age on reproduction of *Tenebrio molitor* (Coleoptera: Tenebrionidae). *Journal of Entomological Science* 47(3), 208–220.

Morales-Ramos, J. A., Kelstrup, H. C., Rojas, M. G. and Emery, V. (2019). Body mass increase induced by eight years of artificial selection in the yellow mealworm (Coleoptera: Tenebrionidae) and life history trade-offs. *Journal of Insect Science* 19(2), 4.

Morales-Ramos, J. A., Rojas, M. G., Kelstrup, H. C. and Emery, V. (2020). Self-selection of agricultural by-products and food ingredients by *Tenebrio molitor* (Coleoptera: Tenebrionidae) and impact on food utilization and nutrient intake. *Insects* 11(12), 827.

Park, J. B., Choi, W. H., Kim, S. H., Jin, H. J., Han, Y. S., Lee, Y. S. and Kim, N. J. (2014). Developmental characteristics of *Tenebrio molitor* Larvae (Coleoptera: Tenebrionidae) in different instars. *International Journal of Industrial Entomology* 28(1), 5–9.

Punzo, F. and Mutchmor, J. A. (1980). Effects of temperature, relative humidity and period of exposure on the survival capacity of *Tenebrio molitor* (Coleoptera: Tenebrionidae). *Journal of the Kansas Entomological Society* 53(2), 260–270.

Ribeiro, N., Abelho, M. and Costa, R. (2018). A review of the scientific literature for optimal conditions for mass rearing *Tenebrio molitor* (Coleoptera: Tenebrionidae). *Journal of Entomological Science* 53(4), 434–454.

Rumbos, C. I., Oonincx, D. G. A. B., Karapanagiotidis, I. T., Vrontaki, M., Gourgouta, M., Asimaki, A., Mente, E. and Athanassiou, C. G. (2022). Agricultural by-products from Greece as feed for yellow mealworm larvae: circular economy at a local level. *Journal of Insects as Food and Feed* 8(1), 9–22.

Chapter 7

Advances in nutrient/substrate sources/composition for rearing yellow mealworms as a protein source

Riccardo Melis and Roberto Anedda, Porto Conte Ricerche S.r.l., Italy

1 Introduction

Edible insect farming is relatively new and currently small, though rapidly expanding, industry. However, there is still insufficient knowledge about the optimal nutritional requirements of each farmed species in order to reach maximum productivity and to scientifically manage mass rearing strategies so as to reach the highest nutritional quality of insects and derived products. The nutritional composition of edible insects has been surveyed in several review publications (Hawkey et al., 2021; Jantzen da Silva Lucas et al., 2020; Osimani et al., 2017), but a wealth of information is still to be produced on this matter, since there is a considerable lack of knowledge on molecular features related to the absorption and metabolism of feed nutrients as well as to process-related effects on insect meals quality. In fact, the effects of rearing substrates on insect metabolism have been less well reviewed so far. The term 'substrate' generically refers to either the materials directly used for insect rearing and

http://dx.doi.org/10.19103/AS.2024.0139.08

formulated growth feeds. Depending on local regulations in force, according to various governmental departments and various levels of government and competent authorities, different countries consider the exploitation of different substrates as being legal or not, and in some countries, where insects are not considered novel foods, specific regulations on insect-rearing substrates are lacking. For example, in some African and Asian states, a large and easily accessible variety of bio-waste products, such as catering waste and manure, are allowed and used as substrates for insect rearing (Kinyuru and Ndung'u, 2020; Kee et al., 2023; Lähteenmäki-Uutela et al., 2021). On the contrary, certain ruminant proteins, catering waste, and meat-and-bone residues are not permitted as feed substrates under EU legislation; therefore only raw materials of vegetable origin should be used. Exceptions mainly consist in some allowed animal-derived products, such as eggs, milk, and related products, rendered fat or blood products, provided they originate from non-ruminant animals (Regulation (EC) No 999/2001). Although new substrates are being continuously tested, EU legislation regulates at different levels (on general rules: Regulation No 178/2002; Regulation No 852/2004. On farmed animal health and disease management: Regulation (EU) No 2016/429) the production, processing and commercialization of insects and derived products. As a consequence, while on one hand, strict regulations are believed to somewhat limit the rapid expansion of the insect rearing sector, on the other hand, high hygiene standards are essential to ensure the safety and quality of marketed products. In any case, insect farming is currently more developed in Asia, with approximately 26 000–28 000 metric tons produced per year, followed by Africa and Middle East, who annually produce together 19 000–21 000 tons (Insects raised for food and feed – global scale, practices, and policy: https://rethinkpriorities.org/publications/insects-raised-for-food-and-feed, accessed March 31, 2023), which means that more than 70% of farmed insect production at global level is concentrated in Asia, Africa, and Middle East.

2 Yellow mealworm: protein quality

Protein content of insects varies according to species, developmental stage, rearing substrate, and method of analysis (Meyer-Rochow et al., 2021; Rumpold and Schlüter, 2013; Payne, 2016; Toviho and Bársony, 2022). Several studies emphasize the nutritional benefits of mealworms, especially in terms of their high protein content (Zhao et al., 2016; van der Fels-Klerx et al., 2018; Costa et al., 2020), which is expected to be able to replace soy proteins in food and animal feed (EFSA Panel on Nutrition, Novel Foods and Food Allergens (NDA): https://www.efsa.europa.eu/en/science/scientific-committee-and-panels/nda, accessed March 31, 2023; Toviho and Bársony, 2022). Amino acid profile makes much of the quality of alternative proteins compared to terrestrial animal meat, fish, and vegetable proteins. According to recent studies (Jensen et al., 2019), the amino acid score of YMW processed by freeze drying, as corrected to consider overall protein digestibility factor (referred to as PDCAAS) and accounting for crude protein fecal digestibility studied in rats, was found to be 76% for 0.5- to 3-year-old children – a slightly better value than the average from vegetable protein sources but lower than what is measured in beef meat and soy. In this regard, the European Food Safety Authority (EFSA) points out that comparing such parameters from different investigations requires particular care to be taken. Apparent protein digestibility of YMW has been evaluated on different species, such as pigs (Yoo et al., 2019; Jin et al., 2016) and chickens (Biasato et al., 2017). Recent studies demonstrated that depending on processing (e.g. drying, defattening, and grinding), digestibility of YMW meal can substantially differ in farmed fish species like trout (Gasco et al., 2022). This is in agreement with the notes released by EFSA and draws attention in the evaluation of results from different techniques (*in vitro*, *in vivo*, fecal digestibility, apparent ileal digestibility and methods of fecal sampling, especially in aqueous environments). As a matter of fact, taking into account current scientific results, YMW meal, if suitably processed, is expected to present higher digestibility of amino acids than other protein-rich meals such as fish or meat meal (Yoo et al., 2019; Lee et al., 2021; Gasco et al., 2022). As far as microbiological quality and safety of YMW meal are taken into account, several countries have established specific regulations (Jo and Lee, 2016; Lee et al., 2022; Sogari et al., 2019) which could expand the perspectives of using insects as novel feed and food.

3 Characteristics of rearing substrates for optimal yellow mealworm growth

Insect producers usually choose insect rearing substrates based on a variety of factors, including nutritional composition, expected effects on weight gain (also known as biomass production), feed efficiency (feed conversion ratio), the time required to reach the harvesting stage, and a consistent year-round supply. Having a significant impact on several of these factors, the nutritional composition of substrate certainly plays a fundamental role for an efficient insect growth.

Some of the key characteristics that determine the quality of insect biomass are protein concentration and amino acid profile, fatty acid profile, micronutrients (such as minerals), the absence of hazards, and ease of removal during harvesting (Pinotti and Ottoboni, 2021; Mancini et al., 2019). High protein and fat contents of the substrate are believed to mainly increase the nutritional value of insect products. In particular, a high content of essential amino acids as well as mono- and poly-unsaturated fatty acids are expected to increase their nutritional quality.

It appears clear that substrate composition has an impact on both the insect nutritional quality and on growth efficiency as well (Fuso et al., 2021; Riekkinen et al., 2022; Pinotti et al., 2021; Bordiean et al., 2022). For this reason, it would be beneficial to formulate the most suitable substrate for each reared insect species. A crucial aspect of optimization of the substrate includes, first of all, protein and fat contents, which should be neither too high nor too low. As previously observed by van Broekhoven et al. (2015), too low protein level in the substrate causes a delay in growth. At the same time, too-high fat content in the substrate may cause the death of YMW in the adult stage, due to substrate agglomeration that hinders breathing and movement of insects (Alves et al., 2016; Dreassi et al., 2017). Additionally, protein and carbohydrates ratio in the substrate has been proven to influence the growth of edible insects. In particular, Morales-Ramos et al. (2020) observed that a macronutrient ratio of 0.23:0.06:0.71 between protein, lipids, and carbohydrate contents, respectively, was optimal for favoring the growth of YMW. Several studies also evidence that the intake of carbohydrates increases the feed efficiency of YMW, whereas the intake of neutral detergent fiber had a negative impact on their growth (van Broekhoven et al., 2015; Bordiean et al., 2022; Prabhat et al., 2023).

4 Yellow mealworm protein content analysis

Some recent investigations suggest that the protein content of insects has been overestimated in several early studies. As for the largest part of food or feed, the total amount of crude protein in edible insect is often assayed by either the Kjeldahl or the combustion (Dumas) methods, where the amount of nitrogen is determined and converted to crude protein by the nitrogen-to-protein conversion factor (generally 6.25). However, in early 2017, some authors pointed out that the protein content of insects has been sometimes overestimated due to the presence of chitin, which is a polysaccharide composed of units of *N*-acetylglucosamine containing a nitrogen atom. Several studies addressed the aforementioned issue. For example, Janssen et al. (2017) argued that the commonly used conversion factor of 6.25 overestimates the protein content due to the presence of non-protein nitrogen. They proposed, as a result of amino acid analysis, a protein conversion factor of 4.76 ± 0.09 for the larvae from YMW. Protein purification finally led to a conversion factor of 5.60 ± 0.39 for the larvae of YMW. After few months, Jonas-Levi and Martinez (2017) suggested that new N-conversion factors should be proposed for all the insect species and according to specific developmental stages, and that nitrogen found in cuticle of insects should be subtracted from the total nitrogen content. In this line, Ritvanen et al. (2020) proposed the use of a reduced nitrogen conversion factor of 5.0 for house cricket and field cricket, and, more recently, Boulos et al. (2020) estimated a conversion factor of 5.41 for YMW.

In general, it can be concluded that overestimation of protein content of insects can be avoided by analyzing non-digestible chitin separately.

5 Factors influencing yellow mealworm protein body content

According to previous studies on the ultrastructure of insects, protein synthesis occurs in a particular anatomical region, named fat body, consisting of a large central fat globule surrounded by numerous peripheral globules (Pemrick and Butz, 1970; Price, 1973; Soltani-Mazouni and Soltani, 1995). In particular, some of these peripheral globules have been classified as protein storage granules (Locke and Collins, 1965, 1967, 1968). It appears that these granules respond to changes in the insect nutritional and hormonal environment which regulate the balance between protein synthesis, release, and sequestration. In particular, Mordue (1965) evidenced that the neurosecretory control of hemolymph protein concentration in YMW responds directly to changes occurring in the hormonal environment. Similar to other Lepidoptera species, it was assumed that a cyclic fluctuation occurs in the rate of incorporation of amino acids into YMW fat body proteins, which could be correlated with cycles of oocyte growth and neuroendocrine activity (Mordue, 1965).

As recently investigated, YMW exhibit different nutrient levels – including total proteins and amino acid composition – according to their developmental phases (Yu et al., 2021). In particular, researchers observed that YMW protein content decreased with the growth of instars larvae, leading to changes in essential amino acids contents. Consequently, the nutritional value of the proteins progressively changed with their physiological status. On the whole, the best YMW protein yield and quality was found at the earlier larval stage (Yu et al., 2021).

Previous studies also looked at different diets for mass production of YMW, evaluated as nutritional supplements to the YMW diet, and when compared to a carbohydrate-rich diet, a diet with a higher protein and lipid content significantly improved most biological parameters measured. In particular, van Broekhoven et al. (2015) showed that YMW growth on high protein and low starch substrates accumulated the highest crude protein and crude fat content. Nevertheless, it was discovered that the protein or fat contents of YMW did not differ significantly depending on substrate, with the exception of the fat content of the YMW fed on low protein-high starch substrate, which was lower than that of YMW fed other substrates (van Broekhoven et al., 2015; Bordiean et al., 2022). Thus, we can conclude that YMW is able to secrete excess protein and to adjust the amount of protein in their body to the optimal level.

6 Use of novel biomasses as substrates to increase the yellow mealworm protein content

Aiming to improve the economic and environmental sustainability of the edible insect production, the replacement of standard rearing substrates with those derived from food loss and agro-industrial waste can lead to the creation of new value chains and employment opportunities (Heckmann et al., 2018; Sangiorgio et al., 2022). It is projected that by 2030 more than 2 billion tons of food waste will be globally available, representing an important source of potentially high-quality nutrients to feed insects (Elleby et al., 2021). A high demand for insect-based proteins is expected to increase in the near future from the livestock sector as well as from aquaculture activities. YMW growth and rearing has been successfully performed on substrates originating from several organic by-products, such as those derived from beer brewing (Chia et al., 2018), bread/cookie baking (Mancini et al., 2019), and bioethanol production (Zhang et al., 2019). It was suggested that the protein level in dietary substrate intended for YMW rearing is able to influence the protein level in insect larvae.

However, protein content of substrates should be limited to the essential supply, since larvae reared on a diet with higher protein levels excrete the excess of protein as uric acid (van Broekhoven et al., 2015; Montalbán et al., 2022). Studies also indicate that even if protein-rich diets could be essential for mealworm larval growth and survival, they do not necessarily result in an increased protein content of larvae (van Broekhoven et al., 2015)

Brewer's spent grain (BSG) is a by-product of the brewing industry that makes up to 85% of brewing waste. BSG is obtained as a mostly solid residue after wort production in the brewing process and presents a high economic potential as a low-cost source of valuable nutrients for animal nutrition (Mussatto et al., 2006; Mussatto, 2014). Recently, BSG has been also positively evaluated as being a very suitable substrate for insect mass production (Chia et al., 2018; Jucker et al., 2020). In particular, it was found that YMW fed on BSGs have higher body protein and carbohydrate contents compared to YMW larvae reared on more traditional substrates such as bread or cookies (Mancini et al., 2019). In another recent study, Melis et al. (2019) observed that the crude fat content was considerably lower in YMW reared on BSG than in mealworms reared on more traditional wheat bran (WB) substrate. It was explained that high-fiber content (cellulose and lignin) of BSG (Li et al., 2016) played a key role in determining different insect body compositions. The same study also reported that rearing YMW on BSG led to a higher feed conversion ratio and efficiency in conversion of ingested food than conventional WB (Melis et al., 2019), in agreement with previous evidences (Kim et al., 2016).

Distillers' grains (DG) are by-products derived from the process of distilling spirits. These grains are marketed in several forms based on moisture content (Böttger and Südekum, 2018; Zeinab et al., 2021). Zhang et al. (2019) tested spirit distillers' grains (SDG) as substrates to rear YMW with good results in terms of protein levels, also observing that YMW fed on fiber-rich substrates had higher protein and lower fat contents than YMW fed on fiber-free substrates.

Olive pomace (OP) is another interesting by-product from the olive oil industry (Akar et al., 2009; Nasopoulou and Zabetakis, 2013) and it has been assessed as a substrate for YMW larvae. YMW reared in 25% OP-enriched diet on a base of wheat middlings exhibited the highest protein and carbohydrate levels with respect to other tested substrate formulations (Ruschioni et al., 2020). In particular, the addition of 25% of OP led to the highest yield on the dry matter (DM), the most interesting protein content (47.58% DM), and an optimized essential/non-essential amino acids ratio (Ruschioni et al., 2020).

Several studies have been carried out to evaluate the efficacy of other promising agro-food by-products and vegetable wastes as YMW dietary substrates and indicated advantages and limitations of each of them and their combinations. For example, Kröncke and Benning (2022) tested 25 different substrates, including by-products and residues (e.g. hulls, husks, bran, cakes, and flours) from the processing of grain, rice and other vegetables, pomace and other plant components, and evidenced a wide variability in the relative consumption of macronutrients by YMW. They concluded that the nutritional requirements are in all growth stages in the series carbohydrates>proteins>lipids>minerals, and that the need for carbohydrates decreases with age, while an increased requirements for proteins and lipids is observed as insect larvae grow up. They also observed that the most consumed substrates were generally high in carbohydrate content, while substrates characterized by a high amount of fibers or proteins showed lower consumption or no consumption at all. The same authors and other similar studies (Morales-Ramos et al., 2013) indicate for YMW the optimal ratio between substrate macronutrients as approximately 10%:20%:70% of lipids, proteins, and carbohydrates, respectively. Use of wheat and rye barn as substrates resulted in 44–48% crude protein (CP) of YMW larvae (Bordiean et al., 2022), use of wheat barn led to 37% CP in larvae (Zielińska et al., 2021), Carob meal substrates allowed obtaining larvae with 45% CP (Antonopoulou et al., 2022), and larvae grown in oat-based substrates had approximately 44% CP (Toviho and Bársony 2022). Varelas and Langton (2017) reviewed the potential of forest biomass by-products for rearing edible insects and provided examples of the negative consequences of unbalanced feeding on insect growth and nutritional values. Moreover, Zhang et al. (2019) investigated the effect of agricultural waste as a feed for YMW by replacing mushroom substrates with wheat bran feed, with similar results in terms of feed efficiency and protein levels. Oonincx et al. (2015), in a work aimed at identifying food by-products that can be used as substrates and be efficiently converted by YMW, evidenced an optimal YMW nitrogen efficiency – higher than that of conventional production animals – especially feeding fresh vegetable like carrot waste.

7 Yellow mealworm rearing on plastic substrates

The need for new strategies to face the global plastic pollution concern is one of the most pressing problems for our society. According to the available data from 2019, the worldwide plastic production reached 368 Mt (megatons) mainly in Asia (51%, of which 31% in China), followed by North American countries (North America, Canada, and Mexico) (19%), Europe (16%), Middle East and Africa (7%), Latin America (4%), and the countries of the Commonwealth of Independent States (3%) (Plastics-the Facts. An analysis of European Plastics Production, Demand and Waste Data. Plastics Europe, Association of Plastics Manufacturers. 2020. Available online: https://www.plasticseurope.org/download_file/force/4261/181, accessed March 31 2023). Only a small percentage of plastics is recovered and recycled, increasing the quantity of waste released into the environment and losing its economic value. At global level, plastic products account for ~120 million tons per year, but only ~10% are recycled and ~12% used for energy recovery through combustion (Kosior and Mitchell, 2020). In European countries, including Norway and Switzerland, it is estimated that, in 2020, 38% of post-consumer plastic waste was discarded in landfills, while only 5% of the over 15 Mt of mixed-waste collected fraction was recycled (Bombelli et al., 2017). In this perspective, the so-called *entomoremediation*, i.e. the use of insects for plastics degradation, has recently opened new interesting opportunities to solve this problem. Several insect species, mainly *Galleria mellonella*, *Zophobas atratus,* and *Tenebrio molitor*, have shown their capability in decomposing plastics such as polystyrene (PS), polyethylene (PE), polypropylene (PP), and polyvinyl chloride (PVC). Among these, YMW can be an potential solution for the bioconversion of plastics, thanks to its particular intestinal microbiota. In particular, Brandon et al. (2018) discovered the eight intestinal microorganisms responsible of PS degradation in YMW, including *Citrobacter freundii*, *Serratia marcescens*, and *Klebsiella aerogenes*, in fact opening a new way for industrial-scale cultivation of specific and efficient microorganisms able to more sustainably discard plastics. The processes of plastic biodegradation and subsequent mineralization start from chewing the materials by the YMW larvae, then shredded plastic comes into contact with unique intestinal microorganisms and extracellular enzymes housed in the larval gut, which cause the decomposition of the long-chain molecules into lower molecular weight metabolites (Yang et al., 2015; An et al., 2023). Several recent investigations have proven that YMW larvae are able to degrade a wide range of inorganic polymeric materials such as PE, PS, PP, polyurethane (PU), polylactic acid (PLA), and PVC (Billen et al., 2020; Bulak et al., 2021; Urbanek et al., 2020; Wu et al., 2019) and even tire crumbs and vulcanized butadiene-styrene elastomer rubber (SBR) (Aboelkheir et al., 2019). The abovementioned studies were also aimed at determining YMW feeding capacities and survival rates on such plastic substrates, also determining the changes in the intestinal microbiota of larvae. One of the main observed detrimental effect of feeding YMW with only plastic substrates for a long time was a noticeable body-weight loss. For example, Bulak et al. (2021) evidenced a remarkable mass reduction efficiency (41.0–69.7%) when YMW were fed PU and PE for 58 days. However, when the plastic-based substrate was supplemented with conventional food for larvae (usually bran), a marked improvement in the plastics degradation and also on the growth performance of larvae was observed. It should be noted that a plastics-only diet lacks sufficient nutrients for protein synthesis to support YMW growth, making growth and even survival very challenging. It was demonstrated that the survival rate of YMW larvae fed PVC as the sole food for 5 weeks was quite high (80%), but it decreased to 39% after 3 months (Peng et al., 2020). Therefore, supplementary nutrients have to be provided to increase survival and to allow completion of the life cycle. To address this issue, for example, Peng et al. (2020) used PVC plus a co-diet (PVC:wheat bran 1:5, w/w) and showed that YMW could complete growth and pupation in 3 months, similarly to what was observed when feeding a bran-only diet. Yu et al. (2021) investigated on YMW biodegradation of PS and PE with supplementary bran as co-diets and showed that a bran and plastic mass ratio of 7:1 (w/w) facilitated the growth of plastic-feeding mealworms. Also, the examined bran-fed YMW gut microbiomes exhibited substantial distinction from those of plastics-fed groups, suggesting that the effect of bran and plastics mixture on the community assembly of gut microbes was distinguished from those of either bran or plastics alone. Despite important achievements have been reached in the knowledge of plastic bioremediation through YMW larvae, still little optimization work has been carried out in order to characterize and maximize the protein content of plastic-reared larvae. In this light, although it is now well recognized that the key to YMW plastic biodegradation is its intestinal microbiome, further research is needed to more deeply characterize YMW intestinal process and related enzymatic activity to upgrade it to full-scale use.

8 Conclusion

The interest in YMW as a source of food and feed has been increasing steadily due to the high nutritional value of mealworms, ease of production, and lower environmental impact in comparison with livestock. On the basis of the evidences reviewed earlier, the following can be summarized:

1 The growth of YMW can be maximized by proper nutritional approaches. Considerable effort has been focused on the development of proper diets that would maximize the YMW biomass production with a concomitant reduction of their development time. Self-selection is an effective method to approximate optimal combination ratios of substrates for YMW. The ratios and combinations of preferred substrates had a significant impact on larval growth, the increase in biomass and feed conversion.
2 YMW chemical composition can be functionally pre-designed: feeding on selected substrates impacted the nutrient composition of YMW, a key finding that calls for further investigations. It seems that the most important factor affecting the YMW nutritional quality is the carbohydrate and protein content of the substrate. In particular, protein content of YWM is strictly influenced by substrate composition and macronutrient uptake, since a higher protein content in the diet can increase the YWM body protein content.
3 Low-value, organic by-products, and agro-food wastes have great potential as substrates for increasing environmental and economic sustainability of YMW production. Agricultural by-products can offer suitable nutrients for YMW, provided they are given in proper ratios and combinations. Previous studies demonstrated that YMW can effectively convert low-energy feeds, substrates, and biomass into products with high nutrient content. Regarding the ability of YMW to degrade plastic waste, more research is needed on the optimization of rearing conditions, dietary approaches and protein content maximization.

At a time when the insect industry is attempting to produce the necessary quantities to cover current and future market demands for edible insects production, the present chapter contributes to the development of diets and dietary formulations that will allow fast YMW growth and development, thereby accentuating this edible insect species as an attractive alternative source of protein.

9 Where to look for further information

- The following articles provide a good overview of the subject:
- Melis, R., Braca, A., Sanna, R., Spada, S., Mulas, G., Fadda, M. L., Sassu, M. M., Serra,G. and Anedda, R. Metabolic response of yellow mealworm larvae to two alternative rearing substrates. *Metabolomics* 15, 113 (2019).
- Edible insects – Future prospects for food and feed security, free available at https://www.fao.org/fsnforum/resources/reports-and-briefs/edible-insects-future-prospects-food-and-feed-security.

Key research in this area can be found at the following organizations:

- Porto Conte Ricerche S.r.l (https://www.portocontericerche.it/en).
- CNR – Research Institute on Terrestrial Ecosystems (https://www.ibe.cnr.it/).

10 References

Aboelkheir, M. G., Visconte, L. Y., Oliveira, G. E., Toledo Filho, R. D. and Souza, F. G. (2019). The biodegradative effect of Tenebrio molitor Linnaeus larvae on vulcanized SBR and tire crumb. *Sci. Total Environ.* 649, 1075–1082. doi: 10.1016/j.scitotenv.2018.08.228.

Akar, T., Tosun, I., Kaynak, Z., Ozkara, E., Yeni, O., Sahin, E. N. and Akar, S. T. (2009). An attractive agro-industrial by-product in environmental cleanup: Dye biosorption potential of untreated olive pomace. *J. Hazard. Mater.* 166(2–3), 1217–1225. doi: 10.1016/j.jhazmat.2008.12.029.

Alves, A. V., Sanjinez-Argandoña, E. J., Linzmeier, A. M., Cardoso, C. A. L. and Macedo, M. L. R. (2016). Food value of mealworm grown on Acrocomia aculeata pulp flour. *PLoS ONE* 11(3), e0151275. doi: 10.1371/journal.pone.0151275.

An, R., Liu, C., Wang, J. and Jia, P. (2023). Recent advances in degradation of polymer plastics by insects inhabiting microorganisms. *Polymers (Basel)* 15(5), 1307. doi: 10.3390/polym15051307.

Antonopoulou, E., Panteli, N., Feidantsis, K., Mastoraki, M., Koutsogeorgiou, E. I., Grivaki, E., Papagrigoriou, T., Christias, S. P., Chatzifotis, S., Lazari, D., Andreadis, S. S. and Krigas, N. (2022). Carob (*Ceratonia siliqua*) as functional feed is beneficial in yellow mealworm (*Tenebrio molitor*) rearing: Evidence from growth, antioxidant status and cellular responses. *Antioxidants* 11(9), 1840. doi: 10.3390/antiox11091840.

Biasato, I., Gasco, L., De Marco, M., Renna, M., Rotolo, L., Dabbou, S., Capucchio, M. T., Biasibetti, E., Tarantola, M., Bianchi, C., Cavallarin, L., Gai, F., Pozzo, L., Dezzutto, D., Bergagna, S. and Schiavone, A. (2017). Effects of yellow mealworm larvae (*Tenebrio molitor*) inclusion in diets for female broiler chickens: Implications for animal health and gut histology. *Anim. Feed Sci. Technol.* 234, 253–263. doi: 10.1016/j.anifeedsci.2017.09.014.

Billen, P., Khalifa, L., Van Gerven, F., Tavernier, S. and Spatari, S. (2020). Technological application potential of polyethylene and polystyrene biodegradation by macro-organisms such as mealworms and wax moth larvae. *Sci. Total Environ.* 735, 139521. doi: 10.1016/j.scitotenv.2020.139521.

Bombelli, P., Howe, C. J. and Bertocchini, F. (2017). Polyethylene bio-degradation by caterpillars of the wax moth *Galleria mellonella*. *Curr. Biol.* 27(8), R292–R293. doi: 10.1016/j.cub.2017.02.060.

Bordiean, A., Krzyżaniak, M., Aljewicz, M. and Stolarski, M. J. (2022). Influence of different diets on growth and nutritional composition of yellow mealworm. *Foods* 11(19), 3075. doi: 10.3390/foods11193075.

Böttger, C. and Südekum, K.-H. (2018). Review: Protein value of distillers dried grains with solubles (DDGS) in animal nutrition as affected by the ethanol production process. *Anim. Feed Sci. Technol.* 244, 11–17. doi: 10.1016/j.anifeedsci.2018.07.018.

Boulos, S., Tännler, A. and Nyström, L. (2020). Nitrogen-to-Protein Conversion Factors for Edible Insects on the Swiss Market: *T. molitor*, *A. domesticus*, and *L.* migratoria. *Front. Nutr.* 7, 89. doi: 10.3389/fnut.2020.00089.

Brandon, A. M., Gao, S. H., Tian, R., Ning, D., Yang, S. S., Zhou, J., Wu, W. M. and Criddle, C. S. (2018). Biodegradation of polyethylene and plastic mixtures in mealworms (larvae of *Tenebrio molitor*) and effects on the gut microbiome. *Environ. Sci. Technol.* 52(11), 6526–6533. doi: 10.1021/acs.est.8b02301.

van Broekhoven, S., Oonincx, D. G. A. B., Van Huis, A. and Van Loon, J. J. A. (2015). Growth performance and feed conversion efficiency of three edible mealworm species (*Coleoptera: Tenebrionidae*) on diets composed of organic by-products. *J. Insect Physiol.* 73, 1–10. doi: 10.1016/j.jinsphys.2014.12.005.

Bulak, P., Proc, K., Pytlak, A., Puszka, A., Gawdzik, B. and Bieganowski, A. (2021). Biodegradation of different types of plastics by *Tenebrio molitor* Insect. *Polymers* 13(20), 3508. doi: 10.3390/polym13203508.

Chia, S. Y., Tanga, C. M., Osuga, I. M., Mohamed, S. A., Khamis, F. M., Salifu, D., Sevgan, S., Fiaboe, K. K. M., Niassy, S., van Loon, J. J. A., Dicke, M. and Ekesi, S. (2018). Effects of waste stream combinations from brewing industry on performance of Black Soldier Fly, *Hermetia illucens* (*Diptera: Stratiomyidae*). *PeerJ* 6, e5885. doi: 10.7717/peerj.5885.

Costa, S., Pedro, S., Lourenço, H., Batista, I., Teixeira, B., Bandarra, N. M., Murta, D., Nunes, R. and Pires, C. (2020). Evaluation of Tenebrio molitor larvae as an alternative food source. *NFS J.* 21, 57–64. doi: 10.1016/j.nfs.2020.10.001.

Dreassi, E., Cito, A., Zanfini, A., Materozzi, L., Botta, M. and Francardi, V. (2017). Dietary fatty acids influence the growth and fatty acid composition of the yellow mealworm *Tenebrio molitor (Coleoptera: Tenebrionidae)*. *Lipids* 52(3), 285–294. doi: 10.1007/s11745-016-4220-3.

Elleby, C., Jensen, H. G., Domínguez, I. P., Chatzopoulos, T. and Charlebois, P. (2021). Insects reared on food waste: A game changer for global agricultural feed markets? *EuroChoices* 20(3), 56–62. doi: 10.1111/1746-692X.12332.

van der Fels-Klerx, H. J., Camenzuli, L., Belluco, S., Meijer, N. and Ricci, A. (2018). Food safety issues related to uses of insects for feeds and foods. *Compr. Rev. Food Sci. Food Saf.* 17(5), 1172–1183. doi: 10.1111/1541-4337.12385.

Fuso, A., Barbi, S., Macavei, L. I., Luparelli, A. V., Maistrello, L., Montorsi, M., Sforza, S. and Caligiani, A. (2021). Effect of the rearing substrate on total protein and amino acid composition in Black soldier Fly. *Foods* 30(8), 1773. doi: 10.3390/foods10081773.

Gasco, L., Caimi, C., Trocino, A., Lussiana, C., Oddon, S. B., Malfatto, V., Anedda, R., Serra, G., Biasato, I., Schiavone, A., Gai, F. and Renna, M. (2022). Digestibility of defatted insect meals for rainbow trout aquafeeds. *J. Insects Food Feed* 8(11), 1385–1399. doi: 10.3920/JIFF2021.0160.

Hawkey, K. J., Lopez-Viso, C., Brameld, J. M., Parr, T. and Salter, A. M. (2021). Insects: A Potential source of protein and other nutrients for feed and food. *Annu. Rev. Anim. Biosci.* 9, 333–354. doi: 10.1146/annurev-animal-021419-083930.

Heckmann, L.-H., Andersen, J. L., Gianotten, N., Calis, M., Fischer, C. H. and Calis, H. (2018). Sustainable mealworm production for feed and food. In: Halloran, A., Flore, R., Vantomme, P. and Roos, N. (Eds). *Edible Insects in Sustainable Food Systems*. Springer International Publishing AG, Midtown Manhattan, New York City, pp. 321–328.

Janssen, R. H., Vincken, J. P., van den Broek, L. A., Fogliano, V. and Lakemond, C. M. (2017). Nitrogen-to-protein conversion factors for three edible insects: *Tenebrio molitor, Alphitobius diaperinus*, and *Hermetia illucens*. *J. Agric. Food Chem.* 65(11), 2275–2278. doi: 10.1021/acs.jafc.7b00471.

Jantzen da Silva Lucas, A., Menegon de Oliveira, L., da Rocha, M. and Prentice, C. (2020). Edible insects: An alternative of nutritional, functional and bioactive compounds. *Food Chem.* 311, 126022. doi: 10.1016/j.foodchem.2019.126022.

Jensen, L. D., Miklos, R., Dalsgaard, T. K., Heckmann, L. H. and Nørgaard, J. V. (2019). Nutritional evaluation of common (*Tenebrio molitor*) and lesser (*Alphitobius diaperinus*) mealworms in rats and processing effect on the lesser mealworm. *J. Insects Food Feed* 5(4), 257–266. doi: 10.3920/JIFF2018.0048.

Jin, X. H., Heo, P. S., Hong, J. S., Kim, N. J. and Kim, Y. Y. (2016). Supplementation of dried mealworm (*Tenebrio molitor* larva) on growth performance, nutrient digestibility and blood profiles in weaning pigs. *Asian-Australas. J. Anim. Sci.* 29(7), 979–986. doi: 10.5713/ajas.15.0535.

Jo, Y. H. and Lee, J. W. (2016). Insect feed for animals under the Hazard Analysis and Critical Control Points (HACCP) regulations. *Entomol. Res.* 46(1), 2–4. doi: 10.1111/1748-5967.12157.

Jonas-Levi, A. and Martinez, J.-J. I. (2017). The high level of protein content reported in insects for food and feed is overestimated. *J. Food Compos. Anal.* 62, 184–188. doi: 10.1016/j.jfca.2017.06.004.

Jucker, C., Leonardi, M. G., Rigamonti, I., Lupi, D. and Savoldelli, S. (2020). Brewery's waste streams as a valuable substrate for black soldier fly *Hermetia illucens* (Diptera: *Stratiomyidae*). *J. Entomol. Acarol. Res.* 51(3). doi: 10.4081/jear.2019.8876.

Kee, P. E., Cheng, Y. S., Chang, J. S., Yim, H. S., Tan, J. C. Y., Lam, S. S., Lan, J. C., Ng, H. S. and Khoo, K. S. (2023). Insect biorefinery: A circular economy concept for biowaste conversion to value-added products. *Environ. Res.* 221(221), 115284. doi: 10.1016/j.envres.2023.115284.

Kim, S. Y. K., Hong, G. K., Kyeong, Y. L., Hyung, J. Y. and Nam, J. K. (2016). Effects of Brewer's spent grain (BSG) on larval growth of mealworms, *Tenebrio molitor* (*Coleoptera: Tenebrionidae*). *Int. J. Indust. Entomol.* 32(1), 41–48. doi: 10.7852/ijie.2016.32.1.41.

Kinyuru, J. N. and Ndung'u, N. W. (2020). Promoting edible insects in Kenya: Historical, present and future perspectives towards establishment of a sustainable value chain. *J. Insects Food Feed* 6(1), 51–58. doi: 10.3920/JIFF2019.0016.

Kosior, E. and Mitchell, J. (2020). Current industry position on plastic production and recycling. In: Letcher, T. M. (Ed.). *Plastic Waste and Recycling Environmental Impact, Societal Issues, Prevention, and Solutions*. Elsevier Inc., Rotterdam, NLE, pp. 133–162.

Kröncke, N. and Benning, R. (2022). Self-selection of feeding substrates by *Tenebrio molitor* larvae of different ages to determine optimal macronutrient intake and the influence on larval growth and protein content. *Insects* 13(7), 657. doi: 10.3390/insects13070657.

Lähteenmäki-Uutela, A., Marimuthu, S. B. and Meijer, N. (2021). Regulations on insects as food and feed: A global comparison. *J. Insects Food Feed* 7(5), 849–856. doi: 10.3920/JIFF2020.0066.

Lee, J. H., Kim, T. K., Cha, J. Y., Jang, H. W., Yong, H. I. and Choi, Y. S. (2022). How to develop strategies to use insects as animal feed: Digestibility, functionality, safety, and regulation. *J. Anim. Sci. Technol.* 64(3), 409–431. doi: 10.5187/jast.2022.e27.

Lee, J. H., Kim, T. K., Jeong, C. H., Yong, H. I., Cha, J. Y., Kim, B. K. and Choi, Y. S. (2021). Biological activity and processing technologies of edible insects: A review. *Food Sci. Biotechnol.* 30(8), 1003–1023. doi: 10.1007/s10068-021-00942-8.

Li, L., Stasiak, M., Li, L., Xie, B., Fu, Y., Gidzinski, D., Dixon, M. and Liu, H. (2016). Rearing *Tenebrio molitor* in BLSS: Dietary fber afects larval growth, development, and respiration characteristics. *Acta Astronaut.* 118, 130–136. doi: 10.1016/j.actaastro.2015.10.003.

Locke, M. and Collins, J. V. (1965). The structure and formation of protein granules in the fat body of an insect. *J. Cell Biol.* 26(3), 857–875. doi: 10.1083/jcb.26.3.857.

Locke, M. and Collins, J. V. (1967). Protein uptake in multivesicular bodies in the molt-intermolt cycle of an insect. *Science* 155(3761), 467–469. doi: 10.1126/science.155.3761.467.

Locke, M. and Collins, J. V. (1968). Protein uptake into multivesicular bodies and storage granules in the fat body of an insect. *J. Cell Biol.* 36(3), 453–483. doi: 10.1083/jcb.36.3.453.

Mancini, S., Fratini, F., Turchi, B., Mattioli, S., Dal Bosco, A., Tuccinardi, T., Nozic, S. and Paci, G. (2019). Former foodstuff products in *Tenebrio molitor* rearing: Effects on growth, chemical composition, microbiological load, and antioxidant status. *Animals* 9(8), 484. doi: 10.3390/ani9080484.

Melis, R., Braca, A., Sanna, R., Spada, S., Mulas, G., Fadda, M. L., Sassu, M. M., Serra, G. and Anedda, R. (2019). Metabolic response of yellow mealworm larvae to two alternative rearing substrates. *Metabolomics* 15(8), 113. doi: 10.1007/s11306-019-1578-2.

Meyer-Rochow, V. B., Gahukar, R. T., Ghosh, S. and Jung, C. (2021). Chemical composition, nutrient quality and acceptability of edible insects are affected by species, developmental stage, gender, diet, and processing method. *Foods* 10(5), 1036. doi: 10.3390/foods10051036.

Montalbán, A., Sánchez, C. J., Hernández, F., Schiavone, A., Madrid, J. and Martínez-Miró, S. (2022). Effects of agro-industrial byproduct-based diets on the growth performance, digestibility, nutritional and microbiota composition of mealworm (*Tenebrio molitor L.*). *Insects* 13(4), 323. doi: 10.3390/insects13040323.

Morales-Ramos, J. A., Rojas, M. G., Kelstrup, H. C. and Emery, V. (2020). Self-selection of agricultural by-products and food ingredients by *Tenebrio molitor (Coleoptera: Tenebrionidae)* and impact on food utilization and nutrient intake. *Insects* 11(12), 827. doi: 10.3390/insects11120827.

Morales-Ramos, J. A., Rojas, M. G., Shapiro-Ilan, D. I. and Tedders, W. L. (2013). Use of nutrient self-selection as a diet refining tool in *Tenebrio molitor* (*Coleoptera: Tenebrionidae*). *J. Entomol. Sci.* 48(3), 206–221. doi: 10.18474/0749-8004-48.3.206.

Mordue, W. (1965). The neuro-endocrine control of oöcyte development in Tenebrio molitor L. *J. Insect Physiol.* 11(4), 505–511. doi: 10.1016/0022-1910(65)90055-7.

Mussatto, S. I. (2014). Brewer's spent grain: A valuable feedstock for industrial applications. *J. Sci. Food Agric.* 94(7), 1264–1275.

Mussatto, S. I., Dragone, G. and Roberto, I. C. (2006). 'Brewers' spent grain: Generation, characteristics and potential applications'. *J. Cereal Sci.* 43(1), 1–14. doi: 10.1016/j.jcs.2005.06.001.

Nasopoulou, C. and Zabetakis, I. (2013). Agricultural and aquacultural potential of olive pomace a review. *J. Agric. Sci.* 5(7), 116. doi: 10.5539/jas.v5n7p116.

Oonincx, D. G., van Broekhoven, S., van Huis, A. and van Loon, J. J. (2015). Feed conversion, survival and development, and composition of four insect species on diets composed of food by-products. *PLoS ONE* 10(12), e0144601. doi: 10.1371/journal.pone.0144601.

Osimani, A., Garofalo, C., Milanović, V., Taccari, M., Cardinali, F., Aquilanti, L., Pasquini, M., Mozzon, M., Raffaelli, N., Ruschioni, S., Riolo, P., Isidoro, N. and Clementi, F. (2017). Insight into the proximate composition and microbial diversity of edible insects marketed in the European Union. *Eur. Food Res. Technol.* 243(7), 1157–1171. doi: 10.1007/s00217-016-2828-4.

Payne, C. L. R., Scarborough, P., Rayner, M. and Nonaka, K. (2016). A systematic review of nutrient composition data available for twelve commercially available edible insects, and comparison with reference values. *Trends Food Sci. Technol.* 47, 69–77. doi: 10.1016/j.tifs.2015.10.012.

Peng, B. Y., Chen, Z., Chen, J., Yu, H., Zhou, X., Criddle, C. S., Wu, W. M. and Zhang, Y. (2020). Biodegradation of Polyvinyl Chloride (PVC) in *Tenebrio molitor* (*Coleoptera: Tenebrionidae*) larvae. *Environ. Int.* 145, 106106. doi: 10.1016/j.envint.2020.106106.

Pemrick, S. M. and Butz, A. (1970). Protein synthesis of the fat body of adult *Tenebrio molitor*. *J. Insect Physiol.* 16(4), 643–651. doi: 10.1016/0022-1910(70)90097-1.

Pinotti, L. and Ottoboni, M. (2021). Substrate as insect feed for bio-mass production. *J. Insects Food Feed* 7(5), 585–596. doi: 10.3920/JIFF2020.0110.

Prabhat, K., Pandey, D., Næss, G., Cabrita, A. R. J., Fonseca, A. J. M., Maia, M. R. G., Timilsina, B., Veldkamp, T., Sapkota, R. and Overrein, H. (2023). Yellow mealworms (Tenebrio molitor) as an alternative animal feed source: A comprehensive characterization of nutritional values and the larval gut microbiome. *J. Clean. Prod.* 389, 136104. doi: 10.1016/j.jclepro.2023.136104.

Price, G. M. (1973). Protein and nucleic acid metabolism in insect fat body. *Biol. Rev. Camb. Philos. Soc.* 48(3), 333–375. doi: 10.1111/j.1469-185X.1973.tb01006.x.

Riekkinen, K., Väkeväinen, K. and Korhonen, J. (2022). The effect of substrate on the nutrient content and fatty acid composition of edible insects. *Insects* 13(7), 590. doi: 10.3390/insects13070590.

Ritvanen, T., Pastell, H., Welling, A. and Raatikainen, M. (2020). The nitrogen-to-protein conversion factor of two cricket species: *Acheta domesticus* and *Gryllus bimaculatus*. *Agric. Food Sci.* 29(1), 1–5. doi: 10.23986/afsci.89101.

Rumpold, B. A. and Schlüter, O. K. (2013). Nutritional composition and safety aspects of edible insects. *Mol. Nutr. Food Res.* 57(5), 802–823. doi: 10.1002/mnfr.201200735.

Ruschioni, S., Loreto, N., Foligni, R., Mannozzi, C., Raffaelli, N., Zamporlini, F., Pasquini, M., Roncolini, A., Cardinali, F., Osimani, A., Aquilanti, L., Isidoro, N., Riolo, P. and Mozzon, M. (2020). Addition of olive pomace to feeding substrate affects growth performance and nutritional value of mealworm (*Tenebrio molitor L.*) larvae. *Foods* 9(3), 317. doi: 10.3390/foods9030317.

Sangiorgio, P., Verardi, A., Dimatteo, S., Spagnoletta, A., Moliterni, S. and Errico, S. (2022). Valorisation of agri-food waste and mealworms rearing residues for improving the sustainability of *Tenebrio molitor* industrial production. *J. Insects Food Feed* 8(5), 509–524. doi: 10.3920/JIFF2021.0101.

Sogari, G., Amato, M., Biasato, I., Chiesa, S. and Gasco, L. (2019). The potential role of insects as feed: A multi-perspective review. *Animals* 9(4), 119. doi: 10.3390/ani9040119.

Soltani-Mazouni, N. and Soltani, N. (1995). Protein synthesis in the fat body of *Tenebrio molitor* (*L.*) during oocyte maturation: Effect of diflubenzuron, cycloheximide and starvation. *J. Stored Prod. Res.* 31(2), 117–122. doi: 10.1016/0022-474X(94)00046-V.

Toviho, O. A. and Bársony, P. (2022). Nutrient composition and growth of yellow mealworm (*Tenebrio molitor*) at different ages and stages of the life cycle. *Agriculture* 12(11), 1924. doi: 10.3390/agriculture12111924.

Urbanek, A. K., Rybak, J., Wróbel, M., Leluk, K. and Mirończuk, A. M. (2020). A comprehensive assessment of microbiome diversity in *Tenebrio molitor* fed with polystyrene waste. *Environ. Pollut.* 262, 114281. doi: 10.1016/j.envpol.2020.114281.

Varelas, V. and Langton, M. (2017). Forest biomass waste as a potential innovative source for rearing edible insects for food and feed: A review. *Innov. Food Sci. Emerg. Technol.* 41, 193–205. doi: 10.1016/j.ifset.2017.03.007.

Wu, Q., Tao, H. and Wong, M. H. (2019). Feeding and metabolism effects of three common microplastics on *Tenebrio molitor L. Environ. Geochem. Health* 41(1), 17–26. doi: 10.1007/s10653-018-0161-5.

Yang, Y., Yang, J., Wu, W. M., Zhao, J., Song, Y., Gao, L., Yang, R. and Jiang, L. (2015). Biodegradation and mineralization of polystyrene by plastic-eating mealworms: Chemical and physical characterization and isotopic tests. *Environ. Sci. Technol.* 49(20), 12080–12086. doi: 10.1021/acs.est.5b02661.

Yoo, J. S., Cho, K. H., Hong, J. S., Jang, H. S., Chung, Y. H., Kwon, G. T., Shin, D. G. and Kim, Y. Y. (2019). Nutrient ileal digestibility evaluation of dried mealworm (*Tenebrio molitor*) larvae compared to three animal protein by-products in growing pigs. *Asian-Australas. J. Anim. Sci.* 32(3), 387–394. doi: 10.5713/ajas.18.0647.

Yu, X., He, Q. and Wang, D. (2021). Dynamic analysis of major components in the different developmental stages of *Tenebrio molitor*. *Front. Nutr.* 8, 689746. doi: 10.3389/fnut.2021.689746.

Zeinab, M. S., Venkitasamy, C. and Wen, Z. (2021). Corn distillers dried grains with solubles: Production, properties, and potential uses. *Cereal Chem.* 98(5), 999–1019. doi: 10.1002/cche.10445.

Zhang, X., Tang, H., Chen, G., Qiao, L., Li, J., Liu, B., Liu, Z., Li, M. and Liu, X. (2019). Growth performance and nutritional profile of mealworms reared on corn stover, soybean meal, and distillers grains. *Eur. Food Res. Technol.* 245(12), 2631–2640. doi: 10.1007/s00217-019-03336-7.

Zhao, X., Vázquez-Gutiérrez, J. L., Johansson, D. P., Landberg, R. and Langton, M. (2016). Yellow mealworm protein for food purposes - extraction and functional properties. *PLoS ONE* 11(2), e0147791. doi: 10.1371/journal.pone.0147791.

Zielińska, E., Zieliński, D., Jakubczyk, A., Karaś, M., Pankiewicz, U., Flasz, B., Dziewięcka, M. and Lewicki, S. (2021). The impact of polystyrene consumption by edible insects *Tenebrio molitor* and *Zophobas morio* on their nutritional value, cytotoxicity, and oxidative stress parameters. *Food Chem.* 345, 128846. doi: 10.1016/j.foodchem.2020.128846.

Chapter 8

Diseases affecting production of yellow mealworms as a protein source

Antoine Lecocq, University of Copenhagen, Denmark; Helen Hesketh and Pascal Herren, UK Centre for Ecology & Hydrology, UK; and Jozsef Takacs and Annette Bruun Jensen, University of Copenhagen, Denmark

1 Introduction

The mass production of edible insects is intensifying. As one of the more promising species produced on an industrial scale designed for livestock feed, the yellow mealworm, *Tenebrio molitor*, is entering a new phase of intensification following its authorization as human food by the European Commission (Turck et al., 2021). As such, the risk of pathogens that could cause diseases in the yellow mealworms and affect productivity is receiving renewed attention. To this day, information on pathogens of insects in mass production systems remains limited. In the past, insect husbandry systems were focused on fewer species, such as the silk moth (*Bombyx mori*) or the honeybee (*Apis mellifera*). Furthermore, the discovery and description of pathogens occurred as a result of epidemic disease outbreaks in insect populations or were based on observations on a few diseased individuals (Andreadis and Weseloh, 1990; Becnel and Andreadis, 2014; Brun, 1984; Majumdar et al., 2008; Valles and Chen, 2006).

The advent of new mass production systems for species such as the black soldier fly (*Hermetia illucens*), the house cricket (*Acheta domesticus*) and the yellow mealworm (*T. molitor*) based on automation and closed rearing systems is challenging methods previously used. There is still much to learn about insects and their pathogens; however, new bioassays testing pathogen susceptibility and a wealth of new molecular techniques have increased the discovery of (insect) pathogens, especially of viruses (De Miranda et al., 2021;

http://dx.doi.org/10.19103/AS.2024.0139.09

Junglen and Drosten, 2013; Liu et al., 2015). Similarly, microbiome studies are shedding light on the complexities of host–microbe interactions in species of interest. In addition, the yellow mealworm has, over the years been considered a storage pest, a model insect species for infection studies and feed for a range of exotic pets. These factors have accelerated the accrued knowledge about mealworm husbandry and its response to a range of pathogenic organisms.

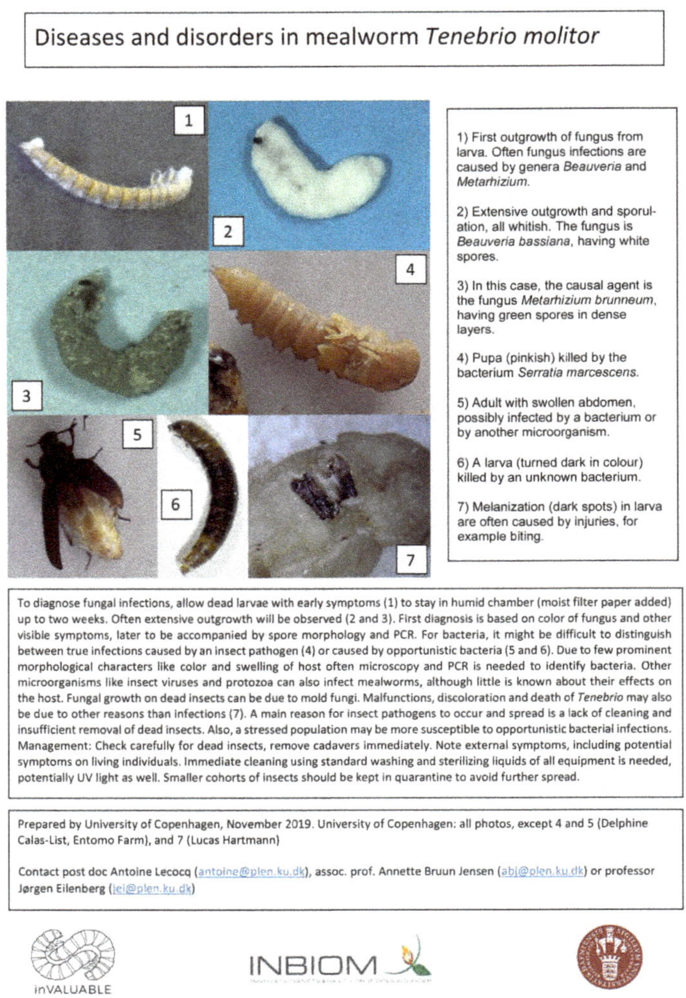

Figure 1 Examples of a factsheet developed in 2019 by research institutions, in collaboration with industrial partners outlining the main diseases and disorders in *T. molitor*. Such factsheets are freely available and regular updates and distribution could inform new and existing yellow mealworm breeders and rearers on emerging threats and common methods of control.

The aim of this chapter is to gather the most up-to-date information on known diseases in the yellow mealworm and to outline the risk for existing, emerging and as-yet unknown pathogens to cause lethal and sub-lethal effects in yellow mealworm production facilities. Moreover, a renewed interest in the mass production of insects and their health has been accompanied by interest in new methods to protect the insects from any future pathogenic threats. To that end, this chapter attempts to summarize some of the most promising ways in which the health and the immune system of *T. molitor* can be improved and better prepared to fight off infections. Finally, we bring attention to the need for coordinated and shared research and resources in an effort to protect this growing industry from future pathogenic threats. Such coordinated efforts have, for example, resulted in the creation of freely available factsheets, outlining the most common threats in yellow mealworm production (Fig. 1), which would benefit from regular updates.

2 Pathogens of *Tenebrio molitor*

2.1 Viruses

Viruses are obligate pathogens, which need the cells of the host to replicate. There are two main groups of viruses which are differentiated based on the type of genetic material, namely DNA (DNA viruses) or RNA (RNA viruses), that encodes the viral genome. These two groups can be subdivided into whether the viral genomes are double stranded or single stranded. Today in total 24 viral taxa are known for their capability to infect invertebrates and they are found within both DNA and RNA viruses. Among these, three double-stranded DNA virus families, Ascoviridae, Baculoviridae and Polydnaviridae, and two single-stranded RNA virus families, Discistroviridae and Iflaviridae, are known only to infect insects (Eilenberg et al., 2015). Insect rearing systems, with high densities of insects in close proximity, are well-suited environments for pathogens, and their rapid spread can lead to epizootic events (Eilenberg et al., 2015). Therefore, viruses represent a major hazard due to their quick infection cycles and capability to eradicate insect stocks entirely (Szelei et al., 2011; Maciel-Vergara and Ros, 2017). The most common route of infection is by oral uptake of viruses by insects (Eilenberg et al., 2015) and with the various feed material utilized by the insect industry, the possibilities for controlling viruses are limited. However, there is only limited information available on viruses that cause diseases in *T. molitor*. The literature only mentions two Iridoviruses capable of causing disease related mortality in *T. molitor*. The first symptom of the invertebrate iridescent virus 6 (IIV-6) is paralysis after 72 h, which is followed by death (Gencer et al., 2020). The invertebrate iridescent virus 29 (IIV-29) is rarely encountered in the larvae of *T. molitor*, but can have a lethal effect upon infection (Kelly et al., 1979; Bertola and Mutinelli, 2021). Some other viruses can be harboured and/or vectored by *T. molitor* and its derived products, but do not have a negative effect on the yellow mealworms. These viruses are the Eastern equine encephalitis virus (Hartley, 1957), the small Iridescent Virus (Kelly et al., 1979), the Acheta domesticus densovirus (Szelei et al., 2011) and the Newcastle disease virus (Turck et al., 2021). Their presence is of importance due to their human or animal pathogen nature, when *T. molitor* is used as food or feed. The viruses present in *T. molitor* can be efficiently eliminated during post-harvest processes, such as boiling the larvae to kill it (Turck et al., 2021). The possibility of new, currently unknown pathogenic viruses also needs to be highlighted, but focus should not only be on viruses causing mortality; viruses of *T. molitor* could be latent or asymptomatic, neither causing mortality nor quantifiable losses in the production. When the rearing conditions are suboptimal, the stressors in the rearing systems (abiotic or biotic) can interact with these present viruses and can result in viral epidemics (Szelei et al., 2011; Maciel-Vergara and Ros, 2017). The permission granted by the European Commission of using *T. molitor* as a food and feed ingredient will likely increase its production volumes which could eventually reveal new viral pathogens of the species.

2.2 Bacteria

The microbiota of mass reared *T. molitor* is shaped by its feed, which can be various agro-industrial byproducts, and its environment (Garofalo et al., 2019, Montalbán et al., 2022). Changes in the gut microbiota have recently been studied in connection to the capacity of *T. molitor* to degrade various kinds of plastics (Brandon et al., 2018, Lou et al., 2021). While several studies have also focused on the detection of human pathogenic microorganisms in the light of food and feed safety (Frentzel et al., 2022; Garofalo et al., 2019; Vandeweyer et al., 2020; Vandeweyer et al., 2021; Stoops et al., 2016), little focus has been given to the insect pathogenic bacteria that can create problems in production. The main infection route of bacteria is through the alimentary canal and thus occurs via oral ingestion of bacteria. The bacteria need to cross or destroy the intestinal barriers in order to reach the hemocoel and enter their pathogenic stage. Infection can also occur if the insects are injured, e.g. by conspecifics, giving pathogenic bacteria direct access to the insect hemocoel (Maciel-Vergara et al., 2018), or they can be vectored by nematodes (Vallet-Gely et al., 2008). Bacteria that are not able to breach the intestinal barrier themselves are considered opportunistic pathogens.

Bacillus thuringiensis is the most commonly used insect pathogen in biological control of pest insects. *Bacillus thuringiensis* var. *tenebrionis*, isolated back in 1982, is the first known insect pathogenic bacteria of *T. molitor* (Krieg et al., 1983). Nowadays, it is commercially used to control coleopteran pest species. *Bacillus thuringiensis* is a gram-positive spore forming bacterium that can cause death of the larval stages of many insects. *Bacillus thuringiensis* normally perforates the insect gut by the action of a pore forming Cry toxin, causing poisoning of the hemolymph. In a recent study by Zanchi et al., (2020) *B. thuringiensis* seemed to be more virulent towards younger *T. molitor* larvae but without the common extensive gut leakage normally observed. In addition, reduced feeding behaviour of the infected individual was observed. Other generalist pathogenic bacteria are known to affect *T. molitor* and recent susceptibility studies have attempted to quantify the risk they pose in the production of yellow mealworms. For example, a feeding exposure experiment with the opportunistic pathogenic bacterium *Serratia marcescens* was found to only cause a slightly increased mortality in *T. molitor* and no interaction with density, starvation or sieving (mechanical) stressors. *Serratia marcescens* however had a high persistence in both the feed and larvae, and could become problematic if larvae are injured (Dupriez et al., 2022). Finally, many other opportunistic bacteria have been recorded in *T. molitor* products for food and feed (Garofalo et al., 2019) and injected directly into the hemocoel, several of these bacteria, e.g. *Pseudomonas aeruginosa*, *Escherichia coli* and *Staphylococcus aureus,* can cause significant mortality (Edosa et al., 2020; Nam et al., 2020; Ham et al., 2021). Some symptoms, such as reduced feeding and decreased movement could be the first indication of a bacterial infection. More severe cases of bacterial proliferation would result in hemolymph poisoning, and eventually *T. molitor* cadavers would change colour, become flaccid, and emit a foul odour (Maciel-Vergara et al., 2021; Dupriez et al., 2022). Bacteria can be difficult to identify morphologically, it is therefore recommended to contact experts in insect diseases or veterinarians in cases of suspected infections (Lecocq et al., 2019).

2.3 Fungi

Insect pathogenic fungi readily infect larvae of *T. molitor*, with most species recorded from the generalist order of the Hypocreales, known to infect a wide range of insect species. Larvae of *T. molitor* are highly susceptible to fungal infection and used extensively as so-called 'bait insects' in environmental samples to isolate fungi, predominantly from the genera *Metarhizium* and *Beauveria* (Bidochka et al., 1998; Kim et al., 2018; Sharma et al., 2018; Steinwender et al., 2014). Most records of entomopathogenic fungi infecting *T. molitor* therefore originate from studies concerned with the ecology and distribution of fungi, primarily from the soil environment (e.g. Bruck, 2010; Korosi et al., 2019; Meyling and Eilenberg, 2006). However, less is known regarding their occurrence in insects being mass reared as a protein source, with only a few studies of fungal isolates collected from rearing facilities (Lee et al., 2014). Fungi infect insects through direct penetration of the integument and do not require ingestion as a route of entry into the host, in comparison to bacteria and viruses. Infection involves a fungal spore attaching to the insect cuticle or other exoskeletal openings, followed by germination, penetration of the cuticle (using a combination of mechanical and enzymatic processes) and dissemination inside the insect (Ortiz-Urquiza and Keyhani, 2013; Vega et al., 2012). Fungi require a high humidity in order to both infect the insect and subsequently to sporulate on the insect cadaver, producing further spores for new infections. Fungal infections could potentially be initiated through direct contact of *T. molitor* larvae or adults with contaminated feed entering the facility, thus requiring attention to hygienic operating procedures for disease management (Van Huis et al., 2021; see Section 3 in this chapter). Identification of the causal fungus in an infection through morphological examination of fungal structures (such as conidia and hyphae/mycelia) can determine the genus of pathogen, but molecular tools are more often required to identify to the species level (Maciel-Vergara et al., 2021).

Most infection studies in *T. molitor* relate to development of entomopathogenic fungi for biological control of the mealworms as a pest species infesting stored grains and grain products. Virulence of fungi varies significantly between strains; for example, in *T. molitor* larvae infected under experimental conditions, isolates of *Metarhizium brunneum* caused 21–100% mortality (Bharadwaj and Stafford III, 2011; Praprotnik et al., 2021) and *Beauveria bassiana* 53–100% mortality (Praprotnik et al., 2021). Similarly, mortality ranged between 20% and 100% in a study using *Metarhizium anisopliae* and *B. bassiana* sprayed either directly to *T. molitor* larvae or applied to grain (Karabörklü et al., 2019). Exposure to soil treated with *M. anisopliae* impregnated granules caused 13–73% mortality in *T. molitor* larvae depending on the sterility of the soil (Shah et al., 2023). However, variability in virulence is difficult to predict, even when strains originate from fungal communities in a limited spatial area (Maistrou et al., 2020). Life stages of *T. molitor* also show differential susceptibility to infection, with pupae generally being more susceptible to infection than larvae, which in turn may be more susceptible, compared to adult beetles (Eski and Gezgin, 2022).

There is evidence that yellow mealworm beetles are able to resist fungal infection through different mechanisms. For example, constitutively expressed antifungal peptides have been shown to have antifungal activity and increase survival of *T. molitor* larvae against infection with *B. bassiana* (Maistrou et al., 2018). It is also known that melanin plays an important role in *T. molitor* immunity with darker adult beetles producing more hemocytes and phenoloxidase which correlates with resistance to pathogen infection (Armitage and Siva-Jothy, 2005). High density populations of adult beetles can lead to darkening of the insect cuticle and demonstrable resistance to infection with *M. anisopliae* (Barnes and Siva-Jothy, 2000). Solutions for preventing disease outbreaks due to fungi in mass reared *T. molitor* could take advantage of enhancing such traits to improve insect resistance to infections (Grau et al., 2017; see Section 3 in this chapter).

2.4 Other pathogens of interest

Without a doubt, viruses, bacteria and fungi together form the greatest threat to *T. molitor* production systems as described and understood by current scientific knowledge. However, other pathogens with sub-lethal effects and poorly understood mechanisms are known to replicate and be present in *T. molitor*. Microsporidia are intracellular obligate parasites, most closely related to fungi (Capella-Gutiérrez et al., 2012). Their transmission most commonly occurs through oral ingestion of infectious spores, although transovarian transmission is known to allow parents to pass on infections to offspring (Becnel and Andreadis, 2014). Presently, there is little evidence to suggest that microsporidians are causing significant problems in *T. molitor* production by increasing mortality or disrupting developmental cycles. The only account of a natural infection where 87% of the beetles in the infected culture carried microsporidia determined that there was no significant harmful effect detected in the insect (Armitage and Siva-Jothy, 2005).

Entomopathogenic nematodes are cylindrical non-segmented organisms that are obligate or sometimes facultative pathogens of insects. While the pathogenic potential of members in the genera *Steinernema* and *Heterorhabditis* and their associated bacteria is obvious based on their adoption as biological control agents against insect pests (Georgis, 1992) no natural infection in *T. molitor* have yet been described. Furthermore, while *in vivo* cultures have been achieved in the yellow mealworm (Shapiro-Ilan et al., 2008), in a study comparing infectivity in various hosts, nematodes generally caused less mortality in *T. molitor* than other model species (Caroli et al., 1996).

Composed of two main orders, eugregarines and neogregarines, the gregarines are a group of protists known to be parasitic to insects. They are mostly non-lethal, but some can lower the insects' fitness. While this lack of obvious effect is particularly true for eugregarines, a neogregarine infection, on the other hand appears to have significant effects on mortality, likely due to having a proliferative phase in the host, which eugregarines do not (Rodriguez et al., 2007; Lange and Lord, 2012). Overall, five species of gregarines are known to infect *T. molitor*: four in the larvae (*Gregarina polymorpha*, *G. cuneata*, *G. steini* and *Mattesia* spp.) and one (*G. niphandrodes*) in the adult (Clopton et al., 1991; Rodriguez et al., 2007). Over the years, studies have highlighted the complex relationship that gregarines form with their host. For example, a 1933 study on the effect of gregarines on *T. molitor* found that gregarines were necessary for normal growth and longevity of the yellow mealworm (Sumner, 1936). Likewise, a study on the relationship between *G. cuneata* and yellow mealworms found that the presence of eugregarines in the larvae increased the host growth rate and reduced the death rate, despite heavy parasitisation (Valigurová, 2012). On the other hand, when studying the impact of *G. niphandrodes* infection in their host, researchers concluded that while population dynamics of *T. molitor* with low and high infection were not significantly different, beetles with low infection lived significantly longer than those with high infection (Rodriguez et al., 2007). In another example of parasitism, *G. cuneata* was shown to destroy *T. molitor* gut epithelial cells (Lipa, 1967; Bessette and Williams, 2022). Overall, microsporidians, nematodes and gregarines do not appear to currently pose a threat to the production of yellow mealworms. However, it is clear that from both fundamental and applied perspectives, there is a need for more research to increase our understanding of those organisms and of their relationships with *T. molitor* larvae and adults.

3 Control of pathogens

3.1 General hygiene practices

As a general rule, especially in the case of closed production systems like those adopted by many European producers, the first step to controlling pathogens in a production facility is to avoid their entry into the rearing system. This can be achieved by having full control of personnel, equipment and substrates entering the facility. Generally, sanitization and sterilization routines (of all equipment and rearing facilities) should be implemented in order to reduce the risk of introducing pathogens into facilities and the risk of transmission within the facilities. Other actions such as separation of rearing and breading stocks could avoid the need for the destruction of the entire production stock before sterilizing and re-starting production with healthy insects (Eilenberg et al., 2017). Furthermore, factors such as density and diet, as well as abiotic factors such as temperature, humidity and photoperiod are integral elements in setting up production facilities and must be considered relative to insect diseases. For example, most entomopathogenic fungi require high relative humidity for their growth (Marcandier and Khachatourians, 1987, Sinha et al., 2016). As such, *T. molitor* has shown more resistance towards fungi from the genera *Beauveria* and *Metarhizium* when reared under relatively dry conditions on diets with low water content, even at higher densities (Barnes and Siva-Jothy, 2000). Methods for disease treatment or prevention in a new livestock species such as *T. molitor*, are much less developed than those within traditional livestock. Apart from a few exceptions, such as for honeybees and silkworms where a number of treatments have existed for decades, insect production systems are devoid of guidelines in dealing with diseases. In the following section, future potential methods for minimizing the risk posed by pathogens in the yellow mealworm are discussed.

3.2 Breeding

The technique of selective breeding to improve immunity against invading pathogens has been used for many years in a number of production systems such as crop production (Nelson et al., 2018), aquaculture (Gjedrem, 2015), poultry, pigs, (Proudfoot et al., 2019) and honeybees (Guichard et al., 2019). In the context of insects as food and feed, selective breeding has been applied to some degree, both voluntarily and involuntarily as a result of close association with humans. Such is the case for the yellow mealworm for which no truly wild populations are thought to remain (Eriksson and Picard, 2021). Reports of selective breeding programmes in other insects as food and feed, like the black soldier fly (*Hermetia illucens*), suggest that more emphasis is placed on body size and other development traits than immunity and protection against pathogens (Facchini et al., 2022). Similarly, in experimental evolution experiments using *T. molitor*, scientists successfully used artificial selection and selective breeding to improve traits including body size, body weight, growth rate (Leclercq, 1963; Morales-Ramos et al., 2019), fecundity and food conversion efficiency (Morales-Ramos et al., 2019). Such selective breeding, focusing on increased production traits can however have negative effects on individual insect health, such as in the case of the study by Morales-Ramos et al. (2019), which resulted in increased mortality of the larvae. However, selective breeding does have potential as a tool to protect yellow mealworm larvae against diseases. For example, a study showed that by selecting for black adult cuticular phenotypes in *T. molitor*, two immune parameters related to pathogen resistance became significantly higher in selection lines, suggesting improved immunity (Armitage and Siva-Jothy, 2005). Large-scale breeding programmes, such as those reported for the black soldier fly, have not yet been carried out, or made publicly available for *T. molitor*. As mentioned, it is likely that such breeding on an industrial scale will initially focus on traits related to size and developmental rate to increase productivity, leaving room for other methods to control pathogens and improve the health of the insects.

3.3 Microbiome manipulation/probiotics

In traditional livestock, probiotics have been confirmed to have a beneficial effect on animal health, particularly in terms of protection against pathogens, stimulation of the immunological response and increased production capacity (Markowiak and Śliżewska, 2018). Their use in yellow mealworm production systems remains undocumented. However, over the last few years, a number of studies have evaluated their effect on the development of yellow mealworm larvae in the laboratory. For example, an indigenous strain, isolated from *T. molitor*, identified as *Pediococcus pentasaceus* was found to accelerate larval development and protect against the entomopathogenic fungus *M. brunneum* compared to larvae reared on a diet of oatmeal with no probiotic addition (Dahal et al., 2022; Lecocq et al., 2021). Furthermore, Rizou et al. (2022) found that the addition of probiotic bacteria, in this case *Bacillus subtilis*, *Bacillus toyonensis* and *Enterococcus faecalis*, resulted in significant improvements in growth, nutritional content and microbial load reduction in *T. molitor* larvae. As mentioned, while yet to be documented in an industrial setting, the addition of probiotics in insect production systems does provide an opportunity to increase the level of protection to the larvae (Savio et al., 2022). Additional benefits could increase the value of the end products, however, it is expected that any industrial application will focus on beneficial effects to the health of the production insects. Further research is expected to focus on the translation of lab scale research to industrial scale before commercialization of insect probiotic products is envisaged.

3.4 Immune priming

Immune priming describes the decreased susceptibility to a pathogen after being previously exposed to a pathogen (i.e. by exposure to pathogen components, a sublethal pathogen dose, or dead or inactivated pathogens) (Little and Kraaijeveld, 2004; Vigneron et al., 2019) or a physical stress such as temperature shock or vibration (Kojour et al., 2022). This memory-like phenomenon of the invertebrate innate immune response was observed as early as the 1930s and the first empirical studies were conducted in the 1980s (Contreras-Garduño et al., 2016). Immune priming is beneficial in environments in which individuals are repeatedly exposed to pathogens and the mechanism of immune priming is most probably the combination of an increased long-term immune response and a stronger or faster immune response after the second exposure (Milutinović et al., 2016).

In *T. molitor*, priming with inactivated gram-positive bacteria (*S. aureus* or *B. thuringiensis*) resulted in protection against the same gram-positive and also against gram–negative bacteria (*E. coli* and *Serratia entomophila*). The protection after priming with gram-negative bacteria, however, was less effective, suggesting that gram-positive bacteria were more important in the evolution of immune priming in *T. molitor* (Dhinaut et al., 2018). Moreover, *T. molitor* larvae exposed to the fungus *M. anisopliae* had an increased survival probability when previously exposed to sublethal doses of either *M. anisopliae* or *B. thuringiensis* but not when previously exposed to the gram-negative *S. marcescens* (Medina-Gómez et al., 2018). This might be explained by the different signal transduction pathways that different pathogens activate. While gram-negative bacteria activate the immunodeficiency (Imd) pathway, gram-positive bacteria and fungi activate the Toll pathway to regulate the synthesis of antimicrobial peptides (AMPs) (Vigneron et al., 2019).

Immune priming also takes place across generations, which is called trans-generational immune priming. In *T. molitor*, trans-generational immune priming has been shown to be effective in eggs (Dubuffet et al., 2015), larvae (Moret, 2006) and adult offspring (Dhinaut et al., 2018). Adult *T. molitor* females challenged with gram-positive (*Arthrobacter globiformis* or *B. thuringiensis*) or gram-negative (*E.coli* or *S. entomophila*) bacteria laid eggs with increased antimicrobial activity against gram-positive bacteria due to a potential maternal transfer of AMPs (i.e. tenecin-1) to the eggs (Dubuffet et al., 2015). Interestingly, the maternal transfer of immunity is limited to only certain eggs, demonstrating that trans-generational immune priming is costly (Zanchi et al., 2012). Moreover, reduced adult body mass (Moret, 2006) and prolonged larval development (Dhinaut et al., 2018) have been reported as consequences of trans-generational immune priming in *T. molitor*.

Immune priming or trans-generational immune priming might be an approach to decrease susceptibility to pathogens of mass-reared *T. molitor* (Grau et al., 2017). Oral priming using heat-inactivated microorganisms, e.g., could offer a safe and efficient way to prime large numbers of insects (González-Acosta et al., 2022). Yet, biological and monetary costs of immune priming and trans-generational immune priming in *T. molitor* need to be assessed thoroughly before industrial application.

4 Conclusion

Upon reading this chapter, it becomes evident that *T. molitor* are subject to infections by a variety of microorganisms, such as viruses, bacteria, fungi, microsporidia, gregarines and nematodes. While diagnosing diseases in an insect involves some of the same methods as applied to more traditional livestock species, such as using morphological or molecular methods, the insect sector lacks certified diagnostic laboratories where samples can be sent for analysis. For this reason, alongside the novelty of the sector and its rapid expansion, it is evident that there is a risk that past, current and future pathogens have or will remain undiagnosed while still causing substantial losses in production facilities. Currently, the only option is to reach out to researchers and hope that they have both the right expertise and the time to provide knowledge and assistance. However, each group of insect pathogens are characterized by their unique biology, infection routes, modes of action, transmission potential, persistency and/or response to environmental conditions and thus will require different approaches for minimizing the risk of entering the production system or for control upon entry. Fundamental and applied research as well as knowledge sharing are therefore vital tools to increase our understanding of insect pathology and fill in remaining gaps. Besides good hygiene practices, research has shown that breeding for improved immunity is a potential method for ensuring future resilient *T. molitor* production. Furthermore, recent laboratory studies have shown that immune priming and trans-generational immune priming can provide a certain protection towards insect pathogens and that probiotics provide added value for production output in addition to disease prevention. Implementation of these novel methods in production systems will however also require more in depth laboratory studies as well as larger scale studies on their potential for upscaling.

5 Where to look for further information

A number of general reviews about diseases of insects in production systems have been written over the last 5 years. These reviews often include a chapter on *T. molitor,* and many were published open access in specialist journals such as *Journal of insects as food and feed*. Additionally the insect doctors website (https://www.insectdoctors.eu/en/insectdoctors.htm) provides additional information on recent projects in insect pathology and general information about insect diseases and disorders.

6 References

Andreadis, T. G. and Weseloh, R. M. 1990. Discovery of *Entomophaga maimaiga* in North American gypsy moth, Lymantria dispar. *Proceedings of the National Academy of Sciences of the United States of America* 87(7), 2461–2465.

Armitage, S. A. O. and Siva-Jothy, M. T. 2005. Immune function responds to selection for cuticular colour in *Tenebrio molitor*. *Heredity* 94(6), 650–656.

Barnes, A. I. and Siva-Jothy, M. T. 2000. Density–dependent prophylaxis in the mealworm beetle *Tenebrio molitor* L.(Coleoptera: Tenebrionidae): Cuticular melanization is an indicator of investment in immunity. *Proceedings of the Royal Society of London. Series B: Biological Sciences* 267(1439), 177–182.

Becnel, J. J. and Andreadis, T. G. 2014. Microsporidia in insects. In Weiss, L. M. and Becnel, J. J. (Eds.), *Microsporidia: Pathogens of Opportunity*, 1st edition. John Wiley and Sons, Hoboken, NJ, 521–570.

Bertola, M. and Mutinelli, F. 2021. A systematic review on viruses in mass-reared edible insect species. *Viruses* 13(11), 2280.

Bessette, E. and Williams, B. 2022. Protists in the insect rearing industry: Benign passengers or potential risk? *Insects* 13(5), 482.

Bharadwaj, A. and Stafford III, K. C. 2011. Potential of *Tenebrio molitor* (Coleoptera: Tenebrionidae) as a bioassay probe for *Metarhizium brunneum* (Hypocreales: Clavicipitaceae) activity against *Ixodes scapularis* (Acari: Ixodidae). *Journal of Economic Entomology* 104(6), 2095–2098.

Bidochka, M. J., Kasperski, J. E. and Wild, G. A. 1998. Occurrence of the entomopathogenic fungi *Metarhizium anisopliae* and *Beauveria bassiana* in soils from temperate and near-northern habitats. *Canadian Journal of Botany* 76(7), 1198–1204.

Brandon, A. M., Gao, S. H., Tian, R., Ning, D., Yang, S. S., Zhou, J., Wu, W. M. and Criddle, C. S. 2018. Biodegradation of polyethylene and plastic mixtures in mealworms (larvae of *Tenebrio molitor*) and effects on the gut microbiome. *Environmental Science and Technology* 52(11), 6526–6533.

Bruck, D. J. 2010. Fungal entomopathogens in the rhizosphere. In: Roy, H. E., Vega, F. E., Chandler, D., Goettel, M. S., Pell, J. and Wajnberg, E. eds., *The Ecology of Fungal Entomopathogens*, 103–112. Dordrecht: Springer.

Brun, G. 1984. Le virus sigma de la Drosophile. *Bulletin de la Société Entomologique de France* 89(1), 674–680.

Capella-Gutiérrez, S., Marcet-Houben, M. and Gabaldon, T. 2012. Phylogenomics supports microsporidia as the earliest diverging clade of sequenced fungi. *BMC biology* 10, 1–14.

Caroli, L., Glazer, I. and Gaugler, R. 1996. Entomopathogenic nematode infectivity assay: Comparison of penetration rate into different hosts. *Biocontrol Science and Technology* 6(2), 227–234.

Clopton, R. E., Percival, T. J. and Janovy, J. 1991. *Gregarina niphandrodes* n. sp. (Apicomplexa: Eugregarinorida) from adult *Tenebrio molitor* (L.) with oocyst descriptions of other gregarine parasites of the yellow mealworm. *The Journal of Protozoology* 38(5), 472–479.

Contreras-Garduño, J., Lanz-Mendoza, H., Franco, B., Nava, A., Pedraza-Reyes, M. and Canales-Lazcano, J. 2016. Insect immune priming: Ecology and experimental evidences. *Ecological Entomology* 41(4), 351–366.

Dahal, S., Jensen, A. B. and Lecocq, A. 2022. Effect of probiotics on *Tenebrio molitor* larval development and resistance against the fungal pathogen *Metarhizium brunneum*. *Insects* 13(12), 1114.

De Miranda, J. R., Granberg, F., Onorati, P., Jansson, A. and Berggren, Å. 2021. Virus prospecting in crickets – discovery and strain divergence of a novel Iflavirus in wild and cultivated *Acheta domesticus*. *Viruses* 13(3), 364. https://doi.org/10.3390/v13030364.

Dhinaut, J., Chogne, M. and Moret, Y. 2018. Immune priming specificity within and across generations reveals the range of pathogens affecting evolution of immunity in an insect. *Journal of Animal Ecology* 87(2), 448–463.

Dubuffet, A., Zanchi, C., Boutet, G., Moreau, J., Teixeira, M. and Moret, Y. 2015. Trans-generational immune priming protects the eggs only against gram-positive bacteria in the mealworm beetle. *PLoS Pathogens* 11(10), e1005178.

Dupriez, F., Rejasse, A., Rios, A., Lefebvre, T. and Nielsen-LeRoux, C. 2022. Impact and persistence of *Serratia marcescens* in *Tenebrio molitor* larvae and feed under optimal and stressed mass rearing conditions. *Insects* 13(5), 458.

Edosa, T. T., Jo, Y. H., Keshavarz, M., Bae, Y. M., Kim, D. H., Lee, Y. S. and Han, Y. S. 2020. Tm Spz6 is essential for regulating the immune response to *Escherichia coli* and *Staphylococcus aureus* infection in *Tenebrio molitor*. *Insects* 11(2), 105.

Eilenberg, J., Vlak, J. M., Nielsen-LeRoux, C., Cappellozza, S. and Jensen, A. B. 2015. Diseases in insects produced for food and feed. *Journal of Insects as Food and Feed* 1(2), 87–102.

Eilenberg, J., Jensen, A. B. and Hajek, A. E. 2017. Prevention and management of diseases in terrestrial invertebrates. In: Hajek, A. E. and Shapiro-Ilan, D. I. (eds),*Ecology of Invertebrate Diseases*, 495–526. John Wiley &Sons: USA.

Eriksson, T. and Picard, C. J. 2021. Genetic and genomic selection in insects as food and feed. *Journal of Insects as Food and Feed* 7(5), 661–682.

Eski, A. and Gezgin, M. M. 2022. Susceptibility of different life stages of *Tenebrio molitor* (Coleoptera: Tenebrionidae) to indigenous entomopathogenic fungi. *Journal of Stored Products Research* 98, 102008.

Facchini, E., Shrestha, K., Van Den Boer, E., Junes, P., Sader, G., Peeters, K. and Schmitt, E. 2022. Long-term artificial selection for increased larval body weight of *Hermetia illucens* in industrial settings. *Frontiers in Genetics* 13, 865490.

Frentzel, H., Kelner-Burgos, Y., Fischer, J., Heise, J., Göhler, A. and Wichmann-Schauer, H. 2022. Occurrence of selected bacterial pathogens in insect-based food products and in-depth characterisation of detected Bacillus cereus group isolates. *International Journal of Food Microbiology* 379, 109860.

Garofalo, C., Milanović, V., Cardinali, F., Aquilanti, L., Clementi, F. and Osimani, A. 2019. Current knowledge on the microbiota of edible insects intended for human consumption: A state-of-the-art review. *Food Research International* 125, 108527.

Gencer, D., Yesilyurt, A., Güllü, M., Demir, İ. and Nalcacioglu, R. 2020. Insecticidal activities of wild type and recombinant invertebrate iridescent viruses on five common pests. *Turkish Journal of Entomology* 44(3), 365–373.

Georgis, R. 1992. Present and future prospects for entomopathogenic nematode products. *Biocontrol Science and Technology* 2(2), 83–99.

Gjedrem, T. 2015. Disease resistant fish and shellfish are within reach: A review. *Journal of Marine Science and Engineering* 3(1), 146–153.

González-Acosta, S., Baca-González, V., Asensio-Calavia, P., Otazo-Pérez, A., López, M. R., Morales-delaNuez, A. and Pérez de la Lastra, J. M. 2022. Efficient oral priming of *Tenebrio molitor* larvae using heat-inactivated microorganisms. *Vaccines* 10(8), 1296.

Grau, T., Vilcinskas, A. and Joop, G. 2017. Sustainable farming of the mealworm *Tenebrio molitor* for the production of food and feed. *Zeitschrift Fur Naturforschung C* 72(9–10), 337–349.

Guichard, M., Neuditschko, M., Fried, P., Soland, G. and Dainat, B. 2019. A future resistance breeding strategy against *Varroa destructor* in a small population of the dark honey bee. *Journal of Apicultural Research* 58(5), 814–823.

Ham, S. Y., Kim, H. S., Jo, M. J., Lee, J. H., Byun, Y., Ko, G. J. and Park, H. D. 2021. Combined treatment of 6-gingerol analog and tobramycin for inhibiting Pseudomonas aeruginosa infections. *Microbiology Spectrum* 9(2), e0019221.

Hartley, C. F. 1957. *Host Induced Alteration of Eastern Equine Encephalomyelitis Virus in Tenebrio Molitor L (The University of Wisconsin-Madison).* Madison ProQuest Dissertations Publishing.

Junglen, S. and Drosten, C. 2013. Virus discovery and recent insights into virus diversity in arthropods. *Current Opinion in Microbiology* 16(4), 507–513.

Karabörklü, S., Altın, N. and Keskin, Y. 2019. Native entomopathogenic fungi ısolated from Duzce, Turkey and their virulence on the mealworm beetle *Tenebrio molitor* L. (Coleoptera: Tenebrionidae). *Philippines Agricultural Scientist* 102(1), 82–89.

Kelly, D. C., Ayres, M. D., Lescott, T., Robertson, J. S. and Happ, G. M. 1979. A small Iridescent Virus (Type 29) isolated from *Tenebrio molitor*: A comparison of its proteins and antigens with six other iridescent viruses. *Journal of General Virology* 42(1), 95–105.

Kim, J. C., Lee, M. R., Kim, S., Lee, S. J., Park, S. E., Nai, Y. S., Lee, G. S., Shin, T. Y. and Kim, J. S. 2018. *Tenebrio molitor*-mediated entomopathogenic fungal library construction for pest management. *Journal of Asia-Pacific Entomology* 21(1), 196–204.

Kojour, M. A. M., Baliarsingh, S., Am Jang, H., Yun, K., Park, K. B., Lee, J. E., Han, Y. S., Patnaik, B. B. and Jo, Y. H. 2022. Current knowledge of immune priming in invertebrates, emphasizing studies on *Tenebrio molitor*. *Developmental and Comparative Immunology* 127, 104284.

Korosi, G. A., Wilson, B. A. L., Powell, K. S., Ash, G. J., Reineke, A. and Savocchia, S. 2019. Occurrence and diversity of entomopathogenic fungi (*Beauveria* spp. and *Metarhizium* spp.) in Australian vineyard soils. *Journal of Invertebrate Pathology* 164, 69–77.

Krieg, A. V., Huger, A. M., Langenbruch, G. A. and Schnetter, W. 1983. *Bacillus thuringiensis* var. tenebrionis: ein neuer, gegenüber Larven von Coleopteren wirksamer Pathotyp. *Zeitschrift für angewandte Entomologie* 96(1–5), 500–508.

Lange, C. E. and Lord, J. C. 2012. Protistan entomopathogens. In Vega, F. E. and Kaya, H. K. (Eds.), *Insect Pathology*, 2nd edition. Elsevier, Amsterdam, 367–394.

Leclercq, J. 1963. Artificial selection for weight and its consequences in *Tenebrio molitor* L. *Nature* 198(4875), 106–107.

Lecocq, A., Jensen, A. B. and Eilenberg, J. 2019. Diseases of insects in European production systems: Diagnosis, prevention and management. *Berliner und Münchener Tierärztliche Wochenschrift* 132(5/6), 245–250.

Lecocq, A., Natsopoulou, M. E., Berggreen, I. E., Eilenberg, J., Heckmann, L.-H. L., Nielsen, H. V., Stensvold, C. R. and Jensen, A. B. 2021. Probiotic properties of an indigenous *Pediococcus pentosaceus* strain on *Tenebrio molitor* larval growth and survival. *Journal of Insects as Food and Feed* 7(6), 975–986.

Lee, S. J., Kim, S. H., Nai, Y. S., Je, Y. H., Parker, B. L. and Kim, J. S. 2014. Management of entomopathogenic fungi in cultures of *Tenebrio molitor* (Coleoptera: Tenebrionidae). *Entomological Research* 44(6), 236–243.

Lipa, J. J. 1967. *Studies on Gregarines (Gregarinomorpha) of Arthropods in Poland*. Instytut Biologii Doświadczalnej im. M. Nenckiego Polskiej Akademii Nauk.

Little, T. J. and Kraaijeveld, A. R. 2004. Ecological and evolutionary implications of immunological priming in invertebrates. *Trends in Ecology and Evolution* 19(2), 58–60.

Liu, S., Chen, Y. and Bonning, B. C. 2015. RNA virus discovery in insects. *Current Opinion in Insect Science* 8, 54–61. https://doi.org/10.1016/j.cois.2014.12.005.

Lou, Y., Li, Y., Lu, B., Liu, Q., Yang, S. S., Liu, B., Ren, N., Wu, W. M. and Xing, D. 2021. Response of the yellow mealworm (*Tenebrio molitor*) gut microbiome to diet shifts during polystyrene and polyethylene biodegradation. *Journal of Hazardous Materials* 416, 126222.

Maciel-Vergara, G. and Ros, V. I. D. 2017. Viruses of insects reared for food and feed. *Journal of Invertebrate Pathology* 147, 60–75.

Maciel-Vergara, G., Jensen, A. B. and Eilenberg, J. 2018. Cannibalism as a possible entry route for opportunistic pathogenic bacteria to insect hosts, exemplified by *Pseudomonas aeruginosa*, a pathogen of the giant mealworm *Zophobas morio*. *Insects* 9(3), 88.

Maciel-Vergara, G., Jensen, A. B., Lecocq, A. and Eilenberg, J. 2021. Diseases in edible insect rearing systems. *Journal of Insects as Food and Feed* 7(5), 621–638.

Maistrou, S., Paris, V., Jensen, A. B., Rolff, J., Meyling, N. V. and Zanchi, C. 2018. A constitutively expressed antifungal peptide protects *Tenebrio molitor* during a natural infection by the entomopathogenic fungus *Beauveria bassiana*. *Developmental and Comparative Immunology* 86, 26–33.

Maistrou, S., Natsopoulou, M. E., Jensen, A. B. and Meyling, N. V. 2020. Virulence traits within a community of the fungal entomopathogen *Beauveria*: Associations with abundance and distribution. *Fungal Ecology* 48, 100992.

Majumdar, A., Boetel, M. A. and Jaronski, S. T. 2008. Discovery of *Fusarium solani* as a naturally occurring pathogen of sugarbeet root maggot (Diptera: Ulidiidae) pupae: Prevalence and baseline susceptibility. *Journal of Invertebrate Pathology* 97(1), 1–8. https://doi.org/10.1016/j.jip.2007.05.003.

Marcandier, S. and Khachatourians, G. G. 1987. Susceptibility of the migratory grasshopper, *Melanoplus sanguinipes* (Fab.)(Orthoptera: Acrididae), to *Beauveria bassiana* (Bals.) Vuillemin (Hyphomycete): Influence of relative humidity. *The Canadian Entomologist* 119(10), 901–907.

Markowiak, P. and Śliżewska, K. 2018. The role of probiotics, prebiotics and synbiotics in animal nutrition. *Gut Pathogens* 10(1), 21.

Medina-Gómez, H., Farriols, M., Santos, F., González-Hernández, A., Torres-Guzmán, J. C., Lanz, H. and Contreras-Garduño, J. 2018. Pathogen-produced catalase affects immune priming: A potential pathogen strategy. *Microbial Pathogenesis* 125, 93–95.

Meyling, N. V. and Eilenberg, J. 2006. Occurrence and distribution of soil borne entomopathogenic fungi within a single organic agroecosystem. *Agriculture, Ecosystems and Environment* 113(1–4), 336–341.

Milutinović, B., Peuß, R., Ferro, K. and Kurtz, J. 2016. Immune priming in arthropods: An update focusing on the red flour beetle. *Zoology* 119(4), 254–261.

Montalbán, A., Sánchez, C. J., Hernández, F., Schiavone, A., Madrid, J. and Martínez-Miró, S. 2022. Effects of agro-industrial byproduct-based diets on the growth performance, digestibility, nutritional and microbiota composition of mealworm (*Tenebrio molitor* L.). *Insects* 13(4), 323.

Morales-Ramos, J. A., Kelstrup, H. C., Rojas, M. G. and Emery, V. 2019. Body mass increase induced by eight years of artificial selection in the yellow mealworm (Coleoptera: Tenebrionidae) and life history trade-offs. *Journal of Insect Science* 19(2), 4.

Moret, Y. 2006. 'Trans-generational immune priming': Specific enhancement of the antimicrobial immune response in the mealworm beetle, *Tenebrio molitor*. *Proceedings of the Royal Society B: Biological Sciences* 273(1592), 1399–1405.

Nam, S., Ham, S. Y., Kwon, H., Kim, H. S., Moon, S., Lee, J. H., Lim, T., Son, S. H., Park, H. D. and Byun, Y. 2020. Discovery and characterization of pure RhlR antagonists against *Pseudomonas aeruginosa* infections. *Journal of Medicinal Chemistry* 63(15), 8388–8407.

Nelson, R., Wiesner-Hanks, T., Wisser, R. and Balint-Kurti, P. 2018. Navigating complexity to breed disease-resistant crops. *Nature Reviews Genetics* 19(1), 21–33.

Ortiz-Urquiza, A. and Keyhani, N. O. 2013. Action on the surface: Entomopathogenic fungi versus the insect cuticle. *Insects* 4(3), 357–374.

Praprotnik, E., Lončar, J. and Razinger, J. 2021. Testing virulence of different species of insect associated fungi against yellow mealworm (Coleoptera: Tenebrionidae) and their potential growth stimulation to maize. *Plants* 10(11), 2498.

Proudfoot, C., Lillico, S. and Tait-Burkard, C. 2019. Genome editing for disease resistance in pigs and chickens. *Animal Frontiers* 9(3), 6–12.

Rizou, E., Kalogiouri, N., Bisba, M., Papadimitriou, A., Kyrila, G., Lazou, A., Andreadis, S., Hatzikamari, M., Mourtzinos, I. and Touraki, M. 2022. Amelioration of growth, nutritional value, and microbial load of *Tenebrio molitor* (Coleoptera: Tenebrionidae) through probiotic supplemented feed. *European Food Research and Technology* 248(3), 727–739.

Rodriguez, Y., Omoto, C. K. and Gomulkiewicz, R. 2007. Individual and population effects of eugregarine, *Gregarina niphandrodes* (Eugregarinida: Gregarinidae), on *Tenebrio molitor* (Coleoptera: Tenebrionidae). *Environmental Entomology* 36(4), 689–693.

Savio, C., Mugo-Kamiri, L. and Upfold, J. K. 2022. Bugs in bugs: The role of probiotics and prebiotics in maintenance of health in mass-reared insects. *Insects* 13(4), 376.

Shah, S., Ash, G. J. and Wilson, B. A. L. 2023. Resporulation of *Metarhizium anisopliae* granules on soil and mortality of *Tenebrio molitor*: Implications for wireworm management in sweet potato. *Annals of Applied Biology* 182(1), 65–76.

Shapiro-Ilan, D., Rojas, M. G., Morales-Ramos, J. A., Lewis, E. E. and Tedders, W. L. 2008. Effects of host nutrition on virulence and fitness of entomopathogenic nematodes: Lipid-and protein-based supplements in *Tenebrio molitor* diets. *Journal of Nematology* 40(1), 13–19.

Sharma, L., Oliveira, I., Torres, L. and Marques, G. 2018. Entomopathogenic fungi in Portuguese vineyards soils: Suggesting a '*Galleria-Tenebrio*-bait method' as bait-insects *Galleria* and *Tenebrio* significantly underestimate the respective recoveries of *Metarhizium (robertsii)* and *Beauveria (bassiana)*. *MycoKeys* 38(38), 1–23.

Sinha, K. K., Choudhary, A. K. and Kumari, P. 2016. Entomopathogenic fungi in ecofriendly pest management for food security. Elsevier Inc., 475–505.

Steinwender, B. M., Enkerli, J., Widmer, F., Eilenberg, J., Thorup-Kristensen, K. and Meyling, N. V. 2014. Molecular diversity of the entomopathogenic fungal *Metarhizium* community within an agroecosystem. *Journal of Invertebrate Pathology* 123, 6–12.

Stoops, J., Crauwels, S., Waud, M., Claes, J., Lievens, B. and Van Campenhout, L. 2016. Microbial community assessment of mealworm larvae (*Tenebrio molitor*) and grasshoppers (*Locusta migratoria migratorioides*) sold for human consumption. *Food Microbiology* 53(B), 122–127.

Sumner, R. 1936. Relation of gregarines to growth and longevity in the mealworm Tenebrio Molitor L. *Annals of the Entomological Society of America* 29(4), 645–648.

Szelei, J., Woodring, J., Goettel, M. S., Duke, G., Jousset, F. X., Liu, K. Y., Zadori, Z., Li, Y., Styer, E., Boucias, D. G., Kleespies,R. G., Bergoin, M. and Tijssen, P. 2011. Susceptibility of North-American and European crickets to Acheta domesticus densovirus (AdDNV) and associated epizootics. *Journal of Invertebrate Pathology* 106(3), 394–399.

Turck, D., Castenmiller, J., De Henauw, S., Hirsch-Ernst, K. I., Kearney, J., Maciuk, A., Mangelsdorf, I., McArdle, H. J., Naska, A., Pelaez, C. and Pentieva, K. 2021. Safety of dried yellow mealworm (*Tenebrio molitor* larva) as a novel food pursuant to Regulation (EU) 2015/2283. *EFSA Journal* 19(1), 6343.

Valigurová, A. 2012. Sophisticated adaptations of *Gregarina cuneata* (Apicomplexa) feeding stages for epicellular parasitism. *PloS One* 7(8), e42606.

Valles, S. M. and Chen, Y. 2006. Serendipitous discovery of an RNA virus from the cricket, *Acheta domesticus*. *Florida Entomologist* 89(2), 282–283.

Vallet-Gely, I., Lemaitre, B. and Boccard, F. 2008. Bacterial strategies to overcome insect defences. *Nature Reviews Microbiology* 6(4), 302–313.

Van Huis, A., Rumpold, B. A., Van der Fels-Klerx, H. J. and Tomberlin, J. K. 2021. Advancing edible insects as food and feed in a circular economy. *Journal of Insects as Food and Feed* 7(5), 935–948.

Vandeweyer, D., Lievens, B. and Van Campenhout, L. 2020. Identification of bacterial endospores and targeted detection of foodborne viruses in industrially reared insects for food. *Nature Food* 1(8), 511–516.

Vandeweyer, D., De Smet, J., Van Looveren, N. and Van Campenhout, L. 2021. Biological contaminants in insects as food and feed. *Journal of Insects as Food and Feed* 7(5), 807–822.

Vega, F. E., Meyling, N. V., Luangsa-ard, J. J. and Blackwell, M. 2012. Fungal entomopathogens. *Insect Pathology*, 171–220.

Vigneron, A., Jehan, C., Rigaud, T. and Moret, Y. 2019. Immune defenses of a beneficial pest: The mealworm beetle, *Tenebrio molitor*. *Frontiers in Physiology* 10, 138.

Zanchi, C., Troussard, J. P., Moreau, J. and Moret, Y. 2012. Relationship between maternal transfer of immunity and mother fecundity in an insect. *Proceedings of the Royal Society B: Biological Sciences* 279(1741), 3223–3230.

Zanchi, C., Lindeza, A. S. and Kurtz, J. 2020. Comparative mortality and adaptation of a Smurf assay in two species of tenebrionid beetles exposed to *Bacillus thuringiensis*. *Insects* 11(4), 261.

Chapter 9

Use of protein from yellow mealworms in poultry feed

M. Mabelebele, E. Malematja, S. D. Kolobe, T. G. Manyelo, N. A. Sebola, and F. Manyeula, University of South Africa, South Africa

1 Introduction

The General Assembly of the United Nations has adopted a resolution to shift the global food production chains onto a sustainable and resilient path (Dicke, 2018). Despite these intensifying efforts, the aim of meeting food security needs means a greater challenge for researchers to implement research that will lead to food production increases – and more is still in pursuit for about 40 years since the establishment of the green revolution (Hobbs, 2007). A trending issue in food security is the production of sufficient, high-quality, and digestible protein sources which hinges mainly on sustainable livestock production (Dicke, 2018). The production of animal protein food relies on the feedstuff for livestock that is either suitable for human consumption (such as cereals and soybean meal) or for natural resources (such as fishmeal) which is becoming scarce due to overfishing of the oceans (Bovera et al., 2016; Dicke,

http://dx.doi.org/10.19103/AS.2024.0139.10

2018). Moreover, the demand for animal protein is significantly higher – and is projected to continue to rise – as the human population increases and its nutritional requirement is unsteady (Selaledi et al., 2020). Therefore, animal production is at the nexus of several of the greatest challenges we faced in the twenty-first century. However, achieving food security by curbing climate change with the same critical resources such as arable land, water, and energy has become increasingly difficult (Oberthür and Groen, 2018). The available land to produce enough feedstuff is shrinking due to increasing urbanisation (Hobbs, 2007). In addition, land use is a major issue when searching for opportunities to plant more feedstuff so as to ensure a steady supply of animal feed ingredients (Bovera et al., 2016). To a large extent, the enormous proportion of land used in livestock farming is occupied mainly by planted crops as feed (Dicke, 2018). Furthermore, there is an increase in the prices of soybean and fishmeal. However, existing competition between humans and livestock for soybean consumption is on the rise (Elahi et al., 2020; Chodová and Tůmová, 2020). As for fishmeal and oil production, the wild capture of forage fish is declining, and high coastal habitat impairment is a reality (Chodová and Tůmová, 2020). Other challenges include the reduction in environmental impacts such as biodiversity loss, land loss, and loss of ecosystem (Smith and Gregory, 2013).

Therefore, conventional feed ingredient production is one of the factors contributing to climate change and it needs to be dealt with at local and international levels. Mitigating greenhouse gas emissions and curbing climate change are top priorities for the Paris Agreement goal achievement (Oberthür and Groen, 2018). It has been projected that climate change in the future will have an impact on food production chains (Smith and Gregory, 2013). Moreover, it will be scientifically impossible for government organisations, the United Nations, and food summits to achieve food security under the Paris Climate agreement to decarbonise the world economy (FAO and WHO, 2022). However, to end this, food security must be achieved while stabilising climate change and reducing the environmental impact of food production (Smith and Gregory, 2013). To successfully meet food security needs to feed the rapidly increasing human population under the Paris Agreement goals, innovative changes in food production are needed (Dicke, 2018). Therefore, alternative and easily accessible feed ingredients as a potential replacement for soybean meal and fishmeal are needed (Chodová and Tůmová, 2020). As a first step towards a sustainable food production approach, various insect species are identified as they provide a great potential alternative and a sustainable protein source for use in sustainable livestock production (Dicke, 2018; Mabelebele et al., 2022). Insects' meals provide an excellent alternative protein source to replace conventional protein sources in poultry and fish diets (Dicke, 2018). Hence, the inclusion of insect meals in poultry diets is expected to lower the price and reduce competition between human and animal feed ingredients, resulting in the cost-effectiveness of small-scale poultry farms (Ballitoc and Sun, 2013; Selaledi et al., 2020). Other benefits of using insect meals relate to health and welfare – with opportunities to reduce the use of synthetic growth promoters in feeding and livestock production (Benzertiha et al., 2020; Elahi et al., 2022). Insects are regarded as the future feed ingredient for poultry, aquaculture, and piggery (Mabelebele et al., 2022; Malematja et al., 2023). Therefore, the aim of this chapter is to outline the significant role played by yellow mealworm larvae as a protein source in the place of soybean, which is expensive. The chapter also explores the significant role of yellow mealworm in improving productivity to feed a billion people sustainably in the context of climate change, with reduced environmental impact.

2 Electronic database accessibility and acquisition of literature

The study was conducted to determine the effects that incorporating insect meals into poultry diets have on growth performance, carcass characteristics, meat quality, and other products. The literature search was performed via electronic databases and was based on peer-revised and published articles that reported the use of insect meals in poultry diets as feed supplements and feed additives. An algorithm search key for recently published articles was conducted using an electronic database such as Directory of Open Access Journals, ResearchGate, ScienceDirect, Google Scholar, and JURN. In the search process, the keywords were 'insect meal', 'growth performance', 'meat quality', 'poultry', and 'yellow mealworm'. In addition, the citations included in articles from the databases were used to search for other relevant articles. Searched articles were evaluated and sorted according to the production parameters, and the results varied according to poultry species and breed, age, sex, level of yellow mealworm inclusion, and the method used for supplementation.

3 The production of conventional protein sources (fishmeal and soybean meal) vs insect larvae

The traditional way of producing animal protein for use in feeds is depicted in Fig. 1. The figure clearly shows that the current feedstuff production has detrimental environmental effects such as biodiversity loss, deforestation, destruction of ecosystems, land degradation, and increasing greenhouse gas emissions resulting in a large carbon footprint (Oberthür and Groen, 2018; Selaledi et al., 2020). As shown in Fig. 2, insect production is environmentally friendly, contrary to soybean production and wild capturing of forage fish for fishmeal production (Chodová and Tůmová, 2020). In addition, the breeding system of insects is simple as they can grow on wild-range substrates such

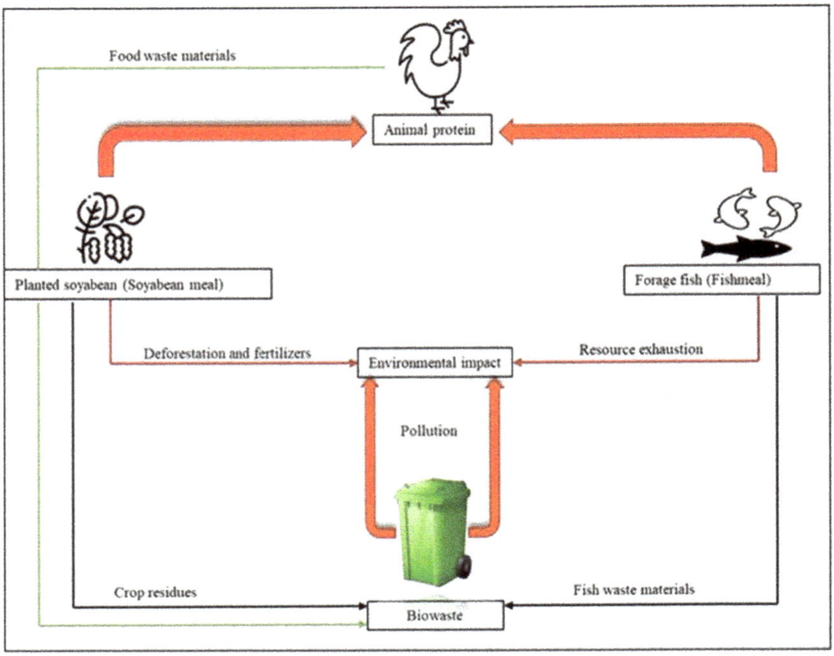

Figure 1 The traditional way of producing animal protein for human consumption.

as cereals, fruit or vegetable residues, decomposing organic matter, animal manure, and industrial by-products that would be environmental problems. Insects are capable of converting low-quality crop residues and animal manure into high-quality protein and fertiliser for growing plants (Elahi et al., 2022). Food waste and other agricultural waste materials such as vegetable residues could be used as substrates to grow insects, rather than being thrown away (Bordiean et al., 2020; Harsányi et al., 2020). Among the possible insect species, yellow mealworm larvae (*Tenebrio molitor*) have been identified as a potential feed ingredient and is regarded as the most bred and traded insect worldwide (Bovera et al., 2015; Benzertiha et al., 2020; Bordiean et al., 2020; Sedgh-Gooya et al., 2022). Hence, the use of yellow mealworm larvae presents a potential solution to improve the sustainability of animal diets and it reduces feed costs. This implies that the insect meal is an acceptable feed ingredient – an affordable, accessible, and appropriate protein source for poultry. The rapid consumption of biomass shown in Fig. 2 reduces methane formation and off-gassing, thus stabilising climate change as one of the Paris Agreement goals (Elahi et al., 2022). Additionally, the consumption of ground mealworms emits no ammonia, hence, they are proving to be biofriendly by causing less harm to the environment (Oonincx, et al., 2010; Hussain et al., 2017).

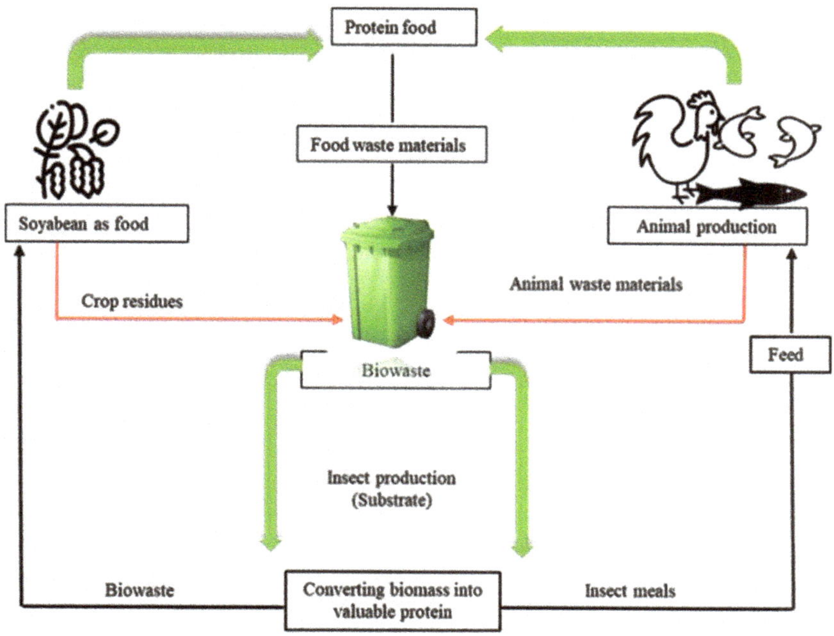

Figure 2 Sustainable way of producing animal protein with the use of insect meals as an alternative protein source for use in livestock production.

3.1 Nutritional benefits of yellow mealworm larvae

Nutrients, especially proteins, are vital in improving the growth health and product quality of poultry (Craig et al., 2017). In the poultry commercial sector, it is important to balance all dietary nutrients (mainly protein) such that they meet the animal nutrition requirements for high productivity and efficiency (Thirumalaisamy et al., 2016). In general, protein is needed for muscle development and health maintenance (Lee et al., 2021). Hence, soybean and fishmeal are the most common ingredients used in the commercial feed industry to provide protein for poultry (Ravindran, 2013). These conventional protein sources are often reported to be expensive and consequently lead to the rise in total costs of feeds (Veldkamp and Bosch, 2015), especially in the poultry industry (Thirumalaisamy et al., 2016). However, edible insects such as yellow mealworms (*Tenebrio molitor* larvae) have been widely documented to have a high nutrient content including mainly protein, essential amino acids, fatty acids, and minerals (Hlongwane et al., 2020; Kröncke et al., 2019; Siemianowska et al., 2013). However, insect species at various stages of development differ in protein, lipids, fatty acid, and amino acid levels (Lee et al., 2021; Hussain et al.,

2017). Hence, the accumulation of these nutrients in edible insects could be beneficial when replacing the conventional protein sources in poultry diets.

Many studies have found that yellow mealworms have nutritional benefits (Rumpold and Schluter, 2013a; Bordiean et al., 2020). At the larvae stage, they have a high crude protein and lipid content (Rumbos et al., 2020; Chodová and Tůmová, 2020). According to various authors, it contains about 49.1–50.8% of dry matter, crude protein which could aid in facilitating growth in poultry diets of mainly broiler chickens (Veldkamp and Bosch, 2015; Hussain et al., 2017; Chodová and Tůmová, 2020). They are also reported to be rich in nitrogen and essential amino acids that are available only in food and cannot be synthesised by humans or livestock (Selaledi et al., 2020). Their fat content is rich in unsaturated fatty acids which have health benefits for humans (Bordiean et al., 2020; Lee et al., 2021). The oil extracted from yellow mealworms is high in essential polyunsaturated fatty acids (mainly linoleic and linolenic acids), which are vital in improving the health status of humans and animals (Heidari-Parsa, 2018). As a result, several authors are exploring the potential of oil-rich insects such as yellow mealworm as a source of fat content and essential fatty acids (Yi et al., 2013; Da Silva Lucas et al., 2020; Kim et al., 2020; Lee et al., 2021).

Mealworm larvae are rich in fibre which helps to improve feed utilisation in the gut (Siemianowska et al., 2013). Anti-nutritional factors such as chitin and uric acid in mealworms improve the health of the chickens (Chodová and Tůmová, 2020). The chitin present in insects could be highly beneficial in improving the health of birds by reducing the accumulation of harmful microbial populations such as *Escherichia coli* and *Salmonella* spp. in the gut, thus reducing the use of antibiotics (prebiotic effect) and posing a minimal threat to animal health and safety when included in poultry diets (Van Huis and Oonaincx, 2017; Bovera et al., 2015). Along with other bioactive compounds, chitin also plays a role in preventing pathogen infections (Motte et al., 2019). Yellow mealworm larvae are more digestible than other chitin – rich edible insects such as crickets, locusts, termites, and stink bugs (Siemianowska et al., 2013). It is known that the type of substrate – including protein and starch – offered to the mealworms has a significant influence on their nutritional composition (Bordiean et al., 2020; Harsányi et al., 2020). Therefore, mealworms are able to produce high-quality nutrients from organic waste materials and by-products of traditional crop residues (Oonincx et al., 2015; Bordiean et al., 2020; Harsányi et al., 2020; Heidari-Parsa, 2018). For example, mealworms could turn wheat bran into a high-quality protein (Selaledi et al., 2020) – therefore, some of the birds have recently been fed diets consisting of 80% wheat bran (Rumbos et al., 2020; Harsányi et al., 2020). Moreover, they could still produce more essential fats even when fed substrates of very low-fat quality (Dreassi et al., 2017).

3.2 Yellow mealworm vs traditionally used protein sources

Soybean and fishmeal have long been used in monogastric diets due to their high crude protein content and amino acid profile which could meet the nutritional requirements of poultry (Veldkamp and Bosch, 2015; Thirumalaisamy et al., 2016). Nevertheless, yellow mealworms as a potential alternative protein source have a higher protein content than soybean but are lower compared to fishmeal (Bovera et al., 2015). However, compared to common housefly and black soldier fly which have been reported to have the highest crude protein content (42.3–62.5%), yellow mealworm larvae contain up to 50.8% (Veldkamp and Bosch, 2015). Again, they have a higher sulphur amino acid index – mainly lysine, methionine, and tryptophan that outcompete soybean but are comparable to that of fishmeal (Elahi et al., 2020; Heidari-Parsa, 2018; Veldkamp and Bosch, 2015), which makes them a suitable insect protein source to replace traditional protein ingredients in poultry diets (Khampakool et al., 2020; Chodová and Tůmová, 2020). Similarly, it has been documented that yellow mealworms contain balanced amino acids (lysine and methionine) than plant-based protein sources. (Rumpold and Schluter, 2013b; Chodová and Tůmová, 2020). In addition, the oil from yellow mealworm and soybean has the same energy value (Kierończyk et al., 2018). However, unlike soybean, yellow mealworms lack calcium and sulphur amino acids (methionine, cysteine, and arginine) which are the limiting amino acids in monogastric animals. Literature reported that sulphur amino acids facilitate the process of protein synthesis (Veldkamp and Bosch, 2015). Hence, there is a need for supplementation of sulphur amino acids when including yellow mealworm as a protein source in poultry diets (Selaledi et al., 2020). Additionally, yellow mealworm is rich in micro minerals such as iron and magnesium, which are responsible for the proper functioning of the cells and also aid in improving the immunity of the animal, hence, improving their health status (Mabelebele et al., 2022). These essential minerals seem to be lacking in both soybean and fishmeal (Table 1).

3.3 Common techniques used in processing yellow mealworm larvae

Feed processing involves manipulating the nutritional composition of the feed, mainly to improve feed utilisation by increasing nutrient digestibility while reducing the anti-nutritional factors (Mabelebele et al., 2022). The various feed processing methods used for food and feed ingredients include defatting, heating, cooking or boiling, baking, frying, solar drying, oven drying, microwave drying, vacuum drying, high-frequency drying, infrared drying, and freezer drying (Keil et al., 2022; Lee et al., 2019). However, methods such as

Table 1 Nutritional comparison of fresh and dried yellow mealworm larvae vs fishmeal and soybean meal

Nutrients	Fresh yellow mealworm larvae	Dried yellow mealworm larvae	Fishmeal	Soybean meal
Moisture %	36.42–56.27	2.43–5.78	–	–
Crude ash %	1.55–3.24	2.51–6.99	15.00–17.15	6.00–7.19
Crude protein %	17.92–52.14	43.96–69.0	64.60–67.53	47.50–50.5
Crude fat %	21.93–28.38	19.12–42.48	10.36–11.76	1.40–1.90
Crude fibre %	24.36	3.97–22.35	0.26–1.20	6.70–7.43
Lysine g/100g	2.41	2.50–4.51	7.91	6.36
Methionine g/100g	0.50	0.52–1.34	3.02	1.40
Threonine g/100g	1.45	1.55–1.64	4.37	3.89
Calcium mg/kg	514.12	382.00–500.12	2.47%	3.20%
Phosphorus mg/kg	950.12	7.00–976.36	2.13%	5.80%
Iron mg/kg	65.36	63.00–68.20	–	–
Zinc mg/kg	96.14	102.00–106.31	–	–
Potassium mg/kg	932.63	8.50–953.20	0.60%	6.66%
Chitin %	–	5.60–8.91	–	–

Source: Adapted from: Banaszkiewicz (2011); Cho and Kim (2011); Siemianowska et al. (2013); Veldkamp and Bosch (2015); Oonicx et al. (2015); Hussain et al. (2017); Heidari-Parsa (2018); Kröncke et al. (2019); Bordiean et al. (2020); Hong et al. (2020); Selaledi et al. (2020).

Table 2 Recent literature on common processing techniques used in yellow mealworms (*Tenebrio molitor* larvae)

Method	Temperature	Duration	Effect on nutrient composition	References
Freeze-drying	–20°C or –50°C	27 h	Had a minimal effect on the fatty acid and amino acid profile and also resulted in high lipid oxidation	Melis et al. (2018); Keil et al. (2022)
Heating	55°C, 75°C, and 95°C	20, 40, and 60 min	Significantly affected protein oxidation and functional properties of *Tenebrio molitor*	Lee et al. (2019)
Rack oven drying	60°C	19 h	Caused darkening of yellow mealworms	Keil et al. (2022)
Infrared drying	130°C	15 min	Caused darkening of yellow mealworms	Keil et al. (2022)
Defatting/lipid extraction	60°C	15 s and 35 s	The hydraulic pressing of yellow mealworms resulted in larval expansion which consequently reduced apparent density with lighter mealworm colour	El Hajj et al. (2023)

defatting or drying through oven, microwave, and freezer are commonly used in the poultry feed industry to process oil-containing feedstuffs such as soybean and edible insects (Veldkamp and Bosch, 2015). Defatting or oil extraction and heating post-harvest has been reported to elevate the crude protein content of insects such as yellow mealworms beyond that of soybean (Veldkamp and Bosch, 2015). However, the extraction of oil heavily influences the protein functionality of insects when taking into consideration the temperature and duration of processing (Lee et al., 2021). The literature has widely reported on different methods and their effects on the nutritional composition of yellow mealworms (Table 2).

3.4 Influence of processing methods on the nutritional values of yellow mealworm larvae

The exposure of yellow mealworms to various processing methods highly influences their nutrient quality, including their physical and chemical characteristics, especially their availability and palatability when utilised as a feed ingredient in poultry diets (Lee et al., 2021; Krzyżaniak et al., 2022;

Bordiean et al., 2020). Hence, the nutritional value of insects depends on the type of processing technique, the temperature, and the processing period (Motte et al., 2019). In addition, Bordiean et al. (2020) reported developmental stages and processing methods as the main factors influencing both protein and amino acid availability in yellow mealworms. According to Heidari-Parsa (2018), processed (dried) yellow mealworms have been reported to contain more nutrients including protein, amino acids, fatty acids, and minerals than fresh mealworms. However, thermal and oven drying are considered as suitable methods to improve the nutritional value of fresh yellow mealworm by elevating its protein content (Bordiean et al., 2020; Selaledi et al., 2020), as well as ensuring safe consumption. This is because the protein content of yellow mealworms has been observed to increase with the rise in temperature (Bjørge et al., 2018). For that, techniques such as heating and freezing are used to humanely slaughter live insects including yellow mealworms prior to drying and grinding (Hong et al., 2020). Moreover, heating and drying techniques could also aid in destroying digestive enzymes in insects, which bind and reduce protein accessibility after grinding (Elahi et al., 2020). In addition, heating mealworms could aid in reducing their microbial population, consequently reducing harmful microbes and promoting the safe consumption of insects (Grau et al., 2017; Krzyżaniak et al., 2022). Finke (2007) also reported techniques such as heating and drying as preferable methods to help reduce antinutrient chitin, which forms the fibre content in insects such as yellow mealworm. High chitin and fat content have been reported to negatively influence performance and nutrient digestibility, respectively, when included in poultry diets (Selaledi et al., 2020). However, insect overheating and drying of yellow mealworms could negatively affect their amino acid availability through denaturation and the Maillard reaction process, consequently destroying their essential nutrients (Kröncke et al., 2019; Bordiean et al., 2020).

Although defatting or fat extraction prior to protein extraction yields even higher protein content while improving the amino acid and fatty acid profiles of yellow mealworms (Mintah et al., 2020; Lee et al., 2021), it could significantly alter the lipid composition of insects by either improving or destroying essential fatty acids. Therefore, it is important to consider the type of lipid extraction technique suitable for a specific insect species (Lee et al., 2021). However, freeze-drying affects the taste of yellow mealworm, hence, reducing intake (Selaledi et al., 2020). Again, the processing of yellow mealworms through dry fractionation (protein extraction method) has been reported to improve their functional properties and nutritional value (Purschke et al., 2018; Lee et al., 2021).

3.5 Influence of processing methods on yellow mealworm storage in poultry diets

Monogastric feed ingredients including insects and insect meals used in poultry diets require storage at room temperature (25°C) for later use (Keil et al., 2022). Therefore, it is of paramount importance to increase and maintain the long shelf life of feed ingredients. However, the quality of insects and insect meal depends on the processing technique to which they were exposed before storage. This is because they are very sensitive to spoilage, especially in a live, fresh, or unprocessed state (Hong et al., 2020), hence, they require good storage conditions. It has been reported that heating insects with salty water prior to oven or solar drying improved their shelf life, although it elevated the amount of sodium chloride in the insects (Musundire et al., 2016; Kewuyemi et al., 2020). According to Keil et al. (2022), it is vital to dry fresh insects prior to consumption or storage, especially since the presence of high microbial load in fresh insects leads to rapid spoilage, thus posing a threat to human and animal health when included in diets (Hong et al., 2020; Bordiean et al., 2020). Hence, drying of yellow mealworms prior to storage has been highly recommended to prolong their shelf life (Krzyżaniak et al., 2022). According to Heidari-Parsa (2018), drying helps to reduce the moisture content, consequently improving handling and storage since the mealworms will weigh less, with prolonged shelf life. Furthermore, fresh and dried yellow mealworms could be safely stored in a refrigerator at −20°C prior to use (Selaledi et al., 2020; Melis et al., 2018). Since yellow mealworms are rich in fats and oils (Da Silva Lucas et al., 2020; Kim et al., 2020; Lee et al., 2021), the accumulation of fats leads to lipid oxidation which reduces their shelf life. Hence, the defatting or oil extraction technique has also been observed to prolong the storage of yellow mealworms (Hong et al., 2020). Therefore, poor storage conditions clearly result in undesirable odour and taste, with an increased number of harmful microbes and eventually spoilage – hence, becoming unsuitable for livestock feeding.

4 Inclusion of yellow mealworm in poultry diets

The inclusion of insect meals in poultry diets has gained popularity mainly because of their high nutritional value and their accessibility (Józefiak and Engberg, 2015; Malematja et al., 2023). Considerable research has shown that incorporating insect meals into poultry diets has positive effects on production parameters and that it is a suitable replacement for standard protein sources (Bovera et al., 2015; Benzertiha et al., 2020). It has been demonstrated that insect meal in poultry diets enhances the palatability and the functional properties of the diet, which in turn increases feed intake and body weight gain as well as other growth parameters (Elahi et al., 2022). The addition of yellow mealworm in chicken diets has positively influenced growth performance through the alteration of gut microbiota (Malematja et al., 2023). There is empirical evidence demonstrating that the inclusion of yellow mealworm in chicken starter diets improved the growth parameters during the starter period (Sedgh-Gooya et al., 2022). Benzertiha et al. (2020) investigated the effects of incorporating meals into chicken diets and observed that the addition of yellow mealworm in a small amount can improve broilers' growth performance. It was, therefore, indicated that improved growth performance is influenced by increased feed intake, which is associated with the palatability of insect meal (Elahi et al., 2022). Similarly, Ballitoc and Sun (2013) reported improved overall feed intake and weight gain in broiler chickens that were fed diets containing yellow mealworm and it was suggested that increased feed intake is associated with insect palatability. Partial replacement of soybean meal with yellow mealworm had positively affected body weight gain in chickens (Vasilopoulos et al., 2023). Zadeh et al. (2019) also demonstrated that increasing yellow mealworm meal inclusion up to 30 g/kg of feed could improve the body weight and feed conversion ratio in quails. The use of yellow mealworm in broiler starter diets has shown positive improvements in body weight, average daily gain, and partial improvement of feed conversion ratio during the starter period (Elahi et al., 2020).

Recently, it has been indicated that insect meals can successfully replace commonly used proteins in poultry diets without causing adverse effects on growth performance and gut development (Dabbou et al., 2020; Malematja et al., 2023). This is in line with the observations made by Sedgh-Gooya et al. (2022) who observed no remarkable effect on the mortality rate and feed intake in chickens that were fed a diet with yellow mealworm. Furthermore, Yildirim et al. (2020) noted that incorporating defatted yellow mealworm into diets as a replacement for fishmeal reduced feed intake in quails. Vasilopoulos et al. (2023) also observed no effect on the mortality rate in broiler chickens when fed diets supplemented with insect meal. Furthermore, Bovera et al. (2015) indicated that the use of meal from yellow mealworm as the main protein source in broiler grower diets had no significant effect on most growth parameters. Nevertheless, it has been found that incorporating yellow mealworm into indigenous chickens' grower diets did not yield desirable results and hence could not be considered an alternative protein source for chickens (Selaledi et al., 2020). In addition, Khan et al. (2022) demonstrated that the supplementation of mealworm could reduce the average feed intake of broiler quails. Table 3 summarises the literature search results regarding the effects that incorporating yellow mealworm into poultry diets has on growth performance. Despite the desirable results obtained from supplementing yellow mealworms, there is a need to identify the recommended levels

Table 3 Effects of yellow mealworm inclusion level in poultry diets on growth performance

Study region	Species/breed	Sex	Inclusion level	Days of feeding	Conclusion	Author
South Korea	Ross 308	Unsexed	1%, 2%, or 10%	35	Dietary treatment increased the average BW (1484, 1470, and 1486 g), respectively	Ballitoc and Sun (2013)
South Korea	Ross 308	Unsexed	5%, 1%, 2%, or 10%	34	Did not affect FI during the starter period. However, insect meal inclusion of 0.5 % improved FI (172 g/day) and FCR (1.7)	Ballitoc and Sun (2013)
Italy	Shaver brown	Male	100%	32	Did not affect the FI, BW, and BWG in male broilers aged 30–62 days. However, the FCR (3.62) was improved. Furthermore, DM (86.42%) and CP (80.20%) digestibility were reduced	Bovera et al. (2015)
Poland	Ross 308	Female	0.2% or 0.3%	35	Did not affect the apparent ileal digestibility of CP and ME.	Benzertiha et al. (2019)
Iran	Japanese quail	Mixed sex	7.5, 15, 22.5, or 30 g/kg	35	Improved FCR (1.90–2.29) and BW (234–269.6 g). However, the daily FI (16.34–17.28 g) was significantly reduced	Zadeh et al. (2019)
China	Ross 308	Male	4%	10	Dietary treatment levels improved FCR (1.41) and BW (664 g) of broiler chickens during the starter period	Elahi et al. (2020)
Turkey	Japanese quail	Mixed sex	Defatted yellow mealworm of 25% or 50%	35	Increased PER (1.135–1.174) and EFR (9.38–9.70). Furthermore, BW (160.56–164.90 g) and BWG (152.35–156.90g) were increased; also improved FRC (3.56–3.68 g: g)	Yildirim et al. (2020)
Poland	Ross 308	Female	0.2% or 0.3%	35	Improved FI (3380 g) and BW (2148 g) throughout experimental period. However, FCR (1.57 was increased	Benzertiha et al. (2020)

(*Continued*)

Table 3 (*Continued*)

Study region	Species/breed	Sex	Inclusion level	Days of feeding	Conclusion	Author
Italy	Label Hubbard hybrid	Female	75 g/kg	54	Did not affect BW at slaughter age	Dabbou et al. (2020)
Iran	Arbor Acres	Mixed sex	2.5% or 5%	42	Did not affect mortality rate throughout the experimental period. However, BWG (178.25–18627 g) was increased during the first 10 days	Sedgh-Gooya et al. (2021)
Iran	Arbor Acres	Mixed sex	2.5%	10	Improved the FCR (1.25), thus increased BWG (186.2 g) during the starter period. However, the FI was not affected	Sedgh-Gooya et al. (2022)
Greece	Ross 308	Male	5% or 10%	35	Increased average FI (3612.80–3823.13 g), BW (2208.25–2346.50 g), and BWG (2162.35–2300.43 g). However, FCR was not affected	Vasilopoulos et al. (2023)
Pakistan	Japanese quail	Mixed sex	1, 2, or 3 g/kg	35	Average FI (478.46–494.17 g) was negative throughout the experimental period. However, average FCR (2.75–2.90) and BWG (164.78–176.25 g) were improved	Khan et al. (2022)

FI, feed intake; BW, body weight; BWG, body weight gain; FCR, feed conversion ratio; PER, protein efficiency ratio; EFR, energy efficiency ratio; CP, crude protein; ME, metabolizable energy; DM, dry matter.

to optimise productivity. The optimal inclusion level will help farmers in estimating the number of specific feed ingredients to be included and mixed in formulated animal diets. There are different incremental levels of up to 100% used by various authors to replace soybean meal or fishmeal with insect meal as an alternative protein source in poultry diets (Veldkamp and Bosch, 2015). However, the optimal insect supplementation levels depend mainly on the nutrient quality of the insect, especially the protein content, in meeting the nutrient requirements of the bird (Thirumalaisamy et al., 2016). It is important to balance the insect meal based on its nutritional profile to avoid oversupply or undersupply of amino acids, hence, negatively affecting the health and growth of the birds (NRC, 1994; Thirumalaisamy et al., 2016; Krzyżaniak et al., 2022). It has been reported that high inclusion levels of yellow mealworm cause toxicity, subsequently leading to the increasing mortality rate in broiler chickens. Therefore, it is important to optimise the inclusion levels of yellow mealworms in poultry diets (Ballitoc and Sun, 2013). According to FAO (2013) the recommended supplementation levels of insects could range up to 100%, depending on the poultry species and the processing methods used. However, the insects can be easily supplemented in mealworm-based diets to improve health and performance (Selaledi et al., 2020).

4.1 Effects on poultry meat quality when feeding insect-based diets

Further processing of poultry meat into meat products calls for higher standards to improve carcass quality and sensory attributes (Bovera et al., 2016; Selaledi et al., 2021). Meat properties such as colour, moisture content, tenderness, and texture are important factors as they drive meat consumers to buy the meat (Malematja et al., 2022). Moisture content is a significant factor in meat storage and desirable moisture content in raw poultry meat should fall within the range of 70–72% (Ballitoc and Sun, 2013). Furthermore, the water-holding capacity has a direct effect on the colour and texture of the meat (Zadeh et al., 2019; Sedgh-Gooya et al., 2021). Nutrition is one of the production-specific factors which influence the quality and shelf life of the meat (Albrecht et al., 2019), therefore it is very important to review the literature on the effects that insect meal have on meat quality and sensory attributes. Zadeh et al. (2019) observed improved carcass, breast meat yields and sensory attributes in Japanese quail when fed with a diet containing insect meal. Likewise, Yildirim et al. (2020) also reported improved hot and carcass weight in quails fed diets with defatted yellow mealworm. These improvements seem to disappear when insect meal is included at 50%, 75%, or 100%. Considerable studies (Shaviklo et al., 2021; Elahi et al., 2022) indicated that yellow mealworm (*Tenebrio molitor* larvae) supplementation in broiler diets improved the meat characteristics, sensory attributes, and meat shelf life. Partial replacement of soybean meal with yellow mealworm larvae in broiler diets positively improved carcass yield, moisture content, and fatty acid composition (Vasilopoulos et al., 2023). Additionally, improved skin and meat colour, abdominal cavity odour, and texture were reported on broilers fed diets containing yellow mealworm (Sedgh-Gooya et al., 2021).

Feeding a chicken's diet containing full-fat yellow mealworm did not prejudice meat quality at slaughter age (Dabbou et al., 2020; Selaledi et al., 2020). Furthermore, vast literature (Bovera et al., 2016; Elahi et al., 2020) indicated a lack of significant effects on meat chemical characterisation when broilers were fed diets supplemented with yellow mealworm. Similarly, Secci et al. (2018) reported no effect on meat quality of Barbary partridge fed diets containing insect meal in place of soybean. Similar meat moisture, colour, and protein content were also reported by Shaviklo et al. (2021) in chickens fed diets with or without insect meal. Indeed, Sedgh-Gooya et al. (2021) demonstrated that incorporating yellow mealworm powder into chickens' diets do not affect the final product such as meat characteristics. Khan et al. (2022) reported no effects on the weights of wings, legs, and necks from quails fed either a basal diet or diets supplemented with mealworms. This implies that the standard protein source can be safely replaced by yellow mealworm without compromising the meat quality. Contrary to these findings, Vasilopoulos et al. (2023) observed negative effects on fat composition in broilers fed diets containing insect meal. Similarly, Zadeh et al. (2019) observed no difference in meat cooking loss from the quails fed diets supplemented with insect meal. Despite the extensive research on the use of insect meal supplementation in poultry diets, there are limited publication works precisely evaluating carcass characteristics and sensory attributes of birds fed diets containing insect meal (Shaviklo et al., 2021). The reports on the effects on meat quality and sensory attributes when feeding chickens insect meals are summarised in Table 4.

4.2 The global stand and direction in relation to the utilisation of yellow mealworm in poultry diets

Although the worldwide interest in insect farming continues, mass rearing of insects is still evolving, and most farmers operate on a smaller scale (Mastoraki et al., 2020). Recently, the European Commission has permitted the use of yellow mealworm in livestock diets (EU, 2017/8993) due to sustainable production (Iaconisi et al., 2017). Based on the current status of yellow mealworm production, the European and Asian countries are actively focused on mass rearing – such that the industry is operating faster in terms of yellow mealworm production through the use of advanced technologies to improve efficiency and output. However, laws and regulations on insects are implemented with regard to health and safety measures for human and livestock consumption (Lahteenmaki-Uutela et al., 2017). According to Ghosh et al. (2017), yellow mealworm larvae increasingly being considered in food-related legislation in Australia, Thailand, and New Zealand. However, Canada has granted no laws and restrictions on the use of insects including yellow mealworm in animal feeds, hence, insects and their by-products are sold in the market (Lahteenmaki-Uutela et al., 2017). In addition, farmers in Thailand have already adopted the mass rearing of yellow mealworm (Fernandez-Cassi et al., 2019). Kolobe et al. (2023) and Babarinde et al. (2021) indicated that there is a lack of insect farming activity in most parts of Africa, hence, there is slow adoption of insect farming on a large scale. Thus, knowledge and awareness on insect production are required across the continent. However, countries such as South Africa, Kenya, and Uganda have recently established the production of yellow mealworm larvae on a small scale (Orinda et al., 2018; Nyangena et al., 2020; Kolobe et al., 2023). Since the European Union has promoted the use of insects' meals in livestock diets, this has tremendously stimulated more research focus on the production of yellow mealworm larvae for poultry species (Van Huis and Oonincx, 2017; Madau et al., 2020). Hence, the mass production of this insect species is expected to expand even further in coming years across insect industries, especially in Europe (Rumbos et al., 2020).

4.3 Challenges that could limit the use of yellow mealworm in poultry diets

Considerable literature has shown that there are still some challenges in maximising the utilisation of insects' meals in poultry diets (Elahi et al. 2020). However, the quality and nutrient profile of insects differ with the growth stage, substrate, rearing environment, and processing method. Insects may also pose the risk of accumulating undesirable substances such as heavy metals and mycotoxins and microbial loads. Since yellow mealworm larvae can grow on various substrates ranging from agricultural crop residues to animal manures, they can accumulate fertiliser and veterinary drug residues in their body (Bordiean et al., 2020). Furthermore, insect larvae are also known to accumulate mycotoxins such as aflatoxin B1, which is one of the most dangerous mycotoxins harmful to both humans and animals (Bordiean et al., 2020). Therefore, the processing method needs to be adjusted and updated regularly for risk reduction of pathogenic and heavy metal contaminations, hence improving the safe use of insects' meals in poultry diets. On the other hand, the absence of large-scale farming of yellow mealworm larvae and a steady supply may delay the transitioning process. However, with the constant supply of insects' larvae, the poultry industry will be able to adopt a resolution to shift the global food production chains onto a sustainable and resilient path.

Table 4 Summary of literature on effects of feeding insect meal on meat quality and sensory attributes of domesticated birds

Study region	Species	Sex	Inclusion level	Days of feeding	Conclusion	Author
South Korea	Ross 308	Mixed sex	0%, 5%, 1%, 2%, or 10%	35	Increased the protein content (21%) and the moisture content (76–78%) of the meat. However, the ash content was not affected	Ballitoc and Sun (2013)
South Korea	Ross 308	Mixed sex	0%, 5%, 1%, or 2%	35	Increased dressed carcass weight (90–92%)	Ballitoc and Sun (2013)
Italy	Shaver brown	Male	100%	32 days	Did not affect carcass yield	Bovera et al. (2016)
Italy	Barbary partridge	Mixed sex	25% or 50%	57	The meat pH, shear force, cholesterol, and colour were not affected	Secci et al. (2018)
Iran	Japanese quail	Male	30 g/kg	35	Improved carcass and breast yields (72.67% and 16.62%) as well as water retention capacity (25.72%), and meat colour (7.63 redness and 12.13 yellowness)	Zadeh et al. (2019)
China	Ross 308	Male	2%, 4%, or 8%	42	Dietary treatments did not affect meat pH, shear force, cooking loss, and meat colour	Elahi et al. (2020)
Italy	Label Hubbard hybrid	Female	75 g/kg	54	Did not influence carcass yields, meat pH and colour and drip loss. Similarly, proximate composition of meat was not affected	Dabbou et al. (2020)
Turkey	Japanese quail	Mixed sex	Defatted yellow mealworm of 25%	35 days	Increased the hot (71.53%) and cold carcass yield (69.90%)	Yildirim et al. (2020)
Iran	Ross 308	Male	1%, 2%, or 3%	24	Did not affect the moisture, protein and fat content of the breast meat. However, the meat pH (5.78–5.96), water holding capacity (71.60–73.93%), and ash content (1.60%) were reduced at 24 days of age	Shaviklo et al. (2021)

Iran	Ross 308	Male	1%, 2%, or 3%	42	Did not affect protein and moisture content and meat colour. However, the meat pH values (5.72–5.86), water holding capacity (52.05–54.06 %), and fat content (1.24–1.55%) were reduced	Shaviklo et al. (2021)
Iran	Arbor Acres	Mixed sex	2.5% or 5%	42	Did not influence carcass yield as well as breast meat, thigh, wing, neck, and back meat weights.	Sedgh-Gooya et al. (2021)
South Africa	Boschveld chickens	Mixed sex	5%, 10%, or 15%	60	Did not affect carcass characteristics such as weight and length of various parts. However, it increased cooking loss (12.23–13.70), meat pH (6.0), meat shear force (55.6), and meat colour lightness before cooking, ranging from 61.7 to 6.9.3 as compared to the control group	Selaledi et al. (2021)
Pakistan	Japanese quail	Mixed sex	1, 2, or 3 g/kg	35	Increased carcass dressing (65.23–68.62%) and breast meat (34.38–39.47%)	Khan et al. (2022)
Greece	Ross 308	Male	5% or 10%	35	Improved carcass yield (1937.58 g), moisture content (73.98–77.54%), and protein content (21.30–24.24%). However, fat content was reduced (1.24–2.67%).	Vasilopoulos et al. (2023)

Although insect meals have shown great potential for use in poultry diets, some countries still consider insect meals as processed animal protein and do not allow it as feed to livestock (Regulation (EC) No. 1069/2009). Nevertheless, due to ecological and economic benefits, overall acceptance by the farmers and consumers, and the fact that poultry consumes various insects' species as part of their natural diet, it is probably possible that the political framework might change in future and promote the usage of insects' meals in poultry diets (Chodová and Tůmová, 2020; Dabbou et al., 2020). Furthermore, in the current environment, considerable studies on the efficacy of yellow mealworm to substitute the standard protein sources presented the most promising results, therefore making yellow mealworm an innovative protein source for use in poultry diets.

5 Conclusion

From a circular economic point of view, introducing the use of yellow mealworm in poultry diets is regarded as an interesting topic in animal nutrition as the production of yellow mealworm larvae has a remarkably lower impact on the environment. Based on nutritional analysis, yellow mealworm larvae are relatively high in protein content (17.92–52.14%) and amino acids which are almost similar to those of conventional proteins. Various studies have shown that yellow mealworm larvae are the most appropriate substitute for soybean meal and fishmeal as they have similar influence on growth performance without having a negative impact on carcass, meat quality, and egg quality. A wide range of literature reported substantial and desirable results on growth performance when including yellow mealworm in the diets and indicated that yellow mealworm can be used safely in poultry feeds without compromising the production parameters and the quality of the produce. However, the level of evidence or results on the effects of yellow mealworm on poultry performance and productivity varies with poultry species, breed, health condition, and age as well as the level of yellow mealworm supplementation and the duration of feeding. In addition, improvements seem to disappear when yellow mealworm is included at 50%, 75%, or 100% in layers and broilers. Furthermore, the method of supplementation is entirely dependent on the type of poultry production, species, age, and the production parameters in question.

6 Recommendation and future direction

Although yellow mealworm has presented great potential as alternative protein source in poultry diets, the larvae contain a significant amount of crude fat and fibre which limit the optimal inclusion level. Hence, the existing information on the effects of yellow mealworm inclusion in poultry diets varies and this prevents yellow mealworm from becoming a primary protein source for certain poultry species or age groups. Therefore, more studies are recommended to ascertain these findings, especially recommended inclusion levels for optimum production.

7 Conflict of interest

The authors declare no conflicts of interest.

8 Ethics approval

This study requires no ethical approval because there is no animal participants.

9 Authors' contribution

ME and KSD contributed to the information gathering and writing framework. MF, ME, MTG, and KSD contributed to the editing and improvement of the manuscript. MM, SNA, and MTG approved the final manuscript.

10 Acknowledgment

The authors would like to thank the College of Agriculture and Environmental Science, University of South Africa, for assisting with resources to conduct this study.

11 Where to look for further information

Data used for the research described in this article were extracted from published papers available in electronic database including; Google scholar, Science Direct, Scopus, JURN, and Research Gate.

12 References

Albrecht, A., Hebel, M., Mittler, M., Hurck, C., Kustwan, K., Heitönig, B., Bitschinski, D. and Kreyenschmidt, J. (2019). Influence of different production systems on the quality and shelf life of poultry meat:a case study in the German sector. *Journal of Food Quality*.

Babarinde, S. A., Mvumi, B. M., Babarinde, G. O., Manditsera, F. A., Akande, T. O. and Adepoju, A. A. (2021). Insects in food and feed systems in sub-Saharan Africa: the untapped potentials. *Int. J. Trop. Insect Sci.* 41(3): 1923–1951.

Ballitoc, D. A. and Sun, S. (2013). Ground yellow mealworms (*Tenebrio molitor* L.) feed supplementation improves growth performance and carcass yield characteristics in broilers. Open sci. Reposit. *Agric. E.* 18.e23050425.

Banaszkiewicz, T. (2011). Nutritional value of soybean meal. *Soybean Nutr.* 12:1–20.

Benzertiha, A., Kierończyk, B., Kołodziejski, P., Pruszyńska–Oszmałek, E., Rawski, M., Józefiak, D. and Józefiak, A. (2020). *Tenebrio molitor* and *Zophobas morio* full-fat meals as functional feed additives affect broiler chickens' growth performance and immune system traits. *Poult. Sci.* 99(1): 196–206.

Benzertiha, A., Kierończyk, B., Rawski, M., Józefiak, A., Kozłowski, K., Jankowski, J. and Józefiak, D. (2019). Tenebrio molitor and Zophobas morio full-fat meals in broiler chicken diets: effects on nutrients digestibility, digestive enzyme activities, and cecal microbiome. *Animals (Basel)* 9(12): 1128.

Bjørge, J. D., Overgaard, J., Malte, H., Gianotten, N. and Heckmann, L. H. (2018). Role of temperature on growth and metabolic rate in the tenebrionid beetles Alphitobius diaperinus and Tenebrio molitor. *J. Insect Physiol.* 107: 89–96.

Bordiean, A., Krzyżaniak, M., Stolarski, M. J., Czachorowski, S. and Peni, D. (2020). Will yellow mealworm become a source of safe proteins for Europe? *Agriculture* 10(6): 233.

Bovera, F., Loponte, R., Marono, S., Piccolo, G., Parisi, G., Iaconisi, V., Gasco, L. and Nizza, A. (2016). Use of *Tenebrio molitor* larvae meal as protein source in broiler diet: effect on growth performance, nutrient digestibility, and carcass and meat traits. *J. Anim. Sci.* 94(2): 639–647.

Bovera, F., Piccolo, G., Gasco, L., Marono, S., Loponte, R., Vassalotti, G., Mastellone, V., Lombardi, P., Attia, Y. A. and Nizza, A. (2015). Yellow mealworm larvae (*Tenebrio molitor*, L.) as a possible alternative to soybean meal in broiler diets. *Br. Poult.* 56(5): 569–575.

Cho, J. H. and Kim, I. H. (2011). Fish meal–nutritive value. *J. Anim. Physiol. Anim. Nutr.* 95(6): 685–692.

Chodová, D. and Tůmová, E. (2020). *Insects in Chicken Nutrition. A Review*. https://doi.org/10.15159/ar.20.003.

Craig, S. R., Helfrich, L. A., Kuhn, D. and Schwarz, M. H. (2017). *Understanding Fish Nutrition, Feeds, and Feeding*. Available at: http://www.ext.vt.edu/.

da Silva Lucas, A.J., Menegon de Oliveira, L., da Rocha, M. and Prentice, C. (2020). Edible insects: an alternative of nutritional, functional and bioactive compounds. *Food Chem.* 311: 126022.

Dabbou, S., Gasco, L., Lussiana, C., Brugiapaglia, A., Biasato, I., Renna, M., Cavallarin, L., Gai, F. and Schiavone, A. (2020). Yellow mealworm (*Tenebrio molitor* L.) larvae inclusion in diets for free-range chickens: effects on meat quality and fatty acid profile. *ReNew* 35(5): 571–578.

Dicke, M. (2018). Insects as feed and the Sustainable Development Goals. *J. Insects Food Feed* 4(3): 147–156.

Dreassi, E., Cito, A., Zanfini, A., Materozzi, L., Botta, M. and Francardi, V. (2017). Dietary fatty acids influence the growth and fatty acid composition of the yellow mealworm *Tenebrio molitor* (Coleoptera: Tenebrionidae). *Lipids* 52(3): 285–294.

El Hajj, R., Mhemdi, H., Besombes, C., Lefrancois, V., Allaf, K. and Vorobiev, E. (2023). Instant controlled pressure-drop assisted pressing for defatting and dewatering of yellow mealworm larvae: process study and optimisation. *J. Insects Food Feed* 9(6): 761–779.

Elahi, U., Wang, J., Ma, Y. B., Wu, S. G., Wu, J., Qi, G. H. and Zhang, H. J. (2020). Evaluation of yellow mealworm meal as a protein feedstuff in the diet of broiler chicks. *Animals* 10(2): 224.

Elahi, U., Xu, C. C., Wang, J., Lin, J., Wu, S. G., Zhang, H. J. and Qi, G. H. (2022). Insect meal as a feed ingredient for poultry. *Anim. Biosci.* 35(2): 332–346.

FAO (2013). Edible insects – future prospects for food and feed security. FAO Forestry Paper 171. IX. Rome. Available at: https://library.wur.nl/WebQuery/wurpubs/fulltext/258042.

FAO and WHO (2022). *CODEX – A Year in Transition*. Rome: Codex Alimentarius Commission. https://doi.org/10.4060/cc3141en.

Fernandez-Cassi, X., Supeanu, A., Vaga, M., Jansson, A., Boqvist, S. and Vagsholm, I. (2019). The house cricket (Acheta domesticus) as a novel food: a risk profile (Acheta domesticus). *J. Insects Food Feed* 5(2): 137–157. https://doi.org/10.3920/JIFF2018.0021.

Finke, M. D. (2007). Estimate of chitin in raw whole insects. *Zoo Biol.* 26(2): 105–115. https://doi.org/10.1002/zoo.20123.

Ghosh, S., Lee, S. M., Jung, C. and Meyer-Rochow, V. B. (2017). Nutritional composition of five commercial edible insects in South Korea. *J. Asia Pac. Entomol.* 20(2): 686–694. https://doi.org/10.1016/j.aspen.2017.04.003.

Grau, T., Vilcinskas, A. and Joop, G. (2017). Sustainable farming of the mealworm Tenebrio molitor for the production of food and feed. *Z. Naturforsch. C. J. Biosci.* 72(9–10): 337–349.

Harsányi, E., Juhász, C., Kovács, E., Huzsvai, L., Pintér, R., Fekete, G., Varga, Z. I., Aleksza, L. and Gyuricza, C. (2020). Evaluation of organic wastes as substrates for rearing Zophobas morio, Tenebrio molitor, and Acheta domesticus Larvae as Alternative Feed Supplements. *Insects* 11(9): 604.

Heidari-Parsa, S. (2018). Determination of yellow mealworm (*Tenebrio molitor*) nutritional value as an animal and human food supplementation. *Arthropods* 7(4): 94.

Hlongwane, Z. T., Slotow, R. and Munyai, T. C. (2020). Nutritional composition of edible insects consumed in Africa: a systematic review. *Nutrients* 12(9): 2786.

Hobbs, P. R. (2007). PAPER PRESENTED AT INTERNATIONAL WORKSHOP ON INCREASING WHEAT YIELD POTENTIAL, CIMMYT, OBREGON, MEXICO, 20–24 MARCH 2006 Conservation agriculture: what is it and why is it important for future sustainable food production? *J. Agric. Sci.* 145(2): 127.

Hong, J., Han, T. and Kim, Y. Y. (2020). Mealworm (*Tenebrio molitor* Larvae) as an alternative protein source for monogastric animal: a review. *Animals* 10(11): 2068.

Hussain, I., Khan, S., Sultan, A., Chand, N., Khan, R., Alam, W. and Ahmad, N. (2017). Meal worm (*Tenebrio molitor*) as potential alternative source of protein supplementation in broiler. *Int. J. Bio.Sci.* 10(4): 225–262.

Iaconisi, V., Marono, S., Parisi, G., Gasco, L., Genovese, L., Maricchiolo, G., Bovera, F. and Piccolo, G. (2017). Dietary inclusion of Tenebrio molitor larvae meal: effects on growth performance and final quality treats of blackspot sea bream (Pagellus bogaraveo). *Aquaculture* 476: 49–58. https://doi.org/10.1016/j.aquaculture.2017.04.007.

Józefiak, D. and Engberg, R. M. (2015). Insects as poultry feed. In: 20th European Symposium on Poultry Nutrition. 24–27.

Keil, C., Grebenteuch, S., Kröncke, N., Kulow, F., Pfeif, S., Kanzler, C., Rohn, S., Boeck, G., Benning, R. and Haase, H. (2022). Systematic studies on the antioxidant capacity and volatile compound profile of yellow mealworm larvae (*T. molitor* L.) under different drying regimes. *Insects* 13(2): 166.

Kewuyemi, Y. O., Kesa, H., Chinma, C E. and Adebo, O. A. (2020). Fermented edible insects for promoting food security in Africa. Review. *Insects* 11(5): 283.

Khampakool, A., Soisungwan, S., You, S. and Park, S. H. (2020). Infrared assisted freeze-drying (IRAFD) to produce shelf-stable insect food from *Protaetia brevitarsis* (white-spotted flower chafer) larva. *Food Sci. Anim. Resour.* 40(5): 813–830.

Khan, S., Jabbar Tanweer, A., Rafiullah, R., Ibrahimullah, I., Abbas, G., Khan, J., Saeed Imran, M. and Ali Kamboh, A. (2022). Effect of supplementation of mealworm scales (*Tenebrio molitor*) on growth performance, carcass traits and histomorphology of Japanese quails. *J. Anim. Health Prod.* 10(3): 273–411.

Kierończyk, B., Rawski, M., Józefiak, A., Mazurkiewicz, J., Swiatkiewicz, S., Siwek, M., Bednarczyk, M., Szumacher-Strabel, M., Cieslak, A., Benzertiha, A. and Józefiak, D. (2018). Effects of replacing soybean oil with selected insect fats on broilers. *Anim. Feed Sci. Technol.* 240: 170–183.

Kim, T. K., Yong, H. I., Chun, H. H., Lee, M. A., Kim, Y. B. and Choi, Y. S. (2020). Changes of amino acid composition and protein technical functionality of edible insects by extracting steps. *J. Asia Pac. Entomol.* 23(2): 298–305.

Kolobe, S. D., Manyelo, T. G., Sebola, N. A. and Mabelebele, M. (2023). Scope and present status of rearing edible insects for animal feeding in Africa. *All Life* 31(16)1:2206542).

Kröncke, N., Grebenteuch, S., Keil, C., Demtröder, S., Kroh, L., Thünemann, A. F., Benning, R. and Haase, H. (2019). Effect of different drying methods on nutrient quality of the yellow mealworm (Tenebrio molitor L.). *Insects* 10(4): 84.

Krzyżaniak, M., Aljewicz, M., Bordiean, A. and Stolarski, M. J. (2022). Yellow mealworm composition after convective and freeze drying–preliminary results. *Agriculture* 12(2): 149.

Lahteenmaki-Uutela, A., Grmelová, N., Hénault-Ethier, L., Deschamps, M. H., Vandenberg, G. W., Zhao, A., Zhang, Y., Yang, B. and Nemane, V. (2017). Insects as food and feed: laws of the European Union, United States, Canada, Mexico, Australia, and China. *Eur. Food Feed L. Rev.* 12: 22.

Lee, H. J., Kim, J. H., Ji, D. S. and Lee, C. H. (2019). Effects of heating time and temperature on functional properties of proteins of yellow mealworm larvae (*Tenebrio molitor* L.). *Food Sci. Anim. Resour.* 39(2): 296–308.

Lee, J. H., Kim, T. K., Jeong, C. H., Yong, H. I., Cha, J. Y., Kim, B. K. and Choi, Y. S. (2021). Biological activity and processing technologies of edible insects: a review. *Food Sci. Biotechnol.* 30(8): 1003–1023.

Mabelebele, M., Kolobe, S. D., Malematja, E., Sebola, N. A. and Manyelo, T. G. (2022). A comprehensive review of the importance of selected trace elements present in edible insects. *Biol. Trace Elem. Res.201*(7): 1–8.

Madau, F. A., Arru, B., Furesi, R. and Pulina, P. (2020). Insect farming for feed and food production from a circular business model perspective. *Sustainability* 12(13): 418. https://doi.org/10.3390/su12135418.

Malematja, E., Manyelo, T. G., Sebola, N. A. and Mabelebel, M. (2023). The role of insects in promoting the health and gut status of poultry. *Comp. Clin. Pathol.* 32(3): 501–513 https://doi.org/10.1007/s00580-023-03447-4.

Malematja, E., Ng'ambi, J. W., Chitura, T., Nemauluma, M. F. D., Kolobe, S. D. and Manyel, T. G. (2022). Onion meal and onion extracts (Allium cepa L.) as natural growth promoters for use in poultry production: a review. *Appl. Ecol. Environ. Res.* 20(1): 383–396.

Mastoraki, M., Ferrándiz, P. M., Vardali, S. C., Kontodimas, D. C., Kotzamanis, Y. P., Gasco, L., Chatzifotis, S. and Antonopoulou, E. (2020). A comparative study on the effect of fish meal substitution with three different insect meals on growth, body composition and metabolism of European sea bass (Dicentrarchus labrax L.). *Aquaculture* 528: 735511. https://doi.org/10.1016/j.aquaculture.2020.735511.

Melis, R., Braca, A., Mulas, G., Sanna, R., Spada, S., Serra, G., Fadda, M. L., Roggio, T., Uzzau, S. and Anedda, R. (2018). Effect of freezing and drying processes on the molecular traits of edible yellow mealworm. *J. Food Process Eng.* 48: 138–149.

Mintah, B. K., He, R., Agyekum, A. A., Dabbour, M., Golly, M. K. and Ma, H. (2020). Edible insect protein for food applications: extraction, composition, and functional properties. *J. Food Process Eng.* 43(4): e13362.

Motte, C., Rios, A., Lefebvre, T., Do, H., Henry, M. and Jintasataporn, O. (2019). Replacing fish meal with defatted insect meal (Yellow Mealworm Tenebrio molitor) improves the growth and immunity of pacific white shrimp (litopenaeus vannamei). *Animals* 9(5): 258.

Musundire, R., Osuga, I. M., Cheseto, X., Irungu, J. and Torto, B. (2016). Aflatoxin contamination detected in nutrient and anti-oxidant rich edible stink bug stored in recycled grain containers. *PLoS ONE* 11(1): e0145914. https://doi.org/10.1371/journal.pone.0145914.

NRC (1994). Poultry nutrition and feeding. *Anim Nutr Handbook Sect.* 12: 410.

Nyangena, D. N., Mutungi, C., Imathiu, S., Kinyuru, J., Affognon, H., Ekesi, S., Nakimbugwe, D. and Fiaboe, K. K. M. (2020). Effects of traditional processing techniques on the nutritional and microbiological quality of four edible insect species used for food and feed in East Africa. *Foods* 9(5): 574.

Oberthür, S. and Groen, L. (2018). Explaining goal achievement in international negotiations: the EU and the Paris Agreement on climate change. *J. Eur. Public Policy* 25(5): 708–727.

Oonincx, D. G. A. B., Van Broekhoven, S., Van Huis, A. and Van Loon, J. J. A. (2015). Feed conversion, survival and development, and composition of four insect species on diets composed of food by-products. *PLoS ONE* 10(12): e0144601.

Oonincx, D. G. A. B., Van Itterbeeck, J., Heetkamp, M. J. W., Van den Brand, H. and Van Loon, J. J. A. (2010). *An Exploration on Green Gas and Ammonia Production by Insects Species Suitable for Animal or Human Consumption. PLoS ONE, 10(12): e0144601.*

Orinda, M., Magara, H., Ayieko, M. and Nyakeri, E. (2018). *GREEiNSECT Technical Brief#2 Report: Insect Production Systems for Food and Feed in Kenya*. Available at: https://www.researchgate.net/publication/327573356.

Purschke, B., Brüggen, H., Scheibelberger, R. and Jäger, H. (2018). Effect of pre-treatment and drying method on physico-chemical properties and dry fractionation behaviour of mealworm larvae (*Tenebrio molitor* L.). *Eur. Food Res. Technol.* 244(2): 269–280.

Ravindran, V. (2013). Poultry feed availability and nutrition in developing countries. *Poult. Dev. Rev.* 2: 60–63. Available at: http://www.fao.org/.

Rumbos, C. I., Karapanagiotidis, I. T., Mente, E., Psofakis, P. and Athanassiou, C. G. (2020). Evaluation of various commodities for the development of the yellow mealworm, *Tenebrio molitor*. *Sci. Rep.* 10(1): 11224.

Rumpold, B. A. and Schlüter, O. K. (2013a). Nutritional composition and safety aspects of edible insects. *Mol. Nutr. Food Res.* 57(5): 802–823.

Rumpold, B. A. and Schlüter, O. K. (2013b). Potential and challenges of insects as an innovative source for food and feed production. *Innov. Food Sci. Emerg. Technol.* 17: 1–11.

Secci, G., Moniello, G., Gasco, L., Bovera, F. and Parisi, G. (2018). Barbary partridge meat quality as affected by Hermetia illucens and *Tenebrio molitor* larva meals in feeds. *Food Res. Int.* 112: 291–298.

Sedgh-Gooya, S., Torki, M., Darbemamieh, M., Khamisabadi, H. and Abdolmohamadi, A. (2022). Growth performance and intestinal morphometric features of broiler chickens fed on dietary inclusion of yellow mealworm (*Tenebrio molitor*) larvae powder. *Vet. Med. Sci.* 8(5): 2050–2058.

Sedgh-Gooya, S., Torki, M., Darbemamieh, M., Khamisabadi, H., Karimi Torshizi, M. A. and Abdolmohamadi, A. (2021). Yellow mealworm, *Tenebrio molitor* (Col: Tenebrionidae), larvae powder as dietary protein sources for broiler chickens: effects on growth performance, carcass traits, selected intestinal microbiota and blood parameters. *J. Anim. Physiol. Anim. Nutr. (Berl)* 105(1): 119–128.

Selaledi, L., Baloyi, J., Mbajiorgu, C., Sebola, A. N., Kock, H. D. and Mabelebele, M. (2021). Meat quality parameters of boschveld indigenous chickens as influenced by dietary yellow mealworm meal. *Foods* 10(12): 3094.

Selaledi, L., Mbajiorgu, C. A. and Mabelebele, M. (2020). The use of yellow mealworm (*T. molitor*) as alternative source of protein in poultry diets: a review. *Trop. Anim. Health Prod.* 52(1): 7–16.

Shaviklo, A. R., Alizadeh-Ghamsari, A. H. and Hosseini, S. A. (2021). Sensory attributes and meat quality of broiler chickens fed with mealworm (*Tenebrio molitor*). *J. Food Sci. Technol.* 58(12): 4587–4597.

Siemianowska, E., Kosewska, A., Aljewicz, M., Skibniewska, K. A., Polak-Juszczak, L., Jarocki, A. and Jedras, M. (2013). Larvae of mealworm (*Tenebrio molitor* L.) as European novel food. *J. Agric. Sci.* 4: 287–291. [Google Scholar].

Smith, P. and Gregory, P. J. (2013). Climate change and sustainable food production. *Proc. Nutr. Soc.* 72(1): 21–28.

Thirumalaisamy, G., Muralidharan, J., Senthilkumar, S., Sayee, R. H. and Priyadharsini, M. (2016). Cost-effective feeding of poultry. *Int. J. Sci. Environ. Technol.* 5(6): 3997–4005.

Van Huis, A. and Oonincx, D. G. A. B. (2017). The environmental sustainability of insects as food and feed. A review. *Agron. Sustain. Dev.* 37(5): 1–14.

Vasilopoulos, S., Giannenas, I., Savvidou, S., Bonos, E., Rumbos, C. I., Papadopoulos, E., Fortomaris, P. and Athanassiou, C. G. (2023). Growth performance, welfare traits and meat characteristics of broilers fed diets partly replaced with whole *Tenebrio molitor* larvae. *Anim. Nutr.* 13: 90–100.

Veldkamp, T. and Bosch, G. (2015). Insects: a protein-rich feed ingredient in pig and poultry diets. *Anim. Front.* 5(2): 45–50.

Yi, L., Lakemond, C. M., Sagis, L. M., Eisner-Schadler, V., van Huis, A. and van Boekel, M. A. (2013). Extraction and characterisation of protein fractions from five insect species. *Food Chem.* 141(4): 3341–3348.

Yildirim, U., Sarica, S. and Kanoglu, B. (2020). Defatted yellow mealworm larvae (*Tenebrio molitor l.*) meal as possible alternative to fish meal in quail diets. *S. Afr. J. Anim. Sci.* 50(3): 481–491.

Zadeh, Z. S., Kheiri, F. and Faghani, M. (2019). Use of yellow mealworm (*Tenebrio molitor*) as a protein source on growth performance, carcass traits, meat quality and intestinal morphology of Japanese quails (Coturnix japonica). *Vet. Anim. Sci.* 8: 100066.

Chapter 10

The use of protein from yellow mealworms in fish feed

Enric Gisbert, Aquaculture Program – Institute of Agrifood Research and Technology (IRTA), Spain; and Alberto Ruiz, Aquaculture Program – Institute of Agrifood Research and Technology (IRTA), Spain and Aquaculture and Fisheries Group – Wageningen University and Research, Wageningen, The Netherlands

1 Introduction

Addressing the global challenge of feeding a growing population within environmental limits requires a shift towards circular food systems (van Riel et al., 2023). Amongst several food production methods, aquaculture has emerged as a crucial solution to meet rising food demands, alleviate poverty, enhance food security, and combat nutritional deficiencies, all with a lower environmental impact compared to alternative animal-source foods (Asche et al., 2022). Currently, aquaculture stands out as the world's fastest-growing food-producing sector, contributing a record 49.2% to the global production of

http://dx.doi.org/10.19103/AS.2024.0139.12

aquatic animals in 2020. The sector's remarkable expansion and intensification are evident across regions, with a total production of 122.6 million tonnes, valued at USD 264.8 billion. Inland waters contributed 54.4 million tonnes, whilst marine and coastal aquaculture yielded 68.1 million tonnes. Notably, most regions, except Africa, have witnessed continued aquaculture growth, driven by key producers like Chile, China, and Norway. However, challenges persist, hindering the sector's pursuit of sustainable outcomes. To ensure the sustainability and economic growth of aquaculture, a focus on innovative and sustainable practices, particularly in aquafeed formulation and feeding practices, is crucial (FAO, 2022).

Feed stands out as the principal input and a constraining element in aquaculture. Commercial aquafeeds are composed of blends of various feedstuffs designed to meet the fundamental nutritional needs of aquaculture species. Predominantly, marine-derived feed ingredients, notably fishmeal (FM) and fish oil (FO), serve as the primary sources of protein and lipids in formulated diets. Fishmeal is a dependable protein source due to its palatability, digestibility, and an optimal nutrient profile that satisfies the dietary requirements of aquatic species. Especially for carnivorous fish species, aquafeeds have heavily relied on FM as the primary protein source due to its superior essential amino acid (AA) profile, high protein content, enhanced nutrient digestibility, and the absence of antinutritional factors (Macusi et al., 2022). Marine-derived oils possess exceptional palatability, a highly digestible energy content, and a favourable nutritional profile rich in long-chain polyunsaturated fatty acids (PUFA), particularly arachidonic acid (ARA, 20:4n-6), eicosapentaenoic acid (EPA, 20:5n-3), and docosahexaenoic acid (DHA, 22:6n-3) (Mozanzadeh et al., 2022). Until 2005, FM and FO were consistently employed as cost-effective sources of protein and lipids in aquafeed formulations for both carnivorous and omnivorous species. Despite aquafeeds constituting only 4% of total industrial feed production (900–1000 Mt in 2018), the aquaculture sector consumes over 70% of global FM (Jannathulla et al., 2019).

However, the decline in fish catches and the persistent growth in FM consumption paint a bleak future for the aquaculture industry, unless there is a paradigm shift towards incorporating non-fish components in fish feed production (Macusi et al., 2022). Given the impracticality of relying on the volume of wild-caught forage fish required to sustain the surge in aquafeed production (projected at 87.1 million tons in 2025) based on existing aquafeed formulations, especially considering that capture fisheries have plateaued, and major fishing areas have reached their maximum potential (Naylor et al., 2021), it is imperative to reduce the dietary inclusion rates of FM and FO to ensure the continued growth of the aquaculture industry (Gephart et al,, 2020).

Reducing reliance on FM and FO in aquafeed is a crucial objective acknowledged by stakeholders as a top priority for the sustainable development of the aquafeed industry. The extent of dependence on FM and FO in feed formulations is largely contingent on the trophic position of the farmed species. Notably, for most omnivorous fish, dependency on FM and FO in diets is nearly negligible, and contemporary diets for carp, tilapia, or catfish are nearly devoid of FM and FO. However, decreasing FM and FO in aquafeeds poses a more significant challenge for carnivorous fish, especially marine species (Oliva-Teles et al., 2015). Over the past two decades, the aquafeed industry has made substantial strides in improving feed formulations and enhancing the efficiency of marine resource utilization for feed formulation. These advancements have been achieved through improvements in feed conversion ratios (FCRs), reductions in FM and FO inclusion levels in aquafeeds, and increased utilization of FM derived from trimmings and alternative raw ingredients (Naylor et al., 2021).

2 Finding alternatives to marine-derived aquafeed

Any alternative feedstuffs to traditional marine-derived ingredients, such as FM and FO, must fulfil a number of specific characteristics. These include nutritional suitability, ready availability, and ease of handling, shipping, storage, and use in feed production. Additionally, the chosen feeds should be evaluated based on their impacts on fish health and performance, consumer acceptance, minimal pollution and ecosystem stress, and potential human health benefits (Naylor et al., 2009). Furthermore, to foster the sustainable use of alternative and eco-friendly aquafeeds, there is a pressing need to opt for ingredients that can be supplied sustainably, grown locally and have a low environmental footprint (Mitra, 2021).

Considerable research has been dedicated to reformulating aquafeeds through the incorporation of novel ingredients and nutritional supplements, such as exogenous enzymes, bioactive compounds, and bioavailable trace metals. These additions complement plant proteins and address the specific needs of aquaculture species. A prominent challenge in the aquaculture industry is the creation of sustainable feeds that enhance fish welfare, maximize growth potential, and remain cost-efficient, especially when feeding costs can represent up to 50–60% of total production costs (Egerton, 2020).

Various nutritional strategies have been employed to replace marine-derived aquafeed ingredients like FM and FO, with approaches varying based on the targeted feedstuff. Plant-derived ingredients stand out as the most abundant and commonly-used alternative protein sources in aquafeeds (Hardy, 2010). Plant feedstuffs serve as major dietary protein sources for omnivorous and herbivorous fish and rank second to FM in diets for carnivorous species (Tacon, 2020). Moreover, plant proteins are frequently employed as replacements for FM in compound diets for aquatic species due to their cost-effectiveness and the negative consumer perception associated with the use of terrestrial animal by-products for fish feed (Egerton, 2020). However, plant feedstuffs come with challenges, including highly variable protein content, inadequate essential AA profiles and anti-nutritional factors, limiting their use in diet formulations (Oliva-Teles et al., 2015; Mitra, 2021). Recent concerns have also arisen about the social and environmental impact of soybean production for aquaculture feed which competes with growing the crop for human consumption or livestock feed (Tzachor, 2019). For these reasons, the aquafeed industry continues to explore new alternative feed protein ingredients to ensure sustainable growth.

Amongst the alternative feed ingredients, efforts have been focused on testing and validating protein sources like

- single-cell proteins (SCP);
- microalgae and macroalgae;
- rendered animal by-products; and
- insects.

Single-cell proteins, regardless of their origin (microalgae, yeast, fungi, or bacteria), offer crucial advantages over conventional protein ingredients, including a good nutritional profile, shorter production time, reduced land use, and independence from seasonal and climatic variations. Bacterial SCP, in particular, are considered popular and sustainable due to their high protein levels (up to 80% on a dry weight basis), AA profile resembling FM, and content in vitamins, phospholipids (PLs), and other functional compounds (Pereira, 2022).

Microalgae, recognized for their nutritional value and beneficial properties, have been acknowledged as an intriguing ingredient for aquafeeds, despite high production costs. However, this limitation is expected to be overcome in the near future (Aragão et al., 2022). Interest in macroalgae as alternative protein sources for FM is primarily due to their potential use as functional ingredients, including bioactive compounds and prebiotic effects (Camacho et al., 2019). Rendered animal by-products, such as poultry by-product meal, blood meal, and feather meal, contribute to the circular economy by offering nutrient profiles competitive in meeting the essential requirements of farmed fish species (Woodgate et al., 2022).

Finally, the last group of potential alternative proteins to be used in aquafeed that has received a lot of attention over the last year are insects. In particular, insects represent a sustainable ingredient with very interesting nutritional profiles and functional properties due to the bioactive compounds they contain (Nogales-Mérida et al., 2019; Gasco et al., 2020; Mousavi et al., 2020; Shafique et al., 2021). Although the use of insects in aquafeeds started only around 40 years ago, the industry has evolved rapidly with remarkable innovations in insect species selection, production systems, nutritional quality, testing, and validation. At least 16 insect species have already been evaluated as alternative protein sources in aquafeeds. Amongst the species tested and used for industrial aquafeed production, eight species are most promising:

1. Silkworms (*Bombyx mori*);
2. Black soldier fly (*Hermetia illucens*);
3. Housefly (*Musca domestica*);
4. Yellow mealworm;
5. Lesser mealworm (*Alphitobius diaperinus*);
6. House cricket (*Acheta domesticus*);
7. Banded cricket (*Gryllodes sigillatus*); and
8. Jamaican field cricket (*Gryllus assimilis*).

These eight species are the most studied species to date in replacing FM in aquafeeds and have also been approved for the production and inclusion in aquafeeds by the European Union (Alfiko et al., 2022).

3 Key advantages of yellow worm meal as an aquafeed

The effect of *Tenebrio molitor* meal (TMM) on fish growth and performance has been a subject of interest in aquaculture and fish nutrition research. The following features of TMM have driven interest in its use in feed for livestock and aquatic species:

1. High protein content: TMM is valued for its high protein content (Li et al., 2022; Tran et al., 2022b), which makes it a nutritious ingredient for fish feed. This is a crucial component in fish diets, essential for growth, muscle development, and overall animal health and condition (Aragão et al., 2022; Barroso et al., 2022).
2. Amino acid profile: the AA profile of TMM is generally well-balanced (Sánchez-Muros et al., 2016; Costa et al., 2020), providing essential AAs that are important for protein synthesis in fish. Amino acids play a key role in various physiological functions, including growth and immunity, and even in the sensory quality of the final product (Li et al., 2009).
3. Lipid content: TMM is also rich in fats (Costa et al., 2020; Hachero-Cruzado et al., 2024). Whilst moderate lipid content is beneficial for energy, excessive levels might need to be considered in diet formulations, especially for species with specific dietary requirements and essential fatty acid contents (Grigorakis, 2015).
4. Palatability: some studies have suggested that fish may find TMM very palatable, which can positively influence feed intake and growth

performance in fish, matching fish fed a FM-based diet (Ng et al., 2001) or even improving key performance indicators (KPIs) (Shafique et al., 2021).

5 Digestibility: the digestibility of TMM meal is an important factor affecting nutrient utilization by fish. High digestibility ensures that the nutrients are efficiently absorbed, contributing to growth and better feed utilization (Mastoraki et al., 2022; Chen et al., 2023).
6 Nutrient enrichment: TMM can serve as a source of various vitamins and minerals (Finke, 2022), contributing to the overall nutritional profile of aquafeeds (Costa et al., 2020; Shafique et al., 2021).
7 Sustainability: TMM is often considered a sustainable protein source (Magnani et al., 2023). The use of TMM in aquafeeds aligns with efforts to find alternative and eco-friendly protein sources for fish farming.

The purpose of this chapter is to provide an overview of the current use of *T. molitor* as alternative source of FM in aquafeeds. It summarizes recent reviews focused on the use of these invertebrates in aquafeeds and their impact on KPIs associated with growth performance and feed efficiency, product nutritional quality, and their immunomodulatory potential and effect on host gut microbiota.

4 Nutritional profile of yellow worm meal for aquafeed

In general terms, the composition of live yellow worm consists of *ca*. 20% protein, 13% fat, 2% fibre, and 62% moisture; in dried form, its composition may reach up approx 53% protein, 28% fat, 6% fibre, and 5% moisture (Mariod, 2020). When yellow worms are transformed into meal, their composition may vary depending on the developmental stage (Shafique et al., 2021) and production systems, rearing conditions, and feedstuffs used for their culture (Jajić et al., 2022) as summarized in Table 1. Figure 1 illustrates how the composition of selected TMM products varies compared to conventional animal-derived ingredients used in aquafeeds.

Looking at the AA profile of yellow worm, although it is generally well-balanced (Sánchez-Muros et al., 2016; Costa et al., 2020, Liland et al., 2021), TMM may have some AA imbalances when compared to FM, particularly in methionine, threonine, lysine, histidine, and cysteine, whilst TMM is rich in valine and tyrosine (Shafique et al., 2021). The AA profile of TMM shows higher lysine, methionine, and threonine content compared to other insect meals (Costa et al., 2020). The AA deficiencies from different TMM also depended on the fish species considered as shown in Figure 2.

For predatory fish species, chitin has the potential to be a significant carbohydrate source. However, chitin has proven to be difficult to digest by

Table 1 Proximate composition of yellow worm meals from different studies based on their development stage

Stage	CP	CF	Fibre	Ash	Moisture	References
Larvae	43.8	35.1	ND	2.9	7.8	Hachero-Cruzado et al. (2024)
Larvae	44.7	42.5	ND	3.7	2.4	Siemianowska et al. (2013)
Larvae	45.3	34.5	ND	4.1	ND	Costa et al. (2020)
D larvae	45.8	34.2	4.0	2.5	5.8	Hussain et al. (2017)
Larvae	46.4	32.7	4.6	2.9	5.3	Ravzanaadii et al. (2012)
D Larvae	47.0	29.6	ND	4.5	ND	Hoffmann et al. (2020)
Larvae	49.1	38.3	ND	4.1	ND	Liu et al. (2020)
Larvae	51.9	21.6	7.2	4.7	ND	Bovera et al. (2015)
FF Larvae	51.9	23.6	4.7	ND	ND	Gasco et al. (2016)
D Larvae	51.3	26.0	8.2	7.1	DD	Gelinçek and Yamaner (2020)
Larvae	51.9	23.6	6.6	4.7	5.1	Iaconisi et al. (2018)
Larvae	52.5	34.1	ND	3.5	3.6	Jeong et al. (2020)
Larvae	52.8	28.5	ND	2.9	6.5	Melenchón et al. (2023)
Larvae	52.8	36.1	ND	3.1	ND	Makkar et al. (2014)
D Larvae	53.0	3.6	3.1	26.8	ND	Khan et al. (2018)
Larvae	53.2	34.5	6.3	4.0	ND	Ghosh et al. (2017)
Larvae	54.9	23.6	6.6	4.4	5.3	Iaconisi et al. (2017)
Larvae	58.4	34.0	ND	4.8	3.5	Sánchez-Muros et al. (2016)
Larvae	58.4	30.1	3.5	8.0	ND	Barroso et al. (2014)
Larvae	60.2	19.1	22.4	4.2	7.3	Heidari-Parsa (2018)
Adult	63.3	7.6	20.0	3.6	3.5	Ravzanaadii et al. (2012)
Larvae	64.7	46.1	ND	5.9	11.6	Li et al. (2022)
Larvae	71.2	7.9	1.2	14.0	3.5	Tran et al. (2022b)

Note: References included in this table are representative of the myriad of studies including data on this issue. Columns fibre, ash, and moisture contents expressed as a percentage in dry weight basis. *Abbreviations:* CP, crude protein; CF, crude fat; D Larvae, dried larvae; FF Larvae, full fat larvae; ND, data not determined or provided by authors.

many fish species (Ringø et al., 2012), even though the inclusion of chitin in compound diets has been described as positive due to its immunomodulatory and prebiotic effects (see later discussion of the use of yellow worm meal as a functional ingredient). The contents of chitin in yellow worm larvae range between 2.9% and 10.1% depending on the stage of development and the methodology used for its quantification (Costa et al., 2020).

Fat and fatty acids constitute the second most abundant nutrient in the composition of yellow worm. As reviewed in Shafique et al. (2021), yellow worm larvae are rich in saturated (21.8–28.3% DM) and monounsaturated (38.1–52.5% DM) fatty acids, with palmitic acid (C16:0) and oleic acid (C18:1 n-9) amongst the most important ones. The content of PUFA varies from 22.7%

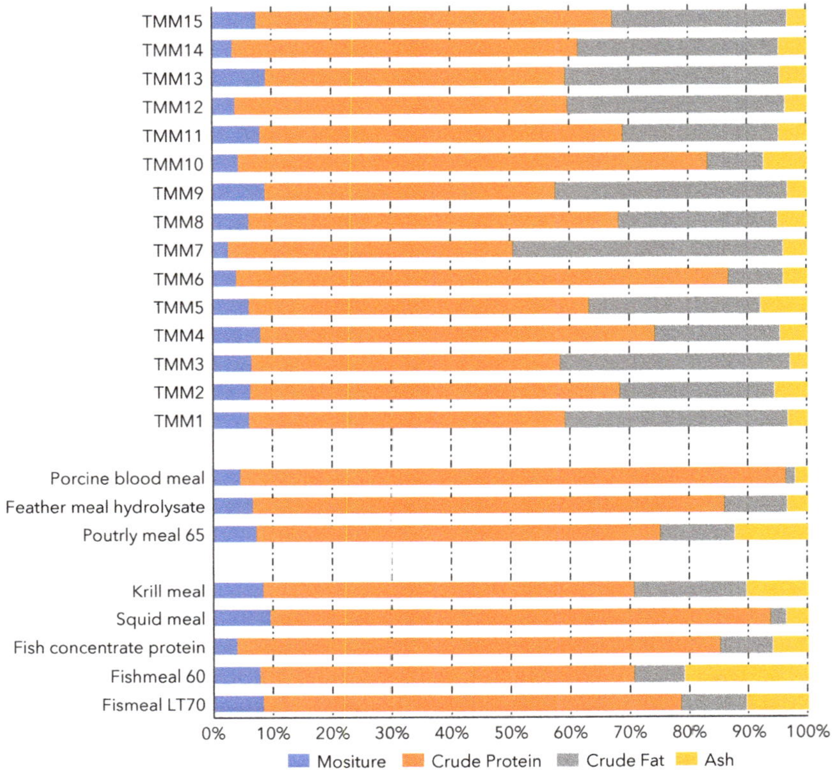

Figure 1 Comparison in terms of proximate composition (percentage in dry matter) of selected *Tenebrio molitor* meals (TMMs) with other marine and terrestrial animal protein sources used in aquafeed formulation. Selected yellow worm meals: TMM1, Gosh et al. (2017); TMM2, Khan et al. (2018); TMM3, Hussain et al. (2017); TMM4, Hoffmann et al. (2020); TMM5, Barroso et al. (2014); TMM6, Tran et al. (2022a); TMM7, Siemianowska et al. (2013); TMM8, Iaconisi et al. (2017); TMM9, Hachero-Cruzado et al. (2023); TMM10, Ravzanaadii et al. (2012); TMM11, Heidari-Parsa (2018); TMM12, Jeong et al. (2020); TMM13, Li et al. (2022); TMM14, Sánchez-Muros et al. (2016); TMN, 15, Melechón et al. (2023). Data from marine and terrestrial protein sources were retrieved from the Technical Booklet: a database of aquaculture feed ingredients (ARRAINA, 2013).

to 33.6%, with linoleic acid (C18:2 n-6) the most abundant one (21.8–31.9% DM). One disadvantage of yellow worm meals is their high lipid content and the lack of n-3 long-chain PUFA (LC-PUFA), such as EPA and DHA (Alfiko et al., 2022). Although defatting yellow worm meals is technically feasible, careful consideration is needed as the processes for fat extraction (i.e. use of organic solvents) and protein purification may reduce the profitability and environmental sustainability of this protein source and may affect its nutritional and functional value (Rawski et al., 2020).

	TMM1	Rainbow trout	Nile tiapia	Common carp	Channel catfish	TMM2	Rainbow trout	Nile tiapia	Common carp	Channel catfish	TMM3	Rainbow trout	Nile tiapia	Common carp	Channel catfish	TMM4	Rainbow trout	Nile tiapia	Common carp	Channel catfish	TMM5	Rainbow trout	Nile tiapia	Common carp	Channel catfish
Arginine	2,2	3,8	4,2	4,3	4,3	4,8	3,8	4,2	4,3	4,3	2,2	3,8	4,2	4,3	4,3	2,4	3,8	4,2	4,3	4,3	6,1	3,8	4,2	4,3	4,3
Histidine	2,8	1,1	1,7	2,1	1,5	3,4	1,1	1,7	2,1	1,5	1,7	1,1	1,7	2,1	1,5	1,4	1,1	1,7	2,1	1,5	3,6	1,1	1,7	2,1	1,5
Isoleucine	2,0	2,1	3,1	2,5	2,6	4,6	2,1	3,1	2,5	2,6	4,5	2,1	3,1	2,5	2,6	2,0	2,1	3,1	2,5	2,6	5,9	2,1	3,1	2,5	2,6
Leucine	3,4	2,6	3,4	3,3	3,5	8,6	2,6	3,4	3,3	3,5	5,3	2,6	3,4	3,3	3,5	3,6	2,6	3,4	3,3	3,5	8,7	2,6	3,4	3,3	3,5
Lysine	2,0	5,1	5,6	5,7	5,1	5,4	5,1	5,6	5,7	5,1	4,5	5,1	5,6	5,7	5,1	2,5	5,1	5,6	5,7	5,1	6,0	5,1	5,6	5,7	5,1
Threonine	1,8	2,6	3,8	3,9	2,2	4,0	2,6	3,8	3,9	2,2	1,6	2,6	3,8	3,9	2,2	1,9	2,6	3,8	3,9	2,2	4,5	2,6	3,8	3,9	2,2
Tryptophan	ND	0,4	1,0	0,8	0,5	0,6	0,4	1,0	0,8	0,5	ND	0,4	1,0	0,8	0,5	ND	0,4	1,0	0,8	0,5	ND	0,4	1,0	0,8	0,5
Valine	2,9	2,4	2,8	3,6	3,0	6,0	2,4	2,8	3,6	3,0	4,4	2,4	2,8	3,6	3,0	3,2	2,4	2,8	3,6	3,0	7,6	2,4	2,8	3,6	3,0
Methionine	ND	1,3	2,7	2,0	2,3	1,5	1,3	2,7	2,0	2,3	1,3	1,3	2,7	2,0	2,3	0,7	1,3	2,7	2,0	2,3	0,6	1,3	2,7	2,0	2,3
Cysteine	3,2	ND	ND	ND	ND	0,8	ND	ND	ND	ND	3,6	ND	ND	ND	ND	ND	ND	ND	ND	ND	ND	ND	ND	ND	ND
Phenylalanine	1,8	2,0	3,8	3,3	2,1	4,0	2,0	3,8	3,3	2,1	1,5	2,0	3,8	3,3	2,1	1,8	2,0	3,8	3,3	2,1	4,3	2,0	3,8	3,3	2,1
Tyrosine	3,5	ND	ND	ND	ND	7,4	ND	ND	ND	ND	2,3	ND	ND	ND	ND	3,0	ND	ND	ND	ND	4,2	ND	ND	ND	ND

Figure 2 Amino acid profile of several *Tenebrio molitor* meals (TMM) compared to the nutritional requirements of juveniles of rainbow trout (*Oncorhynchus mykiss*), Nile tilapia (*Oreochromis niloticus*), common carp (*Cyprinus carpio*) and channel catfish (*Ictalurus punctatus*). A semi-quantitative assessment indicated by different colours showed how close the AA profile of the TMM is from the particular requirement for each AA. For instance, dark green indicates that the AA composition of TMM was higher than 20% of its requirement; light green when the AA composition of the TMM was less than 20% above of its requirement; orange when the AA composition of TMM was deficient in less than 20% of its requirement, and red indicated that the AA composition was deficient in more than 20% of its requirements. Selected *T. molitor* meals: TMM1, Gosh et al. (2017); TMM2, Khan et al. (2018); TMM3, Hussain et al. (2017); TMM4, Hoffmann et al. (2020); TMM5, Barroso et al. (2014). Data on AA requirements for selected fish species has been retrieved from NRC (2011).

Few studies have evaluated the different lipid class composition of yellow worm larvae. Similar to other insects, triglycerides (TAGs) comprise 43.5–56.9% of total lipids. Looking at free fatty acids, percentages of diacylglycerides, monoacylglycerides, are 30.2–34.6%, 2.0–3.0%, and 1.3%, respectively (Costa et al., 2020; Hachero-Cruzado et al., 2024). As mentioned by these authors, TAGs are used for the storage of energy and fatty acids and they are the major lipid class for most insects, representing over 90% of total lipids. Phospholipid levels range from 4.1% to 5.7%, being the second most important lipid class. As PL are structural lipids of cell membranes, their content may considerably vary depending on the developmental stage considered. A wide range of cholesterol content has been described with value ranging from 0.4% to 4.2%, which has been attributed to different feedstuffs used for yellow worm rearing as cholesterol is absorbed from the diet or may result from the metabolism of dietary phytosterols (Costa et al., 2020; Hachero-Cruzado et al., 2024).

Yellow worm larvae are rich in minerals, with phosphorous (7.970 mg/kg) and potassium (8.007 mg/kg) the most abundant. Other minerals are present in smaller quantities such as magnesium (875–2.823 mg/kg), zinc (42–97 mg/kg), iron (38–68 mg/kg), copper (7.8–8.0 mg/kg), and manganese (4.4–11.6 mg/kg) (Costa et al., 2020; Shafique et al., 2021). However, it should be noted that yellow worm larvae may be deficient in calcium, vitamin D3, vitamin A, vitamin B12, thiamine, vitamin E, iodine, and sodium (Finke, 2002); however, this may be corrected by the use of mineral and vitamin premixes when feeds are manufactured.

5 Effects of yellow worm meal on growth and feed efficiency in different fish species

The effect of replacing FM with alternative protein sources not only depends on the target fish species (trophic position and feeding habits of the species), and the type of meal (i.e. full-fat insect meal, defatted or partially defatted insect meal) but also on insect production systems, stage of development, and processing technology used. The current section of this chapter focuses particularly on the evaluation and suitability of TMM as FM substitute, including its fat content, as well as its use in compound feeds for the main groups of farmed marine and freshwater fish species; in particular, salmonids, silurids (catfish), sparids, serranids, and other relevant aquaculture species.

5.1 Salmonids

Looking at salmonids, regardless of the economic importance of Atlantic salmon (*Salmo salar*) in the aquaculture market, most studies conducted in evaluating the feasibility of replacing FM with TMM have been conducted on rainbow trout (*Oncorhynchus mykiss*) and, to a lesser extent, on sea trout

(*Salmo trutta*). In contrast, studies dealing with insect meals in Atlantic salmon have been mainly focused on black soldier fly (*Hermetia illucens*) meals. The replacement of FM with TMM has been successfully tested in many nutritional studies in rainbow trout, even though results in terms of growth performance and composition vary, depending on the use of full-fat or defatted TMM.

The dietary replacement of FM with full-fat TMM up to 50% e.g. did not result in significant differences in growth performance and nutrient composition in rainbow trout (Gasco et al., 2014a). Gasco et al. (2016) found no significant difference in growth when 25% of FM was replaced with full-fat TMM; however, 50% replacement reduced fish growth rate without affecting flesh composition. The full replacement of FM (20%) with defatted TMM did not compromise growth and FCR values in rainbow trout (Chemello et al., 2020). Jeong et al. (2020) found that it was feasible to replace up to 28% of FM with TMM, but the best results in terms of growth performance and protein efficiency ratio (PER) were found at 14% FM replacement levels. Other studies testing the feasibility of replacing FM with defatted TMM showed that graded levels of FM replacement (20%, 30%, 60%, and 100%) with defatted TMM in rainbow trout did not affect body composition, whilst they increased protein and phosphorous retention (Rema et al., 2021). This is of special relevance since efficient phosphorus retention ensures that farmed fish convert the phosphorus in their feed into usable body mass, leading to better feed utilization and cost savings for farmers. In addition, improving phosphorus retention in fish reduces the environmental burden since phosphorus runoff from fish farms can lead to water pollution, causing eutrophication, harmful algal blooms, and oxygen depletion in aquatic ecosystems.

As Shafique et al. (2021) has discussed, levels of TMM required for FM replacement in aquafeeds depend on type of processing of yellow worm larvae. For instance, diets containing defatted TMM from larvae could effectively replace the entire content of FM. Other factors affecting TMM performance in aquafeeds may be associated with its chitin content, digestibility, AA balance, and fatty acid composition, which significantly impact the inclusion levels of TMM in aquafeeds, even though different diet formulation, ingredient selection, and experimental conditions may be also a source of variability when testing this alternative protein source.

5.2 Silurids (catfish)

Amongst other freshwater species, the suitability of TMM has been evaluated in catfish species. For instance, TMM could replace up to 75% of FM in diets containing up to 24% FM in yellow catfish (*Pelteobagrus fulvidraco*) (Su et al., 2017). The study also found that feed conversion efficiency in yellow catfish juveniles fed diets with up to 75% FM replacement by TMM showed similar results. Fishmeal can be replaced up to 40% by TMM without negative effects on growth performance and FCR values in African catfish (*Clarias gariepinus*), whilst fish fed diets with 20% TMM displayed better growth and feed efficiency indices than the control group (Ng et al., 2001). This study concluded that catfish fed diets with up to 80% replacement of FM with TMM still displayed good growth and feed utilization efficiency parameters. This is of special relevance since including TMM in compound diets does not compromise how feed consumed is converted into body mass. However, Roncarati et al. (2015) found a significant decrease in the body weight in common catfish (*Ameiurus melas*) fingerlings fed a diet with 50% of FM replaced by TMM compared to the control group, with no significant changes in the FCR amongst groups, whereas survival was slightly higher in the control diet (79% vs 70%, respectively).

5.3 Sparids

In gilthead seabream (*Sparus aurata*), replacing 25% of FM with TMM had no adverse effects on the growth, crude protein, and fat digestibility, whereas FCR and PER were improved compared to the control group. Considering that feed costs represent the largest operational expense in fish farming, often accounting for 50–70% of total production costs, improving feed efficiency means that fish convert more of the feed they consume into body mass, reducing the overall amount of feed needed. In contrast, at higher level of inclusion (50%), nutrient digestibility was reduced but this did not lead to negative effects on growth performance in comparison to the control group without TMM (Piccolo et al., 2017). Similarly, Fabrikov et al. (2020) demonstrated that FM could be replaced by up to 30% TMM without negative effects on growth performance, weight gain or FCR. Apparent digestibility coefficients (ADCs) of dry matter, organic matter, protein, and energy were not affected by 30% FM replacement by TMM, results that were comparable to those observed at similar replacement levels using other insect meals (Mastoraki et al., 2022). Regarding essential ADC values for essential AA, values (91.9%) were similar to those observed for premium quality FM (90.2%; Davies et al., 2009).

Research on testing TMM in sparids have not only focused on gilthead seabream, the most important species from this group of fishes from both a production and economic point of view. Ido et al. (2019) reported that the full FM replacement with defatted TMM in red seabream (*Pagrus major*) did not compromise growth and feed efficiency, and even promoted growth when compared to the control group fed a compound diet with 65% FM. Another study conducted in blackspot seabream (*Pagellus bogaraveo*) showed that inclusion of TMM larvae meal at 25% and 50% of FM substitution in the diet did not lead to significant effects on growth performance, feed intake or FCR (Iaconisi et al., 2017).

5.4 Serranids

European sea bass (*Dicentrarchus labrax*) is the most important marine fish species in the Mediterranean in terms of economic value and the second in terms of production. Gasco et al. (2014b) found that the dietary inclusion of 25% did not negatively impact somatic growth performance, whereas higher inclusion levels (50%) resulted in a reduction in growth, without compromising PER and feed intake. These results were partly in agreement with those reported by the same group of authors when diets were formulated with 50% full-fat TMM, where no differences were observed in FCR (Gasco et al., 2014b, 2016). Similarly, Reyes et al. (2020) did not find significant differences in FCR and PER, but did find a significant decrease in specific growth rate (SGR) and feed efficiency values in *D. labrax* fed a diet with 50% of FM replaced with TMM. Mastoraki et al. (2022) reported no significant differences in somatic growth and feed intake in *D. labrax* fed a diet with 30% of FM replacement by TMM, even though FCR were slightly worse than for the full-FM diet. Research on black sea bass (*Centropristis striata*), a commercially important species for US aquaculture, found it was possible to replace FM up to 50% with TMM without compromising growth performance, whereas TMM inclusion at 25% enhanced somatic growth without affecting FCR values when compared to the control group (Redman et al., 2019).

5.5 Flatfish and other species

Amongst flatfish species, Hachero-Cruzado et al. (2024) showed that it was feasible to replace up to 10% of FM with full-fat TMM in Senegalese sole (*Solea senegalensis*) with no negative effect on fish growth, FCR, nutrient utilization and survival. Replacing up to 15% of plant protein sources with TMM improved growth rate. In turbot (*Scophthalmus maximus*), FM can be replaced up to 30% TMM without affecting growth and feed efficiency, whereas higher levels of FM replacement (45-75%) resulted in a loss of weight and poor liver condition (Bai et al., 2023). *Tenebrio molitor* meal been used successfully as FM replacement up to 40% in olive flounder (*Paralichthys olivaceus*) without impairing growth, FCR and PER, whilst higher inclusion levels (60–80%) impaired growth and feed efficiency (Jeong et al., 2021).

Considering other carnivorous species, Chen et al. (2023) found that, in largemouth bass (*Micropterus salmoides*), it was possible to replace up to 36% of FM with TMM without affecting growth, whereas inclusion of 24% of TMM enhanced somatic growth compared to the control group. Based on a quadratic dose-response regression model, the authors recommended an optimal TMM inclusion of 19.5%. They determined that ADC of essential AAs in TMM was 93.9%. In perch (*Perca fluviatilis*), another perciform species, somatic growth,

FCR and PER were reduced when fishes were fed 50% and 75% TMM (Tran et al., 2022a). Formulated diets with up to 30% TMM caused no adverse effects on fish performance and feed utilization in mandarin fish (*Siniperca scherzeri*) juveniles, whereas lower inclusion levels (10% and 20%) led to better results in terms of growth and FCR values (Sankian et al., 2018). Regarding tilapiine species, Sánchez-Muros et al. (2016) described a decrease in somatic growth and PER, and an increase in FCR values when FM was replaced with TMM at 50% in Nile tilapia (*Oreochromis niloticus*). Amongst cyprinid species, TMM has been also tested in tench (*Tinca tinca*) (Rema et al., 2021). The replacement of FM at 18%, 28%, and 38% by TMM in diets negatively affected growth performance, regardless of the level of FM replacement considered. These results were also coupled with and increase in feed intake and lower FCR values. In mirror carp (*Cyprinus carpio* var. *specularis*), increasing TMM levels, ranging from 15% to 75%, enhanced somatic growth, reduced FCR and increased PER values without affecting feed intake (Li et al., 2022).

6 Effects of yellow worm meal on fillet quality

The market for farmed fish depends significantly on consumer perceptions of quality, with two crucial parameters at play: the nutritional and technical quality of fish. Nutritional quality relates to the food's nutrient content, specifically the presence of PUFA and their numerous health benefits. This not only contributes to the overall health benefits of fish but also shapes the consumer's perception of its nutritional value. On the other hand, technical quality is primarily an economic consideration, encompassing processing technical losses and edible yields, making it a pivotal factor for both consumers and processors (Grigorakis, 2015).

The relationship between diet and fillet composition and quality is complex and varies depending on the specific nutritional requirements of the fish species, the ingredient composition of fish diets and rearing conditions. The analysis of new ingredients and aquafeed formulations must not only consider their effect on KPIs related to growth and feed performance but also require considering fillet composition since it largely defines fish nutritional value and sensory quality perceived by the consumer.

The inclusion of higher levels than 25% of full-fat TMM has been found to result in reduced levels of n-3 EPA and DHA in fillets and higher levels of n-6 and n-9 fatty acids in rainbow trout (Iaconisi et al., 2018) and blackspot seabream (Iaconisi et al., 2017). The inclusion of TMM also affected the AA profile of the fillet as shown by Iacosini et al. (2017) who reported a reduction in essential AAs like alanine, isoleucine, leucine, and lysine contents in rainbow trout fillets when fed diets replacing 50% FM by full-fat TMM. However, Jeong et al. (2020) showed that the inclusion of 28% TMM in rainbow trout diets had no significant effects on whole body or fillet composition and AA profile compared to the control group (no TMM inclusion). Differences between studies might be due to factors such as age/size of the fish, composition, and nutrient content of the diet, source, or nutritional quality of the TMM, experimental conditions and water temperature, amongst other factors. As Belforti et al. (2015) have indicated, these variations amongst the trials make it difficult to compare the results from different studies and benchmarking studies using TTM from different production systems, stages of development or processing, need to be conducted under the same experimental conditions.

Regarding sparid species, Piccolo et al. (2017) found that the inclusion of 25% and 50% TMM in compound feeds for gilthead seabream did not affect its proximate body composition. Similar results were reported in blackspot seabream fed diets with 25% and 50% of FM replaced with full-fat TMM (Iaconisi et al., 2017). These authors also found an increase in linoleic acid levels with increasing dietary full-fat TMM levels. The same authors also found higher alanine, leucine, lysine, arginine, glycine, and proline levels in gilthead seabream fed diets with 50% of FM replaced with full-fat TMN larvae (Iaconisi et al., 2017). They also found lower levels of taurine, histidine, and phenylalanine in gilthead-seabream-fed diets with 25% and 50% of FM replaced by TMM. However, other authors have found that the dietary inclusion of TMM did not affect the levels of crude lipids in blackspot seabream fillets (Iaconisi et al., 2017).

Gasco et al. (2014b, 2016) found no differences in the overall body composition of European sea bass when fed with diets including 25% and 50% TMM. They also noted a reduction in DHA and EPA in fish given diets with 50% TMM (Gasco et al., 2014b, 2016). Replacement of FM by 80% TMM resulted in an increase in linoleic acid (C18:2) levels in European sea bass fillets but had no effects on EPA and DHA content (Basto et al., 2019). In largemouth bass, Chen et al. (2023) found that replacing FM with TMM at higher levels than 12% reduced body fat content and increased protein levels without impacting on ash and moisture levels.

Regarding flatfish species, Jeong et al. (2021) reported that increasing TMM levels at the expense of FM decreased the levels of crude fat in fillets without changing their protein content nor their essential AA composition in olive flounder. Furthermore, feeding TMM diets increased the levels of C18:1n-9 and C18:2n-6 in the fillet, at the expense of beneficial n-3 LC-PUFA. Concerning the effect of full-fat TM inclusion on the muscle and liver of Senegalese sole, it was observed that including full-fat insect meal in the diet led to a decrease in the total lipid content of the muscle, without changing the total lipid content of the liver (Hachero-Cruzado et al., 2024). Such reduction in muscle lipid content was associated to a decreases triacylglycerols and cholesterol. Lower lipid content in Senegalese sole fillets was thought to be due to the effect of chitin in TMM, which might have affected fat digestibility and lipid absorption.

Results from other species have shown that different levels of FM replacement by TMM did not modify the fillet composition in Nile tilapia (Sanchez-Muros et al., 2016), even though there was an increase in the levels of C18:1n-9 and C18:2n-6, as well as a decrease in n-3 LC-PUFA. In tench, replacement of FM by a combination of TMM with microalgae and macroalgae meals showed a linear increase in ash content with the increase of alternative protein sources to FM, changes that were associated with a decrease in crude lipid content when FM was replaced at 28% and 38% (Rema et al., 2021). In yellow croaker (*Larimichthys crocea*), levels of EPA, saturated fatty acids (SFAs), and the ratio of n-3/n-6 PUFA declined with increasing dietary replacement levels of FM by TMM, but mono-unsaturated fatty acids (MUFA), total PUFA, and n-6 PUFA increased with increasing dietary replacement levels of FM by TMM (Zhang et al., 2022). Overall, fine-tuning of the feed formulation may be required to avoid nutritional deficiencies in terms of essential AA and fatty acid profiles.

7 Effects of yellow worm meal on sensory and nutritional properties of fresh fish

The importance of fish feed on fillet characteristics such as colour, texture, and nutritional quality, as well as functional properties such as shelf life, is well-established. Changes in fish feed ingredients can affect the appearance, smell, and aroma of fish fillets, which may affect the perceived quality of the fillet and consumer acceptance (Magnani et al., 2023).

Amongst other parameters, skin colour plays a major role in the overall acceptability of fish, especially in those species that are sold as whole fish, since colour is commonly used as a key indicator of freshness and quality. In blackspot seabream, the inclusion of TMM did not affect the colour of the dorsal region of the skin, whereas the skin ventral region showed more differences (i.e. lower lightness and hue levels coupled with higher redness values in blackspot seabream fed 50% TMM) (Iaconisi et al., 2017). These changes in the skin pigmentation were associated with the β-carotene content (<200 μg/kg; Finke, 2002) in TMM, which may have enhanced the red colour of the skin (redness index). Amongst cyprinids, the combination of TMM with microalgae and macroalgae led to redder and yellower skin pigmentation in tench but lower lightness values, although the study did not separate out any specific effects of TMM on skin colour (Rema et al., 2021). Similar results regarding the enhancement of skin yellowness when TMM are included in compound diets have been reported in mirror carp and large yellow croaker (Li et al., 2022; Zhang et al., 2022). When evaluating the impact of TMM on skin pigmentation

Iaconisi et al. (2017) pointed out that changes in skin yellowness might be attributed to the riboflavin content in TMM (41–56 mg/kg; Finke, 2002) and the effects of this vitamin on skin chromatophores. The colour of fillets in blackspot seabream, yellowness, and intensity of the colour in the epaxial region of the fillet were also higher when fish were fed diets with TMM (Iaconisi et al., 2018). Similar results in terms of the effect of defatted and full TMM inclusion on fillet colour were reported by Melenchón et al. (2023) in rainbow trout.

Regarding the physical and textural properties of fillets, Iaconisi et al. (2017) found no significant differences in fillet quality parameters like water-holding capacity and texture characteristics (e.g. hardness, cohesiveness, resilience, gumminess, and adhesiveness). pH was lower in blackspot seabream fed with diets containing 50% TMM compared to the control diet (0% TMM) and 25% TMM. In contrast, the inclusion of different levels of TMM (15–75%) did not affect fillet pH values in mirror carp or muscle texture regardless of the level of FM replaced by TMM (Li et al., 2022). Similar results in terms of water-holding capacity and texture of flesh in large yellow croaker-fed diets replacing 15%, 30%, and 45% of FM by TMM were reported by Zhang et al. (2022), confirming the view that inclusion of TMM in aquafeeds does not alter textural properties, as also demonstrated in rainbow trout (Iaconisi et al., 2018).

In terms of consumer perception, Borgogno et al. (2017) reported differences in the perceived intensity of odour, flavour, and texture in rainbow trout fillets fed a diet that included insect meal from black soldier fly. However, the inclusion of TMM in aquafeeds did not affect sensory attributes and physicochemical parameters in fillets from rainbow trout when FM was replaced from 30% to 100% by TMM (Magnani et al., 2023), with similar results reported in blackspot seabream (Iaconisi et al., 2017). In rainbow trout, Magnani et al. (2023) did not find differences in flavour amongst experimental groups, even though numerical scores for earthy, sour, bitter, and for flavour intensity and persistence tended to increase with the full replacement of FM by TMM, which is in the line with other studies conducted with other insect meals (Borgogno et al., 2017). Magnani et al. (2023) found that insect meal-fed products were positively perceived by consumers as being nutritious, natural, and sustainable. However, their willingness to buy products was lower when compared to fish fed diets formulated with FM and plant protein sources. In contrast, Deely et al. (2022) found consumers were willing to pay more for environmentally friendly food products.

8 Effects of yellow worm meal on the host microbiome

The interplay between diet and microbiota in fish is an intricate and dynamic interaction pivotal for the overall health and well-being of aquatic organisms. The dietary choices of fish exert a significant influence on the composition and activity of their gut microbiota, with wide-ranging implications for host physiology, metabolism, immune function, and overall performance. Any disturbance in microbial composition or disruptions in host–microbe interactions, referred to as dysbiosis, can lead to digestive and systemic imbalances, potentially resulting in poor health and increasing disease risk (Sommer et al., 2017; Ruiz et al., 2024). These interactions are reciprocal, as the bacterial community can stimulate the immune system through pathogen-associated molecular patterns (PAMPs) recognized by pattern recognition receptors (PRRs), thereby activating immune signalling pathways. Concurrently, the host plays a role in shaping the microbial composition, regulating the abundance of potentially pathogenic bacterial communities (López-Nadal et al., 2020).

It is therefore important to assess how gut microbes respond to the inclusion of new raw materials replacing FM. One of the first studies evaluating the impact of TMM inclusion in aquafeeds was on rainbow trout where FM was replaced at 60% by TMM (Antonopoulou et al., 2019). They found that gut microbial diversity significantly decreased, indicating a more specialized gut bacterial community after FM replacement. However, Terova et al. (2021) showed no major effects of full FM substitution (100%) with TMM on microbial species richness and diversity of both gut mucosa- and skin mucus-associated bacterial communities in rainbow trout, results that were similar to those found in perch-fed diets containing 25–75% TMM (Tran et al., 2022a) or in sea trout (*Salmo trutta* m. *trutta*)-fed diets containing 40% of hydrolysed TMM (Mikołajczak et al., 2020; Józefiak et al., 2019). Ratios of Proteobacteria:Firmicutes, Proteobacteria:Bacteroidetes, and Firmicutes:Bacteroidetes indicated that replacement of FM at 60% did not affect the composition and functionality of gut microbiome in rainbow trout (Antonopoulou et al., 2019).

In European sea bass and gilthead seabream, Antonopoulou et al. (2019) reported that diversity of bacterial gut communities was not affected by the inclusion of TMM in diets, though effects varied by species. Fishmeal replacement by TMM resulted in an increased number of operational taxonomic units (OTUs), a proxy of bacterial species, in gilthead seabream and European sea bass, whereas the number of OTUs decreased in rainbow trout. Inclusion of TMM was thought to generate new nutritional niches in the intestines of both marine species, resulting in the appearance of new and unique OTUs when compared to the full FM-based diet. Changes in the relative abundance of the main phyla may result in potential gut imbalances that need to be further investigated by using omics approaches and physiological assays.

Józefiak et al. (2019) found that inclusion of 20% TMM in compound diets for rainbow trout resulted in a significant increase in the relative abundance of *Lactobacillus* and *Enterococcus* genera, which was attributed to the presence of chitin in the TMM since this compound has prebiotic properties. Similarly, an increase in *Lactobacillus* was found in the intestinal microbiota of perch-fed diets containing TMM (Tran et al., 2022a). This is of special relevance since the use of TMM may be regarded as beneficial for the host due to its potential probiotic properties (Ringø, 2020), even though its use as a functional ingredient needs to further explored and validated to take account of variables such as production system and feedstuff used.

9 The use of yellow worm meal as a functional ingredient

The effect of alternative feed ingredients or aquafeed formulations has to be addressed holistically, including their effects on host condition, immune and stress responses, and disease resistance against pathogens (Aragão et al., 2022). Insect meal should be regarded not only as a source of safe and high-quality protein for aquafeed but also as a potential source of dietary bioactive compounds. Research has focused on three categories of bioactive compounds found in insects:

- Antimicrobial peptides (AMPs);
- Fatty acids such as lauric acid; and
- Non-digestible polysaccharides like chitin and chitosan (Veldkamp et al., 2022).

Insect meal is recognized as being rich in AMPs, including defensins (effective against Gram-positive bacteria), diptericins, attacins, drosocins, and cecropins (active against Gram-negative bacteria), and drosomycins (which combat fungi) (Singh et al., 2013). These compounds exhibit antimicrobial activities through diverse mechanisms, including membrane pore formation or destruction of bacteria, interference with bacterial intracellular processes, or support for the host's immune system. The antioxidant capacity of insect proteins can also contribute to safeguarding the host against oxidative tissue damage (Veldkamp et al., 2022).

The inclusion of TMM in fish diets has been associated with increased antioxidant content in fish species like yellow catfish (Su et al., 2017), largemouth bass (Gu et al., 2022), mandarin fish (Sankian et al., 2018), and European sea bass (Henry et al., 2018). Antioxidants play roles in eliminating excessive superoxide radicals, contributing to the maintenance of reactive oxygen species (ROS) homeostasis in fish species (Sankian et al., 2018). However, other studies have shown that dietary inclusion of TMM did not promote activity of oxidative stress enzymes in rainbow trout (Jeong et al., 2020), pikeperch (Tran et al., 2022a), Nile tilapia (Sánchez-Muros et al., 2016), or tench (Hidalgo et al., 2022). These differences amongst studies might be due to differences in experimental conditions, species-specific nutritional requirements, and source and quality of the tested TMM. Henry et al. (2018) demonstrated a positive impact of TMM on the immune system in European sea bass. Full-fat TMM included at 24.8% exhibited anti-inflammatory and immunomodulating properties in the host, resulting in improvement of antibacterial and antiparasitic defences. In mandarin fish, an increase in lysozyme was found when fish were fed diets containing 30% full-fat TMM, though other immunological compounds remained unchanged (Sankian et al., 2018). In contrast, no differences in innate immune parameters were found in rainbow trout and tench-fed diets with graded levels of TMM (Valipour et al., 2019; Hidalgo et al., 2022). Although no differences in plasmatic innate immune parameters were found in tench-fed diets containing TMM, there was a trend for lower levels of lymphocytes, monocytes, and neutrophils, coupled with an increase in basophils, which was associated with a higher capacity for controlling bacterial infections, especially in those animals fed 5% TMM (13.8% FM replacement) (Hidalgo et al., 2022). However, other authors have found no differences in haematological indices in fish fed TMM diets (Sankian et al., 2018; Valipour et al., 2019; Mastoraki et al., 2022).

These differences between studies regarding the antioxidant and immunomodulatory properties of TMM may be associated with variables such as nutritional differences, but it is likely that enhancement of antioxidant activities in fish-fed diets containing TMM may be due to their chitin content (Ngo and Kim, 2014) as well as to other bioactive compounds present in yellow worms (Mousavi et al., 2020).

The immunomodulatory properties of feedstuffs and formulated diets may also be analysed by exposing the host to a particular pathogen and evaluating disease resistance mechanisms. Yellow catfish fed a diet containing 27% TMM (75% FM replacement) showed better survival rates than the control diet (0% TMM) when challenged with *Edwardsiella ictaluri* (Su et al., 2017). These results were coupled with an increase in lysozyme in serum and up-regulation of immune-related genes like immunoglobulin M (*igM)*, major histocompatibility complex (*mhc)*, and Interleukin-2 (*il-2*). This study suggests that dietary inclusion of TMM enhanced both innate and adaptive immune responses, allowing fish to better cope with bacterial infection. Similarly, red seabream fed with diets replacing 10% FM by defatted TMM and challenged with *Edwardsiella tarda* showed higher survival rates than those fed diets containing 0% or 5% defatted TMM (Ido et al., 2019). Similarly, pearl gentian grouper (♀ *Epinephelus fuscoguttatus* × ♂ *Epinephelus lanceolatu*) fed diets containing 7.5% TMM (18% FM replacement) and challenged with *Vibrio harveyi* showed higher survival rates than those specimens fed a diet devoid of TMM, whereas higher FM replacement (25% and 31%) compromised disease resistance in this species (Song et al., 2018).

10 Conclusion

The available literature shows that full-fat or defatted TMMs are a safe and sustainable alternative protein source for aquafeed formulations. The inclusion of TMM may reduce the fish-in and fish-out ratio, indicating that a lower quantity of marine forage fish is needed per unit of fish produced. This fact supports the sustainability of TMM with regard to the exploitation of fisheries stocks for marine raw materials for animal feeding. Furthermore, the use of yellow worm meal should not only be considered as a sustainable alternative to FM, but also for its functional characteristics. The immunomodulatory and antioxidant properties of TMM provides added value to this protein source, despite contradictory findings from different studies, which may be associated to its chitin composition and bioactive compound content.

Despite encouraging results in terms of growth and feed performance found in most of freshwater and marine species tested so far, further research needs to focus on yellow worm breeding and production technologies, including feedstuffs, as they have a direct effect on the composition and nutritional value of TMMs, as well as in their digestibility and chitin content (Barroso et al., 2022). Processing technologies used for producing insect meal need further investigation since each operation will have a different impact on the chemical and microbiological properties of the final product. This applies to novel food processing technologies (i.e. high pressure processing, pulsed electric field, ultrasound, and cold plasma) which have shown potential to modify, complement or replace conventional processing steps in insect processing (Ohja et al., 2021). The current heterogeneity in nutrient profiles of TMMs may limit its use in the aquafeed industry, as the feed manufacturer sector relies on the regular supply of ingredients with competitive prices and of homogenous and stable quality.

Although the inclusion of TMM in aquafeeds does not significantly change fillet composition, some changes in their fatty acid profile have been described. When including full-fat TMM in aquafeeds, particular attention should be paid to the final fatty acid profile of the feed, since yellow worm are rich in MUFA, PUFA, and n-6 PUFA and poor in n-3 PUFA, which indicates that FO or oil sources rich in n-3 PUFA may be needed to balance the fatty acid composition of the diet.

Finally, regardless of the advantages of using insect meal in aquafeeds, in order to foster its use for animal nutrition, more emphasis should be placed on the benefits of insect meal in animal feeds to achieve greater awareness and acceptance of the benefits of insect-based aquafeeds amongst consumers and aquaculture practitioners.

11 Where to look for further information

Further information about trends in research, investment, and innovation in insect-based feeds to meet aquafeed demands and maximize its environmental benefits may be found in the works of Fantatto et al. (2024) and of Selvaraj and Won (2024), whereas a detailed review on the effects of insect meals on fish immunity with special focus on mucosal tissues may be found in Islam et al. (2024).

Readers interested in understanding the role of insects in the circular bioeconomy are encouraged to consult Hamam et al. (2024). This study presents a systematic review of the literature, identifying the state-of-the-art and evaluating insects' contributions to closed-loop systems. It also explores opportunities and challenges associated with integrating insects into circular bioeconomy strategies, including applications in aquaculture.

The International Platform of Insects for Food and Feed, an EU non-profit representing the insect production sector to EU policymakers, stakeholders, and citizens, outlines its position on the use of insects in aquafeeds at this link: https://ipiff.org/.

Regarding the legislation involved in the use of insects in aquafeeds, the reader may find more information on the Regulation No. 2001/999 (Annex IV) amended by the Regulation 2017/893 (Annex X) that authorizes the use of insect proteins in aquafeeds from seven insect species: namely black soldier fly (*H. illucens*), common housefly (*M. domestica*), yellow mealworm (*T. molitor*), lesser mealworm (*A. diaperinus*), house cricket (*A. domesticus*), banded cricket (*G. sigillatus*), and field cricket (*G. assimilis*). Furthermore, the EU legislator authorized the use of silkworm (*B. mori*) processed animal proteins in aquaculture through the Regulation (EU) 2021/1925.

With the global aquaculture sector expected to grow in the coming years, research on alternative feed ingredients, such as those derived from insects, will continue to expand. This will lead to new studies, technical reports, and news focused on the use of insect meal and oil in finfish and crustacean nutrition and farming. For this reason, readers are encouraged to regularly look for newer and updated information in technical and scientific open access platforms like PubMed, arXiv, ResearchGate or Google Scholar, or in subscription-based platforms like the Web of Science™, ScienceDirect or SpringerLink among others.

12 References

Alfiko, Y., Xie, D., Astuti, R. T., et al. (2022), 'Insects as a feed ingredient for fish culture: status and trends', *Aquac. Fish.*, 7(2), 166–178.

ARRAINA (2013), Technical booklet. A database of aquaculture feed ingredients. Available at: https://www.sparos.pt/wp-content/uploads/2019/04/BOOKLET-ARRAIANA.pdf

Antonopoulou, E., Nikouli, E., Piccolo, G., et al. (2019). 'Reshaping gut bacterial communities after dietary Tenebrio molitor larvae meal supplementation in three fish species', *Aquaculture*, 503, 628–635.

Aragão, C., Gonçalves, A. T., Costas, B., et al. (2022), 'Alternative proteins for fish diets: implications beyond growth', *Animals*, 12(9), 1211.

Asche, F., Eggert, H., Oglend, A., et al. (2022), 'Aquaculture: externalities and policy options', *Rev. Environ. Econ. Policy*, 16(2), 282–305.

Bai, N., Li, Q., Pan, S. et al. (2023), 'Effects of defatted yellow mealworm (Tenebrio molitor) on growth performance, intestine, and liver health of turbot (*Scophthalmus maximus*)', *Anim. Feed Sci. Technol.*, 302, 115672.

Barroso, F. G., de Haro, C., Sánchez-Muros, M. J., et al. (2014), 'The potential of various insect species for use as food for fish', *Aquaculture*, 422, 193–201.

Barroso, F. G., Trnazado, C. E., Pérz-Jiménez, A., et al. (2022), 'Innovatve protein sources in aquafeeds', In J. M. Lorenzo and J. Simal-Gadara (Eds.), *Sustainable Aquafeeds, Technological Innovation and Novel Ingredients*, CRC Press, Boca Ratón, USA, pp. 139–184.

Basto, A., Maia, M.R., Pérez-Sánchez, J., et al. (2019), *Defatted yellow mealworm (Tenebrio molitor) larvae meal: a promising fishmeal substitute for European seabass*; Aquaculture Europe Conference (EAS), Berlin, Germany. Available at: https://digital.csic.es/handle/10261/202649

Belforti, M., Gai, F., Lussiana, C., et al. (2015). '*Tenebrio molitor* meal in rainbow trout (*Oncorhynchus mykiss*) diets: effects on animal performance, nutrient digestibility and chemical composition of fillets'. *Ital. J. Anim. Sci., 14*(4), 4170.

Borgogno, M., Dinnella, C., Iaconisi, V., et al. (2017), 'Inclusion of *Hermetia illucens* larvae meal on rainbow trout (*Oncorhynchus mykiss*) feed: effect on sensory profile according to static and dynamic evaluations', *J. Sci. Food Agric.*, 97(10), 3402–3411.

Bovera, F., Piccolo, G., Gasco, L., et al. (2015), 'Yellow mealworm larvae (*Tenebrio molitor*, L.) as a possible alternative to soybean meal in broiler diets', *Brit. Poult. Sci.*, 56(5), 569–575.

Camacho, F., Macedo, A. and Malcata, F. (2019), 'Potential industrial applications and commercialization of microalgae in the functional food and feed industries: a short review', *Mar. Drugs*, 17(6), 312.

Costa, S., Pedro, S., Lourenço, H. et al. (2020), 'Evaluation of *Tenebrio molitor* larvae as an alternative food source', *NFS Journal*, 21, 57–64.

Chemello, G., Renna, M., Caimi, C., et al. (2020), 'Partially defatted Tenebrio molitor larva meal in diets for grow-out rainbow trout, *Oncorhynchus mykiss* (Walbaum): Effects on growth performance, diet digestibility and metabolic responses', *Animals*, 10(2), 229.

Chen, H., Yu, J., Ran, X., et al. (2023), 'Effects of yellow mealworm (*Tenebrio molitor*) on growth performance, hepatic health and digestibility in juvenile largemouth bass (*Micropterus salmoides*)', *Animals*, 13, 1389.

Davies, S. J., Gouveia, A., Laporte, J., et al. (2009), 'Nutrient digestibility profile of premium (category III grade) animal protein by-products for temperate marine fish species (European sea bass, gilthead sea bream and turbot)', *Aquac. Res.*, 40, 1759–1769.

Deely, J., Hynes, S., Barquín, J., et al. (2022), 'Are consumers willing to pay for beef that has been produced without the use of uncontrolled burning methods? A contingent valuation study in North-West Spain', *Econ. Anal. Policy*, 75, 577–590.

Egerton, S., Wan, A., Murphy, K., et al. (2020), 'Replacing fishmeal with plant protein in Atlantic salmon (*Salmo salar*) diets by supplementation with fish protein hydrolysate', *Sci. Rep.*, 10(1), 4194.

Fabrikov, D., Sánchez-Muros, M. J., Barroso, F. G., et al. (2020), 'Comparative study of growth performance and amino acid catabolism in *Oncorhynchus mykiss*, *Tinca tinca* and *Sparus aurata* and the catabolic changes in response to insect meal inclusion in the diet', *Aquaculture*, 529, 735731.

Fantatto, R.R., Mota, J., Ligeiro, C., et al. (2024). Exploring sustainable alternatives in aquaculture feeding: The role of insects. *Aquac. Rep.*, *37*, 102228.

FAO (2022), *The State of World Fisheries and Aquaculture 2022. Towards Blue Transformation*. FAO, Rome. Available at: https://www.fao.org/3/cc0461en/cc0461en.pdf

Finke, M. D. (2002), 'Complete nutrient composition of commercially raised invertebrates used as food for insectivores', *Zoo. Biol.*, 21, 269–285.

Gasco, L., Belforti, M., Rotolo, L., et al. (2014a), 'Mealworm (*Tenebrio molitor*) as a potential ingredient in practical diets for rainbow trout (*Oncorhynchus mykiss*)', In *Insect to feed the world*, Wageningen UR, p. 69. Available at: https://flore.unifi.it/handle/2158/865861

Gasco, L., Gai, F., Piccolo, G., et al. (2014b), 'Substitution of fishmeal by *Tenebrio molitor* meal in the diet of *Dicentrarchus labrax* juveniles', In *Insect to feed the world*, Wageningen UR, p. 70. Available at: https://iris.unito.it/handle/2318/158357

Gasco, L., Henry, M., Piccolo, G., et al. (2016), '*Tenebrio molitor* meal in diets for European sea bass (*Dicentrarchus labrax* L.) juveniles: growth performance, whole body composition and in vivo apparent digestibility', *Anim. Feed Sci. Technol.*, 220, 34–45.

Gasco, L., Acuti, G., Bani, P., et al. (2020), 'Insect and fish by-products as sustainable alternatives to conventional animal proteins in animal nutrition', *It. J. Anim. Sci.*, 19, 360–372.

Gelinçek, İ. and Yamaner, G. (2020), 'An investigation on the gamete quality of Black Sea trout (*Salmo trutta labrax*) broodstock fed with mealworm (*Tenebrio molitor*)', *Aquac. Res.*, 51(6), 2379–2388.

Gephart, J. A., Golden, C. D., Asche, F., et al. (2020), 'Scenarios for global aquaculture and its role in human nutrition', *Rev. Fish. Sci. Aquac.*, 29(1), 122–138.

Grigorakis, K. (2015), 'Fillet proximate composition, lipid quality, yields, and organoleptic quality of Mediterranean-farmed marine fish: a review with emphasis on new species', *Crit. Rev. Food Sci. Nutr.*, 57(14), 2956–2969.

Ghosh, S., Lee, S. M., Jung, C., et al. (2017), 'Nutritional composition of five commercial edible insects in South Korea', *J. Asia-Pac. Entomol.*, 20(2), 686–694.

Gu, J., Liang, H., Ge, X., et al. (2022), 'A study of the potential effect of yellow mealworm (*Tenebrio molitor*) substitution for fish meal on growth, immune and antioxidant capacity in juvenile largemouth bass (*Micropterus salmoides*)', *Fish Shellfish Immunol.*, 120, 214–221.

Hamam, M., D'Amico, M., and Di Vita, G. (2024). Advances in the insect industry within a circular bioeconomy context: a research agenda. *Environ. Sci. Eur.*, 36(1), 29.

Hachero-Cruzado, I., Betancor, M. B., Coronel-Dominguez, A. J., et al. (2024), 'Assessment of full-fat *Tenebrio molitor* as feed ingredient for *Solea senegalensis*: effects on growth performance and lipid profile', *Animals*, 14(4), 595.

Hardy, R. W. (2010), 'Utilization of plant proteins in fish diets: effects of global demand and supplies of fishmeal', *Aquac. Res.*, 41(5), 770–776.

Heidari-Parsa, S. (2018), 'Determination of yellow mealworm (*Tenebrio molitor*) nutritional value as an animal and human food supplementation', *Arthropods*, 7(4), 94.

Henry, M. A., Gasco, L., Chatzifotis, S., et al. (2018), 'Does dietary insect meal affect the fish immune system? The case of mealworm, *Tenebrio molitor* on European sea bass, *Dicentrarchus labrax*', *Dev. Comp. Immunol.*, 81, 204–209.

Hidalgo, M. C., Morales, A. E., Pula, H. J., et al. (2022), Oxidative metabolism of gut and innate immune status in skin and blood of tench (*Tinca tinca*) fed with different insect meals (*Hermetia illucens* and *Tenebrio molitor*)', *Aquaculture*, 558, 738384.

Hoffmann, L., Rawski, M., Nogales-Merida, S., et al. (2020), 'Dietary inclusion of *Tenebrio molitor* meal in sea trout larvae rearing: Effects on fish growth performance, survival, condition, and GIT and liver enzymatic activity', *An. Anim. Sci.*, 20(2), 579–598.

Hussain, I., Khan, S., Sultan, A., et al. (2017), 'Meal worm (*Tenebrio molitor*) as potential alternative source of protein supplementation in broiler', *Int. J. Biosci.*, 10(4), 225–262.

Ido, A., Hashizume, A., Ohta, T., et al. (2019), 'Replacement of fish meal by defatted yellow mealworm (*Tenebrio molitor*) larvae in diet improves growth performance and disease resistance in red seabream (*Pargus major*)', *Animals*, 9(3), 100.

Iaconisi, V., Marono, S., Parisi, G., et al. (2017), 'Dietary inclusion of *Tenebrio molitor* larvae meal: effects on growth performance and final quality treats of blackspot sea bream (*Pagellus bogaraveo*)', *Aquaculture*, 476, 49–58.

Iaconisi, V., Bonelli, A., Pupino, R., et al. (2018), 'Mealworm as dietary protein source for rainbow trout: body and fillet quality traits', *Aquaculture*, 484, 197–204.

Islam, S. M., Siddik, M. A., Sørensen, M., et al. (2024). Insect meal in aquafeeds: a sustainable path to enhanced mucosal immunity in fish. *Fish Shellfish Immunol.* 150, 109625.

Jajić, I., Krstović, S., Petrović, M., et al. (2022), 'Changes in the chemical composition of the yellow mealworm (*Tenebrio molitor* L.) reared on different feedstuffs', *J. Anim. Feed Sci.*, 31(2), 191–200.

Jannathulla, R., Rajaram, V., Kalanjiam, R., et al. (2019), 'Fishmeal availability in the scenarios of climate change: inevitability of fishmeal replacement in aquafeeds and approaches for the utilization of plant protein sources', *Aquac. Res.*, 50(12), 3493–3506.

Jeong, S. M., Khosravi, S., Mauliasari, I. R., et al. (2020). Dietary inclusion of mealworm (*Tenebrio molitor*) meal as an alternative protein source in practical diets for rainbow trout (*Oncorhynchus mykiss*) fry', *Fish. Aquat. Sci.*, 23(1), 1–8.

Jeong, S. M., Khosravi, S., Yoon, K. Y., et al. (2021), 'Mealworm, *Tenebrio molitor*, as a feed ingredient for juvenile olive flounder, *Paralichthys olivaceus*', Aquac. *Rep.*, 20, 100747.

Józefiak, A., Nogales-Mérida, S., Mikołajczak, Z., et al. (2019), 'The utilization of full-fat insect meal in rainbow trout (*Oncorhynchus mykiss*) nutrition: the effects on growth performance, intestinal microbiota and gastrointestinal tract histomorphology', *Ann. Anim. Sci.*, 19, 747–765.

Khan, S., Khan, R. U., Alam, W., et al. (2018), 'Evaluating the nutritive profile of three insect meals and their effects to replace soya bean in broiler diet', *J. Anim. Physiol. Anim. Nutr.*, 102(2), e662–e668.

Li, H., Hu, Z., Liu, S., et al. (2022), 'Influence of dietary soybean meal replacement with yellow mealworm (*Tenebrio molitor*) on growth performance, antioxidant capacity, skin color, and flesh quality of mirror carp (*Cyprinus carpio* var. *specularis*), *Aquaculture*, 561, 738686.

Li, P., Mai, K., Trushenski, J., et al. (2009), 'New developments in fish amino acid nutrition: towards functional and environmentally oriented aquafeeds', *Amino Acids*, 37, 43–53.

Liland, N. S., Araujo, P., Xu, X. X., et al. (2021), 'A meta-analysis on the nutritional value of insects in aquafeeds', *J. Insects Food Feed*, 7, 743–759.

Liu, C., Masri, J., Perez, V., et al. (2020), 'Growth performance and nutrient composition of mealworms (*Tenebrio molitor*) fed on fresh plant materials-supplemented diets', *Foods*, 9(2), 151.

Li, H., Hu, Z., Liu, S., et al. (2022), 'Influence of dietary soybean meal replacement with yellow mealworm (*Tenebrio molitor*) on growth performance, antioxidant capacity, skin color, and flesh quality of mirror carp (*Cyprinus carpio* var. *specularis*)', *Aquaculture*, 561, 738686.

López Nadal, A., Ikeda-Ohtsubo, W., Sipkema, D., et al. (2020), 'Feed, microbiota, and gut immunity: using the zebrafish model to understand fish health', *Front. Immunol.*, 11, 114.

Macusi, E. D., Cayacay, M. A., Borazon, E. Q., et al. (2022), 'Protein fishmeal replacement in aquaculture: a systematic review and implications on growth and adoption viability', *Sustainability*, 15(16), 12500.

Magnani, M., Claret, A., Gisbert, E., et al. (2023), 'Consumer expectation and perception of farmed rainbow trout (*Oncorhynchus mykiss*) fed with insect meal (*Tenebrio molitor*)', *Foods*, 12(23), 4356.

Makkar, H. P., Tran, G., Heuzé, V., et al. (2014), 'State-of-the-art on use of insects as animal feed', *Anim. Feed Sci. Technol.*, 197, 1–33.

Mariod, A. A. (2020), 'Nutrient composition of mealworm (*Tenebrio molitor*)', In A. A. Mariod (Ed.), *African edible insects as alternative source of food, oil, protein and bioactive components*, Springer Nature, Switzerland, pp. 275–280.

Mastoraki, M., Panteli, N., Kotzamanis, Y. P., et al. (2022), 'Nutrient digestibility of diets containing five different insect meals in gilthead sea bream (*Sparus aurata*) and European sea bass (*Dicentrarchus labrax*)', *Anim. Feed Sci. Technol.*, 292, 115425.

Melenchón, F., Larrán, A. M., Sanz, M. Á., et al. (2023), 'Different diets based on yellow mealworm (*Tenebrio molitor*)–part A: facing the decrease in omega‒ 3 fatty acids in fillets of rainbow trout (*Oncorhynchus mykiss*)', *Fishes*, 8(6), 286.

Mikołajczak, Z., Rawski, M., Mazurkiewicz, J., et al. (2020), 'The effect of hydrolyzed insect meals in sea trout fingerling (*Salmo trutta* m. *trutta*) diets on growth performance, microbiota and biochemical blood parameters', *Animals*, 10(6), 1031.

Mitra, A. (2021), 'Thought of alternate aquafeed: conundrum in aquaculture sustainability?', *Proc. Zoo. Soc.*, 74(1), 1–18.

Mousavi, S., Zahedinezhad, S., Loh, J. Y. (2020), 'A review on insect meals in aquaculture: the immunomodulatory and physiological effects', *Int. Aquat. Res.*, 12, 100–115.

Mozanzadeh, M. T., Hetmatpour, F., Gisbert, E. (2022), In J. M. Lorenzo and J. Simal-Gadara (Eds.), *Sustainable Aquafeeds, Technological Innovation and Novel Ingredients*, CRC Press, Boca Ratón, USA, pp. 185–292.

Naylor, R. L., Hardy, R. W., Bureau, D. P., et al. (2009), 'Feeding aquaculture in an era of finite resources', *Proc. Natl. Acad. Sci. USA*, 106(36), 15103–15110.

Naylor, R. L., Hardy, R. W., Buschmann, A. H., et al. (2021), 'A 20-year retrospective review of global aquaculture', *Nature*, 591(7851), 551–563.

Ng, W. K., Liew, F. L., Ang, L. P., et al. (2001), 'Potential of mealworm (*Tenebrio molitor*) as an alternative protein source in practical diets for African catfish, *Clarias gariepinus*', *Aquac. Res.*, 32, 273–280.

Ngo, D. H. and Kim, S. K. (2014), 'Antioxidant effects of chitin, chitosan, and their derivatives', *Adv. Food Nutr. Res.*, 73, 15–31.

Nogales-Mérida, S., Gobbi, P., Józefiak, D., et al. (2019), 'Insect meals in fish nutrition', *Rev. Aquac.*, 11, 1080–1103.

NRC (National Research Council) (2011), *Nutrient Requirements of Fish and Shrimp*, The National Academies Press, Washington, USA.

Ojha, S., Bußler, S., Psarianos, M., et al. (2021), 'Edible insect processing pathways and implementation of emerging technologies', *J. Insects Food Feed*, 7(5), 877–900.

Oliva-Teles, A., Enes, P., Peres, H. (2015), 'Replacing fishmeal and fish oil in industrial aquafeeds for carnivorous fish', In D. A. Davis (Ed.), *Feed and Feeding Practices in Aquaculture, Woodhead Publishing Series in Food Science, Technology and Nutrition*, Woodhead Publishing, Sawston, Waltham, pp. 203–233.

Pereira, A. G., Fraga-Corral, M., Garcia-Oliveira, P., et al. (2022), 'Single-cell proteins obtained by circular economy intended as a feed ingredient in aquaculture', *Foods*, 11(18), 2831.

Piccolo, G., Iaconisi, V., Marono, S., et al. (2017), 'Effect of *Tenebrio molitor* larvae meal on growth performance, in vivo nutrients digestibility, somatic and marketable indexes of gilthead sea bream (*Sparus aurata*)', *Anim. Feed Sci. Technol.*, 226, 12–20.

Ravzanaadii, N., Kim, S. H., Choi, W. H., et al. (2012), 'Nutritional value of mealworm, *Tenebrio molitor* as food source', *Int. J. Indus. Entomol.*, 25(1), 93–98.

Rawski, M., Mazurkiewicz, J., Kierończyk, B., et al. (2020), 'Black soldier fly full-fat larvae meal as an alternative to fish meal and fish oil in Siberian sturgeon nutrition: the effects on physical properties of the feed, animal growth performance, and feed acceptance and utilization', *Animals*, 10(11), 2119.

Redman, D. H., Nelson, D. A., Roy, J., et al. (2019), 'A pilot study using graded yellow mealworm (*Tenebrio molitor*) meal in formulated diets for growth performance of black sea bass (*Centropristis striata*)', *NOAA Technical Memorandum NMFS-NE-253*, 1–20.

Rema, P., Jorge, S., Leite, F., et al. (2021), 'Efficacy of microalgae (*Chlorella* and *Spirulina*) and insect (*Tenebrio molitor*) meals as fishmeal replacers in feed for juvenile tench (*Tinca tinca*)', *Revista Portuguesa de Zootecnia*, 6(1), 37–48.

Reyes, M., Rodríguez, M., Montes, J., et al. (2020), 'Nutritional and growth effect of insect meal inclusion on seabass (*Dicentrarchuss labrax*) feeds', *Fishes*, 5(2), 16.

Ringø, E., Zhou, Z., Olsen, R. E., et al. (2012), 'Use of chitin and krill in aquaculture–the effect on gut microbiota and the immune system: a review', *Aquac. Nutr.*, 18(2), 117–131.

Ringø, E. (2020), 'Probiotics in shellfish aquaculture', *Aquac. Fisher.*, 5(1), 1–27.

Roncarati, A., Gasco, L., Parisi, G., et al. (2015), 'Growth performance of common catfish (*Ameiurus melas* Raf.) fingerlings fed mealworm (*Tenebrio molitor*) diet', *J. Insects Food Feed*, 1(3), 233–240.

Ruiz, A., Gisbert, E., Andree, K. B. (2024), 'Impact of the diet in the gut microbiota after an inter-species microbial transplantation in fish', *Sci. Rep.*, 14(1), 4007.

Sánchez-Muros, M., De Haro, C., Sanz, A., et al. (2016), 'Nutritional evaluation of *Tenebrio molitor* meal as fishmeal substitute for tilapia (*Oreochromis niloticus*) diet', *Aquac. Nutr.*, 22(5), 943–955.

Sándor, Z. J., Banjac, V., Vidosavljević, S., et al. (2022), 'Apparent digestibility coefficients of black soldier fly (*Hermetia illucens*), yellow mealworm (*Tenebrio molitor*), and blue bottle fly (*Calliphora vicina*) insects for juvenile African catfish hybrids (*Clarias gariepinus* × *Heterobranchus longifilis*)', *Aquac. Nutr.*, 2022, 4717014.

Sankian, Z., Khosravi, S., Kim, Y. O., et al. (2018), Effects of dietary inclusion of yellow mealworm (*Tenebrio molitor*) meal on growth performance, feed utilization, body composition, plasma biochemical indices, selected immune parameters and antioxidant enzyme activities of mandarin fish (*Siniperca scherzeri*) juveniles', *Aquaculture*, 496, 79–87.

Selvaraj, V., Won, E. (2024). Transforming aquaculture with insect-based feed: restraining factors, *Animal Frontiers*, 14(4), 24–7.

Shafique, L., Abdel-Latif, H. M. R., Hassan, F. U., et al. (2021), 'The feasibility of using yellow mealworms (*Tenebrio molitor*): towards a sustainable aquafeed industry', *Animals*, 11, 811.

Siemianowska, E., Kosewska, A., Aljewicz, M., et al. (2013), Larvae of mealworm (*Tenebrio molitor* L.) as European novel food', *Agric. Sci.,* 4(6), 336935.

Singh, A., Phougat, N., Kumar, M., et al. (2013), 'Antifungal proteins: potent candidate for inhibition of pathogenic fungi', *Curr. Bioact. Compd.*, 9(2), 101–112.

Sommer, F., Anderson, J. M., Bharti, R., Raes, et al. (2017), 'The resilience of the intestinal microbiota influences health and disease'. *Nat. Rev. Microbiol.*, 15(10), 630-8.

Song, S. G., Chi, S. Y., Tan, B. P., et al. (2018), 'Effects of fishmeal replacement by *Tenebrio molitor* meal on growth performance, antioxidant enzyme activities and disease resistance of the juvenile pearl gentian grouper (*Epinephelus lanceolatus* ♂ × *Epinephelus fuscoguttatus* ♀)', *Aquac. Res.*, 49(6), 2210–2217.

Su, J., Gong, Y., Cao, S., et al. (2017), 'Effects of dietary *Tenebrio molitor* meal on the growth performance, immune response and disease resistance of yellow catfish (*Pelteobagrus fulvidraco*)', *Fish Shellfish Immunol.*, 69, 59–66.

Tacon, A. G. J. (2020), 'Trends in global aquaculture and aquafeed production: 2000–2017', *Rev. Fish. Sci. Aquacult.*, 28, 43–56.

Terova, G., Gini, E., Gasco, L. et al. (2021), 'Effects of full replacement of dietary fishmeal with insect meal from Tenebrio molitor on rainbow trout gut and skin microbiota', *J. Animal Sci. Biotechnol.*, 12, 30.

Tran, H. Q., Nguyen, T. T.2, Prokešová, M., et al. (2022a), 'Systematic review and meta-analysis of production performance of aquaculture species fed dietary insect meals', *Rev. Aquac.*, 14(3), 1637–1655.

Tran, H. Q., Prokešová, M., Zare, M., et al. (2022b), 'Production performance, nutrient digestibility, serum biochemistry, fillet composition, intestinal microbiota and environmental impacts of European perch (*Perca fluviatilis*) fed defatted mealworm (*Tenebrio molitor*)', *Aquaculture*, 547, 737499.

Tzachor, A. (2019), 'The future of feed: integrating technologies to decouple feed production from environmental impacts', *Ind. Biotechnol.*, 15(2), 52–62.

van Riel, A. J., Nederlof, M. A., Chary, K., et al. (2023), 'Feed-food competition in global aquaculture: current trends and prospects', *Rev. Aquac,* 15, 1142–1158.

Valipour, M., Oujifard, A., Hosseini, A., et al. (2019), 'Effects of dietary replacement of fishmeal by yellow mealworm (*Tenebrio molitor*) larvae meal on growth performance, hematological indices and some of non-specific immune responses of juvenile rainbow trout (*Oncorhynchus mykiss*)', *Iran. Sci. Fish. J.*, 28, 13–26.

Veldkamp, T., Dong, L., Paul, A., et al. (2022), 'Bioactive properties of insect products for monogastric animals–a review', *J. Insects Food Feed*, 8(9), 1027–1040.

Woodgate, S. L., Wan, A. H., Hartnett, F., et al. (2022), 'The utilisation of European processed animal proteins as safe, sustainable and circular ingredients for global aquafeeds', *Rev. Aquac.*, 14(3), 1572–1596.

Zhang, Z., Yuan, J., Tian, S., et al. (2022), 'Effects of dietary vitamin E supplementation on growth, feed utilization and flesh quality of large yellow croaker *Larimichthys crocea* fed with different levels of dietary yellow mealworm *Tenebrio molitor* meal', *Aquaculture*, 551, 737954.

Chapter 11

Advances in production of house flies as food and animal feed: key challenges and solutions

Nathaniel A. Sibinga, Katholieke Universiteit (KU) Leuven, Belgium; Eugene Won and Vimal Selvaraj, Cornell University, USA

1 Introduction

The ubiquity of the common housefly (*Musca domestica*) in all climate zones has made it a frequent topic of study in a variety of scientific fields. The earliest published study on the use of housefly as an animal feed ingredient appears to be from 1969 (Calvert et al., 1969), though the idea is quite a bit older: Calvert credits a 1919 German paper for the idea of rearing fly larvae (Lindner, 1919). The impulse to cite very old studies of houseflies appears itself to be something of a tradition: Lindner, drawing upon what by 1919 was evidently already an ancient and extensive body of literature on fly biology, cites a calculation by Antonie van Leeuwenhoek (1632–1723) on their extraordinary fecundity.

The basic arguments made by Lindner and many after him – that fly larvae are easy to propagate, grow rapidly on a wide range of existing organic waste products, and generally have a more desirable nutritional profile than their feed inputs – remain sound 100 years later. It must be acknowledged, however,

http://dx.doi.org/10.19103/AS.2024.0139.13

that despite their extensively studied biology and increasingly favorable economic conditions for insect production, housefly larvae still do not have a clear pathway to commercial viability as a feed ingredient at any significant scale. The question this chapter seeks to address is why houseflies have not caught on as a production species and to what extent that failure is likely to be reversible through further study.

2 The evolution of research on houseflies

Historically, the rearing of houseflies has been a solution in search of the appropriate problem. Lindner (Lindner, 1919), for example, made explicit reference to the devastating food shortages of the First World War in proposing farmed insects as a source of edible lipids for human consumption. By the late 1960s, food production was increasingly industrialized, and concern had shifted to the challenges posed by the quantities of manure in large-scale poultry facilities. In the 1970s, Miller and Calvert each independently proposed the use of housefly larvae to break down poultry manure and conducted several studies to optimize rearing conditions and establish the utility of poultry manure-reared housefly larvae and pupae as a poultry feed supplement (Calvert et al., 1970; Miller et al., 1974; Teotia and Miller, 1974). Ultimately, separation of larvae/pupae from the processed manure appears to have been too impractical at large scale – according to a 1991 review, in a practical application on an egg farm, 'the final product after drying would be a mixture of fly pupae, larvae and manure residue', which would be 'richer as a feedstuff than the untreated manure' alone (El Boushy, 1991). This rather noncommittal assessment of housefly larvae as a feedstuff underscores the practical challenge of isolating the desirable insect component from the unwanted and potentially microbe-rich feed residue in a housefly production system.

A series of controlled experiments in the 1990s by Hogsette – who developed the Gainesville laboratory diet for houseflies (Hogsette, 1992b) – documented housefly larval growth under various conditions in manure substrates from dairy cows (Hogsette, 1996), poultry and swine (Farkas et al., 1998). Hogsette also designed a system to take advantage of the 'self-harvesting' behavior of prepupal larvae in the wandering stage (Hogsette, 1992a), during which they seek a dry location prior to pupation. It is critical to note, however, that while this work has clear relevance for housefly production, the primary research interest of these studies was always the control of wild pest fly populations. Indeed, laboratory-scale rearing methods of houseflies far pre-date both Hogsette and the recent interest in insect production as a feedstuff: no fewer than 12 published studies documenting use of a wide range of substrates are referenced in the introduction of Hafez's 1948 method using cotton pads soaked in milk (Hafez, 1948).

The scientific literature on the intentional rearing of houseflies is dwarfed by (and, until recently, was usually a subset of) the literature on how to suppress them. Adult houseflies are generally recognized as a pest animal, capable of spreading a wide variety of human and animal pathogens (Nayduch et al., 2023), which is likely one of the main reasons their industrial production has not been more extensively pursued. Research conducted on houseflies under the aegis of pest control can be found for many topics with clear relevance for production:

- Metabolism (Heslop et al., 1963; Wakayama et al., 1985)
- Nutrition (Chang and Wang, 1958; House and Barlow, 1958; Bridges, 1971)
- Reproduction (Hampton, 1952; Adams and Nelson, 1990)
- Disease (Güssow, 1917; Moussa, 1978)
- Parasitism (Nagel and Pimentel, 1964; Georgis et al., 1987)
- Microbiota (Nicoll, 1911; Steinhaus, 1940)
- Spread of foodborne pathogens (Greenberg et al., 1970)
- Genetics (Hiroyoshi, 1961, 1977; Scott et al., 2014)
- Chemical susceptibility (Soloway, 1976; Singh and Singh, 1991)

The scope and depth of this literature is too extensive to adequately review in a single chapter, and, as will be discussed in greater detail below, its relevance does not extend to answering the most fundamental outstanding questions about the viability of large-scale housefly production for use as an ingredient for farmed animal diets. Still, for all the ways that insect farming is a novel field, it is worth remembering that flies are one of the most-studied invertebrates on Earth – the existing literature is full of useful insights and intriguing hypotheses.

3 The housefly life cycle

Houseflies are holometabolous insects found worldwide, often in close association with human settlements. Recent research suggests that coevolution of the human–housefly relationship dates back millions of years and that adult houseflies preferentially associate with primates (Gogarten et al., 2019). Both larval and adult houseflies feed on decaying organic matter, with the larvae being prodigious consumers of a wide variety of substrates.

Houseflies have a temperature-dependent life cycle with generation times ranging from 14 to 27 days between 20°C and 30°C (Black, 1985) (Figure 1):

- 1 day in the egg stage
- 5–9 days across three instars in the larval stage
- 6–11 days in the pupal stage

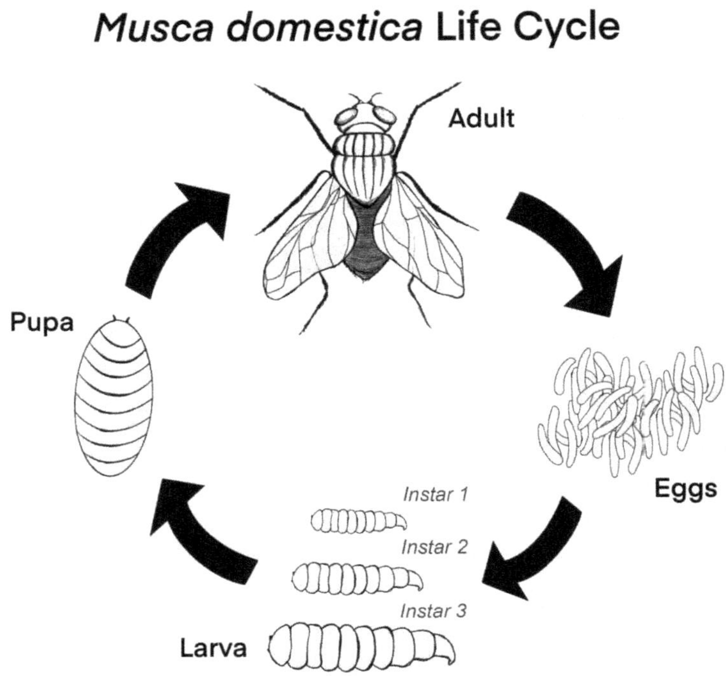

Figure 1 Life cycle of the common housefly (*Musca domestica*). Adult flies lay batches of eggs that can hatch in 8–24 hours to yield larvae. Housefly larvae pro-gress through three developmental instars of growth before pupating and the eclosion of adult flies. (Illustration prepared by Madeleine Wenger.)

- 3–6 days before sexual maturity in the adult stage

This facilitates rapid adaptation. Houseflies display both regional (Maria Alves and Bélo, 2002) and seasonal (Black and Krafsur, 1986) adaptation in the wild, as well as the capacity for rapid recombination and evolution of traits such as pesticide resistance when subjected to selective pressures (Scott and Georghiou, 1985). Sequencing of the housefly genome in 2014 (Scott et al., 2014) has accelerated understanding of the genetic basis of several unique characteristics and gene flow among geographically diverse housefly populations. Both the extant diversity of subtypes and rapid life cycle suggest an almost entirely unexplored potential for creating desirable production phenotypes.

4 Production parameters and issues: temperature, stocking density, and propagation

4.1 Temperature

A number of studies have established that the growth and development rate of housefly larvae are highly dependent on temperature (Lysyk and Axtell, 1987; Wang et al., 2018). The fastest growth rates – approximately 5–7 days from egg to the prepupal stage, when they are currently assumed ready for harvest – are typically observed between 30°C and 35°C, although there appears to be a tradeoff with lower survival at temperatures above 30°C (Barnard and Geden, 1993). Furthermore, higher rearing temperatures have been shown to increase the ratio of saturated:unsaturated fatty acids in larvae (Robb et al., 1972), which could have significant implications for feed applications (Alao and Balnave, 1985; Naghshineh et al., 2010).

Historically, studies of temperature have been conducted by controlling ambient temperature, which, depending on rates of decomposition, evaporative cooling, metabolic activity, and larval density, may not accurately reflect the temperature of the immediate larval microenvironment (Larsen and Thomsen, 1940). Evidence from other fly species suggests that housefly larvae may have the capacity to increase the temperature of their microenvironment via swarming or aggregation behaviors (Johnson and Wallman, 2014; Kortsmit et al., 2023); and the capacity to thermoregulate by migrating towards microenvironments at optimal development temperatures (Aubernon et al., 2016).

Finally, housefly resilience and response to temperature stress vary by population, with genetic variation playing an important role (Kjærsgaard et al., 2015; Delclos et al., 2021). This suggests the possibility of genetic selection for diverse thermal optima to meet the production requirements of different climatic zones.

4.2 Stocking density

In lab experiments, higher densities of housefly larvae are associated with smaller individual sizes and lower survival (Barnard and Geden, 1993; Hussein et al., 2017). However, in a production context, individual growth must be balanced with total biomass production, which is increased at higher densities. Increasing stocking density from 1 egg:1 g dairy cow manure to 16 eggs:1 g decreased individual larval size by about 50% and individual survival from ~75% to ~60%, but resulted in an approximately 4.5-fold increase in total biomass of housefly larvae (Hussein et al., 2017). This falls toward the high end of the range of reported optimal stocking densities at larger scale.

Other studies have suggested differing optimal stocking densities: Comparing housefly larvae reared on poultry manure and kitchen waste, Zi-zhe et al. (2017) found that total biomass production in both substrates peaked at 8000 eggs:1 kg; Using a substrate of food waste and wheat bran, Niu et al. (2017) concluded that the optimal stocking density was 24 000 eggs:1 kg; Using solid kitchen waste, Fan et al. (2023) used 10 000 eggs:1 kg (but did not report testing other densities); and Čičková et al. (2012b) reported optima of 11 000 eggs:1 kg of swine manure without sawdust and 4400 eggs:1 kg of swine manure with sawdust.

Many studies at larger scales report stocking density by egg mass rather than number, making it difficult to normalize such comparisons given that reports of average housefly egg mass vary across the literature: 55 μg (Painter and Kilgore, 1967), 59 μg (Zi-zhe et al., 2017), 62.5 μg (Niu et al., 2017), and 110 μg (Pastor et al., 2011). Using the median value of these studies (60.75 μg/egg), 16 eggs:1 g manure corresponds to 0.97 g:1 kg, which aligns reasonably well with available data for industrial-scale production of housefly larvae using manure as a substrate, including egg:manure ratios of 0.55 g:1 kg (Yang et al., 2014); 0.75 g:1 kg (El Boushy, 1991); and 1 g:1 kg (Cortes Ortiz et al., 2016).

Egg quality is reported to be independent of egg size for houseflies (McIntyre and Gooding, 2000), thus stocking densities should be reported by (estimated) egg number, rather than mass or volume.

As suggested by the variability between studies, optimal stocking density depends on the qualities of the feed substrate that will be transformed into insect biomass, as well as a variety of environmental conditions, including temperature, moisture content, substrate depth, aeration, etc. Interestingly, some studies utilizing higher-nutrient substrates appear to favor relatively low densities (though it should be noted that these studies do not explicitly optimize larval density). For example, using a diet of bran and full-fat milk powder, Shipp and Osborn recommended ~3.8 eggs:1 g media (Shipp and Osborn, 1967). Similarly, recent studies that compared multiple protein-rich substrates reported favorable egg:media densities of ~3.7:1 g (Leyo et al., 2023) and ~9:1 g (Ganda et al., 2022). The high variation in stocking density across studies represents a major knowledge gap in the field. In the absence of reliable models to estimate how many larvae can be produced on a given volume of substrate, the cost and design of production systems are subject to a potentially untenable degree of uncertainty.

A final consideration is that in black soldier fly (BSF) (*Hermetia illucens*) production, larval density has been shown to have different effects on growth at different production scales (Yang and Tomberlin, 2020; Yakti et al., 2022). At higher densities and larger scales, housefly larvae may exert significant effects on the substrate, potentially influencing physical (temperature, evaporation rate, and aeration) and biological (fermentation rate, fungal growth, communal digestion, and competition) parameters. More research is clearly needed to refine and model optimal stocking densities for feed substrates of varying qualities and at larger scales that mimic industrial conditions.

4.3 Propagation

Adult houseflies typically achieve sexual maturity and begin to lay eggs within the first week after eclosion, though this depends on genetic strain, temperature, and nutrition (Francuski et al., 2020). Fecundity peaks in the second week but generally remains strong through the third week and potentially beyond. Under optimal conditions, female houseflies lay 300–500 eggs in their lifetime, though this number drops by as much as 50% in cages at production densities (Pastor et al., 2011) and at higher temperatures (Francuski et al., 2020). However, high temperatures (above 30°C) also have the benefit of accelerating the onset of laying and increasing average daily production at the cost of a shorter laying period and reduced lifetime fecundity (Francuski et al., 2020).

This suggests that in a production context, there is a tradeoff between maximizing lifetime egg production and maximizing peak daily egg production. Broodstock maintained at 25°C in mesh cages at densities as high as 40 000 flies per m^3 (25 cm^3 per fly) have been reported, with average daily yields of ~130 350 eggs from 25 000 adults between days 7 and 22 post-eclosion (0.70 m^3 cages) (Čičková et al., 2012b) or ~96 000 eggs from 5000 adults (0.125 m^3 cages) over an unspecified duration (Niu et al., 2017). Comparison of average daily egg yields can be complicated by differences in the length and timing of the sampling period, thus age of the flies and collection period should be specified in fecundity values in future studies.

Unlike BSF broodstock that lay their eggs in the easily accessible crevasses of collection devices, housefly egg collection from food substrates can be labor-intensive, and oviposition is subject to a variety of factors (Machtinger et al., 2014). Housefly broodstock requires adequate nutrition, and providing high-quality protein (in addition to sugar) has been consistently linked to increased egg production (Shipp and Osborn, 1967; Pastor et al., 2011; Ganda et al., 2020). Nutrition during the larval stage can also influence fecundity, independent of the food available to adults (Khan et al., 2012).

Houseflies can be selective about where they lay their eggs, with some substrates strongly preferred even when all available options represent suitable food for larvae (Amano, 1985). There is ample data to support the possible utility of chemical (Richardson, 1916; Larsen et al., 1966; Tang et al., 2016) and biological (Jiang et al., 2002; Lam et al., 2007) attractants to stimulate oviposition, as well as evidence for the importance of behavioral/environmental cues (Pastor et al., 2011; Kortsmit et al., 2023). Given these physiological and environmental variables, it is important to continue optimizing broodstock husbandry protocols to efficiently generate the substantial numbers of eggs necessary for industrial larval production.

5 Production parameters and issues: feed substrate and harvesting/collection

5.1 Feed substrate

Selection of feed substrate can have a substantial impact on insect larvae growth, nutritional value, and economics, and can also be limited at the national or regional level by regulatory restrictions. Broadly speaking, there are three major categories of feed substrates in the housefly literature:

- Livestock manure
- Kitchen/restaurant food waste
- Agricultural or food processing byproducts

Among these substrates, livestock manure is the most extensively studied, though there is evidence that manure is suboptimal for growth when compared either to laboratory diets (Amano, 1985) or any of several agricultural byproducts (Ganda et al., 2019, 2022; Mafwila Kinkela et al., 2019; Leyo et al., 2023). However, given that industrial or agricultural waste processing is generally considered an added benefit of insect production, optimizing growth is not the only relevant paradigm. The foundation of industrial BSF production thus far has been the use of agricultural byproducts which have a more consistent composition than kitchen waste and fewer regulatory and safety concerns than manure. Pre-processing of kitchen waste in modern solid waste facilities in China also holds promise as a consistent ingredient suitable for large-scale insect production, with housefly larvae shown to grow well on the solid phase fraction (Fan et al., 2023).

It is important to remember that outside the laboratory environment, housefly larvae are not reared on formulated feed. Thus, physical properties of the substrate (e.g. moisture level, density, and particle size) and microbiological components may be as predictive of larval performance as proximate composition (Leyo et al., 2023). It has been established that housefly larvae feed on, and in some cases require, live microorganisms (Zurek et al., 2000). Therefore, the benefits of a diverse bacterial community and the anticipated temporal build-up of bacterial mass in the substrates used for housefly larvae ought to be an active consideration. Substrate microbiology has not been a notable experimental factor in any study evaluating the suitability of a waste material for the commercial production of housefly larvae. However, early results in BSF suggest a critical role for the substrate's microbial community in normal development (Pei et al., 2022; Gorrens et al., 2023).

Combining substrates can help improve physical profiles and balance nutritional components, which in some cases dramatically improves larval yields (Ganda et al., 2019; Mafwila Kinkela et al., 2019). The selection of feed substrates will always require balancing suitability for fly larvae growth with local availability, along with the potential ancillary benefits of mitigating the disposal of organic waste products. Finally, substrate selection should be informed by the downstream application of the produced larvae as a source of nutrition for farm animals. Larvae meal nutritional profiles are strongly influenced by rearing substrate, especially fatty acid profile and ash content, and specific nutritional profiles of insect meal can be targeted by selecting appropriate substrates (Lobina and Mkupasi, 2020; Oonincx and Finke, 2021). The nutritional profile of larvae meal can impact the health and performance of the livestock that consumet it, as well as the quality of the animal products ultimately consumed by humans.

5.2 Harvest/collection

Methods for efficiently separating housefly larvae from their rearing substrate remain a significant bottleneck. Practical larval harvesting techniques have yet to be fully developed for mass production. Such techniques are a major issue considering that a commercial harvest would entail on the order of thousands of tons comprised of <1 g larvae, embedded in moist substrate with high water tension. In addition to the quality of the feed substrate, the timing of harvest can affect the quality of the larval product (Aniebo and Owen, 2010).

Separation from substrate based on larval behavioral cues (i.e. self-harvesting) in theory represents the least intrusive option and has shown potential at the lab scale (Hogsette, 1992a), but its scalability at the industrial scale has yet to be demonstrated. Housefly larvae can be induced to exit rearing substrate by spreading over a sieve and exposing to bright light (Eby and Dendy, 1978; Miranda et al., 2023), though this requires significant manipulation and may not be practical for all substrate types. Reducing oxygen content can also drive larvae to migrate from the substrate; an elegant method of achieving this through use of lids placed on the surface of trays of swine manure substrate was described by Čičková et al. (2012a). However, this technique has not been widely adopted – possibly because reducing oxygen content requires at least a semi-enclosed production system and is furthermore likely to result in reduced survival. A final consideration of behavioral separation methods is that they are unlikely to be 100% effective, meaning that substrate residue must be sterilized (typically by heating or freezing) before disposal to avoid generating adult pests from unharvested larvae.

Physical separation of housefly larvae from rearing substrate is challenging – though not impossible – due to their relatively small size and preference for moist substrates. Older studies often report separation by flotation or washing with water (Morgan, 1981a; Hogsette, 1992b; Tobin and Pitts, 1999) but this has the downside of increasing water use and increasing the aqueous volume of the spent substrate (Čičková et al., 2012a), or the energy spent drying the frass if it is to be used as fertilizer. An alternative method is to reduce substrate moisture content immediately prior to harvesting by addition of dry absorptive material, such as wheat bran (Zi-zhe et al., 2017) or rice husk powder (Fan et al., 2023), to facilitate easier mechanical separation, though this also increases the volume of residue.

The rapid development cycle of housefly larvae presents an additional challenge for harvesting, as they spend only a short time in the third instar before pupating, and exact synchronization of development time across a population is not possible. As a result, some production methods opt to collect pupae, which are more stable and can be separated from dried media by sieving or flotation (Čičková et al., 2012b). Due to their relative ease of harvesting compared to larvae, housefly pupae have also been commonly investigated as a feed ingredient (Teotia and Miller, 1973; Ido et al., 2015). Housefly pupae tend to have a higher protein:lipid ratio than larvae (Pieterse et al., 2013), though the crude fiber content of pupae is increased due to their higher chitin levels (Sánchez et al., 2021) and the pupation process represents a net expenditure of energy that would ideally be preserved in the final feed product.

6 Nutritional value of houseflies as a feedstuff

The lack of current large-scale commercial producers of housefly-derived feed ingredients remains a hindrance for the kind of all-encompassing, industry-supported animal nutrition trials that are now possible with BSF and mealworms. That said, data from many pilot-scale trials support the viability of housefly meals made from both larvae and pupae. Other than its reputation as a pest, there is little to suggest that the housefly should be eliminated from consideration alongside other insect cultivar candidates, assuming it can be produced economically.

Meals produced from housefly larvae and pupae have appealing nutrient profiles, high digestibility, and have been successfully incorporated into the diets of several production animal species at a research scale (Makinde, 2015). From a purely nutritional standpoint, they are eminently suitable for use as feed ingredients. Given the competitive market for feed commodities and the comparatively high cost of production of insect ingredients, however, the key long-term question is whether they have the potential to be more than simply suitable and can command premium prices or otherwise compete with existing feedstuff alternatives (Tavares et al., 2022). There are perhaps reasons for optimism on this front, but data from feeding trials is admittedly limited.

Houseflies have high levels of the essential amino acids lysine and methionine relative both to BSF and soy protein (St-Hilaire et al., 2007; Hussein et al., 2017) and provide a high overall ratio of indispensible:dispensible amino acids relative to soybean meal or other insect species (Nogales-Mérida et al., 2019). The apparent amino acid digestibility of housefly meal fed to pigs was also found to be higher than that of soybean meal, fishmeal, or BSF meal (Tan et al., 2020). Even with minimal processing, housefly larvae meal has excellent digestibility in poultry (Hwangbo et al., 2009; Elahi et al., 2020), perhaps reflecting the significant role of fly larvae in the natural diet of ground-hunting birds. In fish, where the digestibility of terrestrial plant ingredients is a persistent challenge (Zhuo et al., 2016), housefly ingredients have been reported to have higher digestibility than BSF but lower than mealworm (Gasco et al., 2018; Mastoraki et al., 2022a, 2022b). Like all protein ingredients, their amino acid should be optimized for the target species via combination with complementary protein sources.

As with BSF, there is also significant potential for modulation of the housefly nutritional profile via choice of substrate to enrich, for example, to target fatty acids or minerals (Oonincx and Finke, 2021; Van Huis et al., 2021). Some studies have suggested that the array of bioactive compounds associated with houseflies could provide protection against infection (Ido et al., 2015; Gasco et al., 2021). A final consideration is that improved processing of insect raw materials can potentially increase their value as feed ingredients (Ravi et al., 2020), through, for example, the removal of antinutrients or indigestible components analogous to those found in some plant-derived feed ingredients (Krogdahl et al., 2010). Further research, ideally at the production scale, is needed to establish the marginal value of housefly ingredients relative to other available protein sources.

7 Unresolved issues facing commercial housefly production

7.1 What are the optimal conditions for houseflies under realistic production conditions?

Farmed animals, including insects, exist in a unique biological setting. Some limitations that occur in natural settings (e.g. food availability, predation, toxicity, and environmental fluctuations) are reduced or eliminated, while others (e.g. stress, disease risk, and lack of biological diversity) could be exacerbated at high density and need to be managed. As detailed above, much of the research done on housefly biology has been in the context of pest control. This work largely focuses on the survival of houseflies at the extremes of their biological tolerance, with the goal of pest control being explicitly to discover how to create conditions beyond this tolerable range. A farm environment, by contrast, would strive to maintain optimal conditions for houseflies to thrive; limited research on housefly larvae has been done in this context. As a result, the extremes of housefly biology are often better described than the optimal conditions.

For example, extensive work was conducted in the 1950s to establish the minimum nutritional requirements of housefly larvae (Brookes and Fraenkel, 1958; Chang and Wang, 1958; House and Barlow, 1958). This research turns out to be of surprisingly limited utility in a production context. Housefly larvae were shown to be extremely adaptable and to have only a few fixed requirements – the standard animal essential amino acids, some B vitamins, choline, and sterols – all of which are likely to be present in decaying organic matter. Critically, minimum requirements (and, at the other end of the scale, maximum tolerances) do not necessarily allow extrapolation of optimal conditions, and significant empirical work remains to determine the nutritional suitability of practical substrates for housefly production.

Where production-relevant parameters have previously been systematically optimized – for example, adult diets to maximize fecundity (Shipp and Osborn, 1967) or a system to mass-rear pupae as live feed for Hymenopeteran parasites (Morgan, 1981b) – these have often been pursued at laboratory scale with limited regard for the cost of labor or inputs. A final challenge for determining optimum conditions is that many recent housefly production studies have prioritized comparisons of multiple feed substrates rather than optimization of conditions for a single substrate.

Furthermore, the most commonly tested substrates for housefly production remain livestock manures, with little recent critical examination of how realistic this model is in practice. Manure is undoubtedly cheap and widely available, but it presents a variety of biosafety and logistical challenges, in addition to regulatory concerns (Lähteenmäki-Uutela et al., 2021). The use of animal manure is also sub-optimal when compared to other plentiful waste streams: a recent comparative study of cow manure versus plant-based side streams (e.g. wheat or millet bran, spent brewers grain) showed that manure produced smaller yields of housefly larvae with inferior nutritional content (Leyo et al., 2023). In BSF, the early literature also focused on livestock manure (Zhou et al., 2013), but it has adapted in recent years to the reality that industrial production has thus far favored non-manure substrates (Parodi et al., 2020). After the initial phase of identification of high-performing, cheap, consistent, and abundant substrates for housefly, it is hoped that research will shift to systematic optimization of production conditions.

7.2 What has gone wrong in previous attempts at industrial scale production and to what extent are these problems correctable?

To the best of our knowledge, there are currently no notable industrial producers of housefly larvae. While multiple other insect species have seen an explosion in interest and investment over the past ten years, houseflies have moved in the opposite direction. In 2014, Agriprotein (South Africa) and the Hangzhou Tianyuan Agriculture Company (China) received coverage for ambitious plans for large-scale production of housefly protein meal from slaughter waste and swine manure, respectively. Neither has any discernible presence in the market today. Agriprotein quietly shifted production to BSF and became a major player before going bankrupt in 2021, and Hangzhou Tianyuan Agriculture Company does not appear to have a website or to have received English-language media coverage since 2015. Around the same time, BioFlyTech (Spain) was involved in studies of housefly production on swine manure (Roffeis et al., 2015); BioFlyTech still exists but also appears to have entirely transitioned to BSF production.

There is only limited publicly available information on the specific problems that doomed these and other efforts, though the update provided by Pastor et al. (2015) is potentially enlightening. Although this report deals with farmed insect production generally, some key challenges specific to houseflies are mentioned, including the preference of housefly larvae to feed only in the top few centimeters of substrate – as previously reported by Morgan (1981b); and the high ratio of eggs (32%) needed to maintain the broodstock population.

A further datapoint comes from a media interview given by an Agriprotein founder. Larvae were reportedly being reared at 37°C to accelerate the life cycle to 24 hours per instar, but at the expense of survival and size (Sserunjogi, 2014). These are substantial issues, though all of them should be possible to ameliorate through further research and development. This is not a comprehensive list, however, and more data on unsuccessful industrial-scale production of houseflies would be of enormous utility to future efforts.

The preference for shallow substrate creates a demand for large surface area, e.g. using numerous shallow-tiered larvae-raising beds or constantly rotating substrate. Managing a large number of modular rearing units could be expected to add infrastructure and/or labor costs. Research also indicates that the reduction in mass of some substrates is not substantial with stocking densities that approach the maxima for efficient larvae rearing (Hussein et al., 2017). Therefore, there needs to be a path for continuous waste disposal or alternative use/applications for the spent substrates. Improved aeration of substrates could allow for more space-efficient production, and automation of production facilities may help to alleviate the challenges associated with high turnover.

The high ratio of eggs required to maintain the broodstock population would appear to be a fundamental challenge of housefly biology. but one that should be possible to abate with enhanced nutrition and rearing conditions to optimize fecundity, as well as potential genetic improvement of production strains. Selection for adults with higher fecundity is clearly desirable, but so is selection for larger larvae. Not only will larger larvae result in larger adults, which should improve fecundity (Khan et al., 2012), but larger larvae would also reduce required stocking densities and hopefully allow for the maintenance of smaller broodstock populations.

Finally, while rearing housefly at 37°C is possible, most strains are unlikely to be healthy and would be expected to have decreased size and survival rates (Barnard and Geden, 1993). Accelerating the life cycle so dramatically also results in a need for extremely fast turnover of the substrate and would be expected to exacerbate the challenges of broodstock production. Lower temperatures are optimal for fecundity (Francuski et al., 2020), and the reduced survival rate at 37°C would make keeping larvae intended for use as broodstock at such a high temperature untenable. However, using a lower temperature would result in mismatched generation time between broodstock and production populations and a larger broodstock population.

Ultimately, however, these explanations are based on informed speculation. More data on industrial-scale operations – successful and otherwise – is needed to draw firm conclusions and set future research priorities.

7.3 Is there a viable model for small-scale or integrated production?

Housefly larvae are relatively simple to farm with minimal capital investment, even relative to other production insect species (Roffeis et al., 2017). Housefly adults are easy to catch in the wild, and mating and egg collection are straightforward (Basden, 1947). The small size of the larvae makes them easier to dry than those of BSF, and their reduced carapace requires minimal processing to convert larvae into digestible feed ingredients (Kenis et al., 2014). Their global distribution, adaptation to a wide range of temperatures and climatic conditions, and ability to thrive on available organic waste make production flexible and straightforward to implement without substantial overhead.

Produced larvae can be sold or incorporated directly into farm operations – for example, as poultry feed (Leyo et al., 2023). Given the abundance of housefly adults in the farm environment, some authors have suggested that, in appropriate settings, production can be accomplished via natural oviposition

(Koné et al., 2017), greatly reducing the labor associated with egg collection. These ideas echo existing proposals for integrated production of, for example, farmed fish and poultry (Shoko et al., 2019). The production of housefly larvae represents a low-cost way to augment the productivity of extant smallholder farm operations via on-farm upcycling of organic waste (Bloukounon-Goubalan et al., 2020). Indeed, one survey in Benin found that cultivation and use of housefly larvae by small-scale poultry farmers was associated with higher income levels and larger flocks (Pomalégni et al., 2017).

The counterargument to this line of reasoning is that low-capital approaches risk potentially exacerbating the existing problems with wild housefly populations in the farm environment: transmission of human and animal pathogens (Nayduch et al., 2023), the spread of antimicrobial resistance genes (Bertelloni et al., 2023), and irritation to farm animals and workers alike (Malik et al., 2007). Given their rapid lifecycle, uncontained housefly farming can quickly turn into housefly infestation. The production of houseflies without proper containment and controlled oviposition may pose significant safety risks. Furthermore, production outside of an intensive facility is likely to be difficult to optimize and could quickly fall behind improving industrial production methods without access to, for example, improved genetic stocks. Models for small-scale and integrated farming of houseflies hold promise but must adequately address these concerns to have sustained success.

7.4 Is there a reason to choose houseflies instead of black soldier flies?

It is undeniable at this point that the BSF is the favored fly species for industrial processing of organic side streams into feed ingredients. What a decade or two ago was largely a speculative proposal based on a handful of scientific studies (Sheppard et al., 1994) is now a reality. There are several large-scale commercial BSF producers, as well as a rapidly growing research community and body of literature focused on production. By comparison, very little progress has been made toward the production of housefly larvae over the same period. There are compelling arguments in favor of BSF – these are briefly summarized below – but from a housefly perspective, the question is whether those arguments are comprehensive and decisive, or are there as yet unrealized production niches in which the housefly has comparative utility?

Although houseflies and BSF share a similar ecological niche and often compete for substrate in the wild (Bradley and Sheppard, 1984), their biology does differ in important ways that have been well-reviewed in recent years by van Huis et al. (2020) and Kortsmit et al. (2023). The life cycle of BSF is longer than that of the housefly, and BSF larvae achieve approximately ten times larger size than their housefly equivalents. Ten times as many individual housefly larvae as BSF larvae are thus required for a given amount of insect biomass. A greater number of eggs and a larger broodstock population are required in turn, as individual BSF and housefly females lay a similar number of eggs (Chia et al., 2018). The greater numbers of individuals per rearing cycle and the increased frequency of rearing cycles may result in significantly more labor. Both types of larvae can grow on a wide range of substrates – including various kinds of animal manure, food scraps, and organic byproducts – but BSF larvae appear to have a more active and diverse array of digestive enzymes (Kim et al., 2011). These features, along with the housefly larva's smaller size, which restricts its ability to handle larger particle sizes, may lead to less complete bioconversion of manure and other organic substrates (van Huis et al., 2020).

Perhaps the most important and intractable consideration, however, is the difference between the adult stages: housefly adults are well-known pests with a richly documented capacity to transmit human diseases (Khamesipour et al., 2018), while BSF adults do not eat, tend to be reclusive, are short-lived, and pose almost no risk to people. While in theory insects will be contained in a production facility, strong biosecurity measures are needed to ensure that adult flies do not escape. Furthermore, since a fraction of the larval population tends to avoid harvest, safety measures must realistically extend to post-harvest treatment of feed residues. Containment of adult houseflies during routine production is achievable, though subject to failure in the event of natural disasters or human error. An escape could potentially lead to liability issues in the event of a disease outbreak traceable to a housefly production facility. The benign nature of the BSF adult stage means that, even in cases where larval production of the two species is comparable, BSF is likely to be favored.

A perspective shift is therefore necessary. Future research on the production of non-BSF fly larvae, including housefly, should acknowledge the leading position of BSF in the insect production scheme and seek to explicitly demonstrate comparative advantages of diversification. There is every reason to think that such advantages might exist – experience with traditional livestock shows that there are niches for a variety of production species – but researchers should think deliberately about what those advantages are. It is no longer likely to be scientifically significant to show that housefly larvae can be produced on a given substrate or incorporated into animal feed without in some way addressing the comparison to BSF as the current standard.

7.5 Can houseflies get past regulatory considerations?

Of the broad range of substrates that housefly larvae can consume, relatively few that could today realistically be used in Europe or North America for food or feed applications have been studied. In Africa and Asia, where feed regulations are generally less restrictive, manure could potentially constitute a viable substrate for housefly larvae used as animal feed (Koné et al., 2017). However, manure is explicitly prohibited as a feed substrate for farmed animals (including insects) under EU law, and its regulatory status in much of the rest of the world remains ambiguous (Sogari et al., 2019; Lähteenmäki-Uutela et al., 2021). While there is momentum to modernize aspects of feed legislation with regard to insect production and application, approving the use of manure as a growth medium does not appear to be a priority.

Furthermore, as the production and knowledge gap between insect species grows, there is a risk that housefly production will be left out of new legislation entirely. Houseflies are included among the seven insect species currently authorized for animal feed in the EU, but BSF already enjoys a favored regulatory status in the United States and Canada as the only insect species recognized for livestock feed by the Association of American Feed Control Officials (Lähteenmäki-Uutela et al., 2021) and the Canadian Food Inspection Agency (Sogari et al., 2019), respectively.

Research on consumer perception of insects has found they are generally viewed favorably as animal feed ingredients (Sogari et al., 2022), but this research does not generally make explicit mention of the use of houseflies, and, critically, it has not yet addressed the question of locally unwanted land use (Popper, 1983): consumers may be in favor of insect ingredients in theory while simultaneously steadfastly opposing their production locally. Concerns about the potential of escaped houseflies seem likely to translate to decreased social licensing for housefly production facilities compared to less-familiar insect species and result in regulatory challenges at a local level (i.e. permitting). Ultimately, regulatory concerns must be addressed with a combination of strong biosafety measures, clear communication, and persistence.

8 Conclusion and future trends

Despite the challenges, there is growing interest in insect-based protein sources, including housefly larvae, as sustainable ingredients for animal feed due to their high protein content and potential environmental benefits. Researchers and entrepreneurs are working to address these challenges and develop efficient and cost-effective production methods for scaling up housefly larva production.

However, as highlighted by their conspicuous absence among insect startups in the past decade, houseflies occupy a non-obvious niche for commercial production. At present, there do not appear to be clear competitive advantages for houseflies relative to BSF at the industrial scale. There are few, if any, documented examples of substrates that are unsuitable for BSF but suitable for houseflies. Many of their advantages relative to BSF come in the form of being easier to rear and use at small scale: easier propagation, faster life cycle, global distribution, wider temperature range, and smaller larvae that require less energy to dry and less processing prior to use in feed applications. The problem is that the ease and convenience of rearing are at odds with the higher need for biosecurity and containment of adults to prevent the spread of disease. While there are some differences in the nutritional profile of housefly- and BSF-derived ingredients, there is limited evidence to date that housefly-derived ingredients offer clear performance advantages in animal feeds. Identification of specific use cases of houseflies is thus a priority for future development.

The strongest argument for the continued development of housefly production may revolve around the dangers of monoculture BSF production. Little is known about infectious diseases that could pose a risk to BSF production, though putative pathogens have been isolated (Klüber et al., 2022), and it would be naïve to assume that there is zero risk of significant disruption. As with other industries, diversification can be a precaution against total loss resulting from biological bottlenecks or shifts in industry preferences.

There is also a potentially unexplored opportunity for housefly larvae to be grown sequentially with BSF and other insect cultivars in integrated rearing systems (Miranda et al., 2019). Research on houseflies and BSF are also complementary; fundamental insights into fly biology have been explored in houseflies (as well as fruit flies), and researchers studying BSF would be wise to avail themselves of this abundant resource. At the same time, research on BSF now provides a valuable template for production parameters and technologies with relevance for houseflies, and many insights are likely to be transferable. Nevertheless, a candid examination of the challenges and opportunities confronting housefly production reveals an uncertain future.

9 References

Adams, T. S. and Nelson, D. R. (1990). The influence of diet on ovarian maturation, mating, and pheromone production in the housefly, Musca domestica. *Invertebrate Reproduction & Development* 17, 193–201. https://doi.org/10.1080/07924259.1990.9672111.

Alao, S. J. and Balnave, D. (1985). Nutritional significance of different fat sources for growing broilers. *Poultry Science* 64, 1602–1604. https://doi.org/10.3382/ps.0641602.

Amano, K. (1985). Breeding of the house fly, musca domestica (Diptera : Muscidae), in fresh dung of cattle fed on pasture grass. *Applied Entomology and Zoology* 20, 143–150. https://doi.org/10.1303/aez.20.143.

Aniebo, A. O. and Owen, O. J. (2010). Effects of age and method of drying on the proximate composition of housefly larvae (Musca domestica Linnaeus) meal (HFLM). *Pakistan Journal of Nutrition* 9, 485–487.

Aubernon, C., et al. (2016). Thermoregulation in gregarious dipteran larvae: evidence of species-specific temperature selection. *Entomologia Experimentalis et Applicata* 160, 101–108. https://doi.org/10.1111/eea.12468.

Barnard, D. R. and Geden, C. J. (1993). Influence of larval density and temperature in poultry manure on development of the house fly (Diptera: Muscidae). *Environmental Entomology* 22, 971–977. https://doi.org/10.1093/ee/22.5.971.

Basden, E. B. (1947). Breeding the house-fly (*Musca domestica*, L.) in the Laboratory: I. Introduction and technique. *Bulletin of Entomological Research* 37, 381–387. https://doi.org/10.1017/S0007485300030443.

Bertelloni, F., et al. (2023). House flies (*Musca domestica*) from swine and poultry farms carrying antimicrobial resistant enterobacteriaceae and salmonella. *Veterinary Sciences* 10, 118. https://doi.org/10.3390/vetsci10020118.

Black, W. C. (1985). Seasonal breeding structure in the housefly (*Musca domestica* L.) populations. *Heredity* 56, 289.

Black, W. C. and Krafsur, E. S. (1986). Seasonal breeding structure in house fly, Musca domestica L., populations. *Heredity* 56, 289–298. https://doi.org/10.1038/hdy.1986.49.

Bloukounon-Goubalan, A. Y., et al. (2020). Physical and chemical properties of the agro-processing by-products decomposed by Larvae of Musca domestica and Hermetia illucens. *Waste and Biomass Valorization* 11, 2735–2743. https://doi.org/10.1007/s12649-019-00587-z.

Bradley, S. W. and Sheppard, D. C. (1984). House fly oviposition inhibition by larvae ofHermetia illucens, the black soldier fly. *Journal of Chemical Ecology* 10, 853–859. https://doi.org/10.1007/BF00987968.

Bridges, R. G. (1971). Incorporation of fatty acids into the lipids of the housefly, Musca domestica. *Journal of Insect Physiology* 17, 881–895. https://doi.org/10.1016/0022-1910(71)90105-3.

Brookes, V. J. and Fraenkel, G. (1958). The nutrition of the larva of the housefly, Musca domestica L. *Physiological Zoology* 31, 208–223. https://doi.org/10.1086/physzool.31.3.30157842.

Calvert, C. C., et al. (1969). House fly pupae as food for poultry. *Journal of Economic Entomology* 62, 938–939. https://doi.org/10.1093/jee/62.4.938.

Calvert, C. C., et al. (1970). House fly Larvae: Biodegradation of Hen Excreta to useful products. *Poultry Science* 49, 588–589. https://doi.org/10.3382/ps.0490588.

Chang, J. T. and Wang, M. Y. (1958). Nutritional requirements of the common housefly, Musca domestica vicina Macq. *Nature* 181, 566–566. https://doi.org/10.1038/181566a0.

Chia, S. Y., et al. (2018). Threshold temperatures and thermal requirements of black soldier fly Hermetia illucens: Implications for mass production. *PLoS One* 13, e0206097. https://doi.org/10.1371/journal.pone.0206097.

Čičková, H., et al. (2012a). A behavioral method for separation of house fly (Diptera: Muscidae) Larvae from processed pig manure. *Journal of Economic Entomology* 105, 62–66. https://doi.org/10.1603/EC11202.

Čičková, H., et al. (2012b). Biodegradation of pig manure by the housefly, Musca domestica: a viable ecological strategy for pig manure management. *PLoS One* 7, e32798. https://doi.org/10.1371/journal.pone.0032798.

Cortes Ortiz, J. A., et al. (2016). Chapter 6 - Insect mass production technologies. In *Insects as Sustainable Food Ingredients*, eds. A. T. Dossey, et al., San Diego:

Academic Press, pp. 153–201. https://doi.org/10.1016/B978-0-12-802856-8.00006-5.

Delclos, P. J., et al. (2021). Thermal tolerance and preference are both consistent with the clinal distribution of house fly proto-Y chromosomes. *Evolution Letters* 5, 495–506. https://doi.org/10.1002/evl3.248.

Eby, H. J. and Dendy, W. L. (1978). An attempt to mechanize nutrient recovery from animal excreta. *Transactions of the ASAE* 21, 395–398. https://doi.org/10.13031/2013.35311.

El Boushy, A. R. (1991). House-fly pupae as poultry manure converters for animal feed: a review. *Bioresource Technology* 38, 45–49. https://doi.org/10.1016/0960-8524(91)90220-E.

Elahi, U., et al. (2020). Growth performance, carcass characteristics, meat quality and serum profile of broiler chicks fed on housefly maggot meal as a replacement of soybean meal. *Journal of Animal Physiology and Animal Nutrition* 104, 1075–1084. https://doi.org/10.1111/jpn.13265.

Fan, X., et al. (2023). Optimization of substrate intensity and feeding method for the bioconversion of kitchen waste by housefly larvae. *Biomass Conv. Bioref.* https://doi.org/10.1007/s13399-023-04426-z.

Farkas, R., et al. (1998). Development of Hydrotaea aenescens and Musca domestica (Diptera: Muscidae) in poultry and pig manures of different moisture content. *Environmental Entomology* 27, 695–699. https://doi.org/10.1093/ee/27.3.695.

Francuski, L., et al. (2020). Effect of temperature on egg production in the common housefly. *Entomologia Experimentalis et Applicata* 168, 513–522. https://doi.org/10.1111/eea.12912.

Ganda, H., et al. (2020). Influence of adult diet on biological parameters of the housefly, Musca domestica L. (Diptera: Muscidae). *JoBAZ* 81, 46. https://doi.org/10.1186/s41936-020-00181-z.

Ganda, H., et al. (2022). Effect of four rearing substrates on the yield and the chemical composition of housefly larvae, Musca domestica L. 1758 (Diptera: Muscidae). *International Journal of Tropical Insect Science* 42, 1331–1339. https://doi.org/10.1007/s42690-021-00651-z.

Ganda, H., et al. (2019). Potentials of animal, crop and agri-food wastes for the production of fly larvae. *Journal of Insects as Food and Feed* 5, 59–67. https://doi.org/10.3920/JIFF2017.0064.

Gasco, L., et al. (2021). Beyond the protein concept: health aspects of using edible insects on animals. *Journal of Insects as Food and Feed* 7, 715–741. https://doi.org/10.3920/JIFF2020.0077.

Gasco, L., et al. (2018). Apparent digestibility of insect protein meals for rainbow trout. *Journal of Insects as Food and Feed* (Wageningen Academic Publishers), S39. https://doi.org/10.3920/JIFF2018.S1.

Georgis, R., et al. (1987). Survival and movement of insect parasitic nematodes in poultry manure and their infectivity against Musca domestica. *Journal of Nematology* 19, 292–295.

Gogarten, J. F., et al. (2019). Tropical rainforest flies carrying pathogens form stable associations with social nonhuman primates. *Molecular Ecology* 28, 4242–4258. https://doi.org/10.1111/mec.15145.

Gorrens, E., et al. (2023). The use of probiotics during rearing of hermetia illucens: potential, caveats, and knowledge gaps. *Microorganisms* 11, 245. https://doi.org/10.3390/microorganisms11020245.

Greenberg, B., et al. (1970). Factors affecting the transmission of salmonella by flies: natural resistance to colonization and bacterial interference. *Infection and Immunity* 2, 800–809. https://doi.org/10.1128/iai.2.6.800-809.1970.

Güssow, H. T. (1917). Empusa muscae versus Musca domestica L. *Annals of Applied Biology* 3, 150–158.

Hafez, M. (1948). A simple method for breeding the house-fly, Musca domestica, L., in the laboratory. *Bulletin of Entomological Research* 39, 385–386. https://doi.org/10.1017/S0007485300022483.

Hampton, U. M. (1952). Reproduction in the housefly (Musca domestica L.). *Proceedings of the Royal Entomological Society of London, A* 27, 29–32.

Heslop, J. P., et al. (1963). Anaerobic metabolism in the housefly, Musca domestica L. *Biochemical Journal* 87, 35–38.

Hiroyoshi, T. (1961). The linkage map of the house fly, Musca domestica L. *Genetics* 46, 1373–1380.

Hiroyoshi, T. (1977). Some new mutants and revised linkage maps of the housefly, Musca domestica L. *Japanese Journal of Genetics* 52, 275–288. https://doi.org/10.1266/jjg.52.275.

Hogsette, J. A. (1992a). Autoseparation method for harvesting house fly (Diptera: Muscidae) Pupae of known age. *Journal of Economic Entomology* 85, 2295–2297. https://doi.org/10.1093/jee/85.6.2295.

Hogsette, J. A. (1992b). New diets for production of house flies and stable flies (Diptera: Muscidae) in the laboratory. *Journal of Economic Entomology* 85, 2291–2294. https://doi.org/10.1093/jee/85.6.2291.

Hogsette, J. A. (1996). Development of house flies (Diptera: Muscidae) in sand containing varying amounts of manure solids and moisture. *Journal of Economic Entomology* 89, 940–945. https://doi.org/10.1093/jee/89.4.940.

House, H. L. and Barlow, J. S. (1958). Vitamin requirements of the house fly, Musca domestica L. (Diptera: Muscidae). *Annals of the Entomological Society of America* 51, 299–302. https://doi.org/10.1093/aesa/51.3.299.

Hussein, M., et al. (2017). Sustainable production of housefly (Musca domestica) larvae as a protein-rich feed ingredient by utilizing cattle manure. *PLoS One* 12, e0171708. https://doi.org/10.1371/journal.pone.0171708.

Hwangbo, J., et al. (2009). Utilization of house fly-maggots, a feed supplement in the production of broiler chickens. *Journal of Environmental Biology* 30, 609–614.

Ido, A., et al. (2015). Dietary effects of housefly (Musca domestica) (Diptera: Muscidae) pupae on the growth performance and the resistance against bacterial pathogen in red sea bream (Pagrus major) (Perciformes: Sparidae). *Applied Entomology and Zoology* 50, 213–221. https://doi.org/10.1007/s13355-015-0325-z.

Jiang, Y., et al. (2002). Semiochemicals from ovaries of gravid females attract ovipositing female houseflies, Musca domestica. *Journal of Insect Physiology* 48, 945–950. https://doi.org/10.1016/S0022-1910(02)00162-2.

Johnson, A. P. and Wallman, J. F. (2014). Effect of massing on larval growth rate. *Forensic Science International* 241, 141–149. https://doi.org/10.1016/j.forsciint.2014.05.006.

Kenis, M., et al. (2014). Insects used for animal feed in West Africa. *Entomologia* 2. https://doi.org/10.4081/entomologia.2014.218.

Khamesipour, F., et al. (2018). A systematic review of human pathogens carried by the housefly (Musca domestica L.). *BMC Public Health* 18, 1049. https://doi.org/10.1186/s12889-018-5934-3.

Khan, H. A. A., et al. (2012). Effect of livestock manures on the fitness of house fly, Musca domestica L. (Diptera: Muscidae). *Parasitol Res* 111, 1165–1171. https://doi.org/10.1007/s00436-012-2947-1.

Kim, W., et al. (2011). Biochemical characterization of digestive enzymes in the black soldier fly, Hermetia illucens (Diptera: Stratiomyidae). *Journal of Asia-Pacific Entomology* 14, 11–14. https://doi.org/10.1016/j.aspen.2010.11.003.

Kjærsgaard, A., et al. (2015). Plasticity in behavioural responses and resistance to temperature stress in Musca domestica. *Animal Behaviour* 99, 123–130. https://doi.org/10.1016/j.anbehav.2014.11.003.

Klüber, P., et al. (2022). Isolation of bacterial and fungal microbiota associated with hermetia illucens Larvae reveals novel insights into entomopathogenicity. *Microorganisms* 10, 319. https://doi.org/10.3390/microorganisms10020319.

Koné, N., et al. (2017). Production of house fly larvae for animal feed through natural oviposition. *Journal of Insects as Food and Feed* 3, 177–186. https://doi.org/10.3920/JIFF2016.0044.

Kortsmit, Y., et al. (2023). Behaviour of two fly species reared for livestock feed: optimising production and insect welfare. *Journal of Insects as Food and Feed* 9, 149–169. https://doi.org/10.3920/JIFF2021.0214.

Krogdahl, Å., et al. (2010). Important antinutrients in plant feedstuffs for aquaculture: an update on recent findings regarding responses in salmonids. *Aquaculture Research* 41, 333–344. https://doi.org/10.1111/j.1365-2109.2009.02426.x.

Lähteenmäki-Uutela, A., et al. (2021). Regulations on insects as food and feed: a global comparison. *Journal of Insects as Food and Feed* 7, 849–856. https://doi.org/10.3920/JIFF2020.0066.

Lam, K., et al. (2007). Proliferating bacterial symbionts on house fly eggs affect oviposition behaviour of adult flies. *Animal Behaviour* 74, 81–92. https://doi.org/10.1016/j.anbehav.2006.11.013.

Larsen, E. B. and Thomsen, M. (1940). The influence of temperature on the development of some species of díptera. *Vidensk. Medd. fra Dansk naturh. Foren.* 104. Available at: https://www-cabdirect-org.proxy.library.cornell.edu/cabdirect/abstract/19462901908 [Accessed June 20, 2023].

Larsen, J. R., et al. (1966). Olfactory and oviposition responses of the house fly to domestic manures, with notes on an autogenous strain1. *Journal of Economic Entomology* 59, 610–615. https://doi.org/10.1093/jee/59.3.610.

Leyo, I. H., et al. (2023). Optimal substrates for producing housefly larvae with high nutritional composition for sustainable poultry feed in Niger. *Journal of Insects as Food and Feed* 9, 193–203. https://doi.org/10.3920/JIFF2022.0013.

Lindner, P. (1919). Extraction of fat from small animals. *Zeitschrift fr Technische Biologie (Journal of Technical Biology)* 7, 213.

Lobina, A. A. and Mkupasi, E. M. (2020). Enhancing performance and fatty acid composition of Nile tilapia (Oreochromis niloticus) from housefly maggot (Musca domestica) feeds. *Asian Journal of Tropical Biotechnology* 17. Available at: https://smujo.id/bbs/article/view/12218 [Accessed August 4, 2023].

Lysyk, T. J. and Axtell, R. C. (1987). A simulation model of house fly (Diptera Muscidae) development in poultry manure. *The Canadian Entomologist* 119, 427–437. https://doi.org/10.4039/Ent119427-5.

Machtinger, E. T., et al. (2014). Development and oviposition preference of house flies and stable flies (Diptera: Muscidae) in six substrates from florida equine facilities. *Journal of Medical Entomology* 51, 1144–1150. https://doi.org/10.1603/ME14040.

Mafwila Kinkela, P., et al. (2019). Optimization of housefly larvae production on pig wastes and brewers' grains for integrated fish and pig farms in the tropics. *Livestock Research for Rural Development*. Available at: https://orbi.uliege.be/handle/2268/232535 [Accessed June 22, 2023].

Makinde, O. J. (2015). Maggot meal: a sustainable protein source for livestock production-a review. *Advances in Life Science and Technology* 31, 35.

Malik, A., et al. (2007). House fly (Musca domestica): a review of control strategies for a challenging pest. *Journal of Environmental Science and Health, Part B* 42, 453–469. https://doi.org/10.1080/03601230701316481.

Maria Alves, S. and Bélo, M. (2002). Morphometric variations in the housefly, Musca domestica (L.) with latitude. *Genetica* 115, 243–251. https://doi.org/10.1023/A:1020685727460.

Mastoraki, M., et al. (2022a). Insect meals in feeds for juvenile gilthead seabream (Sparus aurata): effects on growth, blood chemistry, hepatic metabolic enzymes, body composition and nutrient utilization. *Aquaculture* 561, 738674. https://doi.org/10.1016/j.aquaculture.2022.738674.

Mastoraki, M., et al. (2022b). Nutrient digestibility of diets containing five different insect meals in gilthead sea bream (Sparus aurata) and European sea bass (Dicentrarchus labrax). *Animal Feed Science and Technology* 292, 115425. https://doi.org/10.1016/j.anifeedsci.2022.115425.

McIntyre, G. S. and Gooding, R. H. (2000). Egg size, contents, and quality: maternal-age and -size effects on house fly eggs. *Can. J. Zool.* 78, 1544–1551. https://doi.org/10.1139/z00-086.

Miller, B. F., et al. (1974). Digestion of poultry manure by Musca domestica. *British Poultry Science* 15, 231–234. https://doi.org/10.1080/00071667408416100.

Miranda, C. D., et al. (2019). Interspecific competition between the house fly, Musca domestica L. (Diptera: Muscidae) and black soldier fly, hermetia illucens (L.) (Diptera: Stratiomyidae) when reared on poultry manure. *Insects* 10, 440. https://doi.org/10.3390/insects10120440.

Miranda, C. D., et al. (2023). Large-scale production of house fly, Musca domestica (Diptera: Muscidae), larvae fed 3 manure types. *Journal of Economic Entomology*, toad099. https://doi.org/10.1093/jee/toad099.

Morgan, P. B. (1981a). Mass production of Musca domestica L. *Status of Biological Control of Filth Flies. USD A, ARS* 212, 189–191.

Morgan, P. B. (1981b). *Status of Biological Control of Filth Flies: Proceedings of a Workshop, February 4-5, 1981, University of Florida, Gainesville*. Agricultural Research (Southern Region), Science and Education Administration, U.S. Department of Agriculture.

Moussa, A. Y. (1978). A new virus disease in the housefly, Musca domestica (Diptera). *Journal of Invertebrate Pathology* 31, 204–216. https://doi.org/10.1016/0022-2011(78)90009-5.

Nagel, W. P. and Pimentel, D. (1964). The intrinsic rate of natural increase of the pteromalid parasite nasonia vitripennis (Walk.) on its muscoid host Musca domestica L. *Ecology* 45, 658–660. https://doi.org/10.2307/1936125.

Naghshineh, M., et al. (2010). Effect of saturated/unsaturated fatty acid ratio on physicochemical properties of palm olein–olive oil blend. *Journal of the American Oil Chemists' Society* 87, 255–262. https://doi.org/10.1007/s11746-009-1495-z.

Nayduch, D., et al. (2023). House flies are underappreciated yet important reservoirs and vectors of microbial threats to animal and human health. *Microorganisms* 11, 583. https://doi.org/10.3390/microorganisms11030583.

Nicoll, W. (1911). On the varieties of bacillus coli associated with the house-fly (Musca Domestica). *Epidemiology & Infection* 11, 381–389. https://doi.org/10.1017/S0022172400016831.

Niu, Y., et al. (2017). A novel bioconversion for value-added products from food waste using Musca domestica. *Waste Management* 61, 455–460. https://doi.org/10.1016/j.wasman.2016.10.054.

Nogales-Mérida, S., et al. (2019). Insect meals in fish nutrition. *Reviews in Aquaculture* 11, 1080–1103. https://doi.org/10.1111/raq.12281.

Oonincx, D. G. A. B. and Finke, M. D. (2021). Nutritional value of insects and ways to manipulate their composition. *Journal of Insects as Food and Feed* 7, 639–659. https://doi.org/10.3920/JIFF2020.0050.

Painter, R. R. and Kilgore, W. W. (1967). Some physical and chemical characteristics of normal eggs, Larvae, and Chorions of the house fly, Musca domestica1. *Annals of the Entomological Society of America* 60, 1163–1166. https://doi.org/10.1093/aesa/60.6.1163.

Parodi, A., et al. (2020). Black soldier fly larvae show a stronger preference for manure than for a mass-rearing diet. *Journal of Applied Entomology* 144, 560–565. https://doi.org/10.1111/jen.12768.

Pastor, B., et al. (2011). Effect of the size of the pupae, adult diet, oviposition substrate and adult population density on egg production in Musca domestica (Diptera: Muscidae). *EJE* 108, 587–596. https://doi.org/10.14411/eje.2011.076.

Pastor, B., et al. (2015). Conversion of organic wastes into fly larval biomass: bottlenecks and challenges. *Journal of Insects as Food and Feed* 1, 179–193. https://doi.org/10.3920/JIFF2014.0024.

Pei, Y., et al. (2022). Bacillus velezensis EEAM 10B strengthens nutrient metabolic process in black soldier fly Larvae (Hermetia illucens) via changing gut microbiome and metabolic pathways. *Frontiers in Nutrition* 9. Available at: https://www.frontiersin.org/articles/10.3389/fnut.2022.880488 [Accessed September 9, 2023].

Pieterse, E., et al. (2013). Nutritional evaluation of dried larvae and pupae meal of the housefly (Musca domestica) using chemical- and broiler-based biological assays. *Animal Production Science* 54, 347–355. https://doi.org/10.1071/AN12370.

Pomalégni, S. C. B., et al. (2017). Traditional use of fly larvae by small poultry farmers in Benin. *Journal of Insects as Food and Feed* 3, 187–192. https://doi.org/10.3920/JIFF2016.0061.

Popper, F. J. (1983). LULUs: locally unwanted land uses. *Resources; (United States)* 73. Available at: https://www.osti.gov/biblio/5869734 [Accessed September 11, 2023].

Ravi, H. K., et al. (2020). Larvae mediated valorization of industrial, agriculture and food wastes: biorefinery concept through bioconversion, processes, procedures, and products. *Processes* 8, 857. https://doi.org/10.3390/pr8070857.

Richardson, O. H. (1916). A chemotropic response of the house fly (Musca domestica L.). *Science* 43, 613–616. https://doi.org/10.1126/science.43.1113.613.

Robb, R., et al. (1972). Temperature-dependent changes in sphingosine composition and composition of fatty acids of glycerophosphatides from Musca domestica larvae. *Insect Biochemistry* 2, 131–136. https://doi.org/10.1016/0020-1790(72)90044-3.

Roffeis, M., et al. (2017). Life cycle inventory analysis of prospective insect based feed production in West Africa. *Sustainability* 9, 1697. https://doi.org/10.3390/su9101697.

Roffeis, M., et al. (2015). Pig manure treatment with housefly (*Musca domestica*) rearing – an environmental life cycle assessment. *Journal of Insects as Food and Feed* 1, 195–214. https://doi.org/10.3920/JIFF2014.0021.

Sánchez, M., et al. (2021). House fly (Musca domestica) larvae meal as an ingredient with high nutritional value: microencapsulation and improvement of organoleptic characteristics. *Food Research International* 145, 110423. https://doi.org/10.1016/j.foodres.2021.110423.

Scott, J. G. and Georghiou, G. P. (1985). Rapid development of high-level permethrin resistance in a field-collected strain of the house fly (Diptera: Muscidae) under laboratory selection. *Journal of Economic Entomology* 78, 316–319. https://doi.org/10.1093/jee/78.2.316.

Scott, J. G., et al. (2014). Genome of the house fly, Musca domestica L., a global vector of diseases with adaptations to a septic environment. *Genome Biol* 15, 466. https://doi.org/10.1186/s13059-014-0466-3.

Sheppard, C. D., et al. (1994). A value added manure management system using the black soldier fly. *Bioresource Technology* 50, 275–279. https://doi.org/10.1016/0960-8524(94)90102-3.

Shipp, E. and Osborn, A. W. (1967). The effect of protein sources and of the frequency of egg collection on egg production by the housefly (Musca domestica L.). *Bull World Health Organ* 37, 331–335.

Shoko, A., et al. (2019). The role of fish-poultry integration on fish growth performance, yields and economic benefits among smallholder farmers in sub-Saharan Africa, Tanzania. *African Journal of Aquatic Science* 44, 15–24. https://doi.org/10.2989/16085914.2018.1555512.

Singh, D. and Singh, A. K. (1991). Repellent and insecticidal properties of essential oils against housefly, Musca domestica L. *Int J Trop Insect Sci* 12, 487–491. https://doi.org/10.1017/S1742758400011401.

Sogari, G., et al. (2019). The potential role of insects as feed: a multi-perspective review. *Animals* 9, 119. https://doi.org/10.3390/ani9040119.

Sogari, G., et al. (2022). Consumer perceptions and acceptance of insects as feed and food: current findings and future outlook. *Novel Foods and Edible Insects in the European Union*, 147–169.

Soloway, S. B. (1976). Naturally occurring insecticides. *Environmental Health Perspectives* 14, 109–117. https://doi.org/10.1289/ehp.7614109.

Sserunjogi, E. M. (2014). Housefly could be the star of 21st century. *Monitor*. Available at: https://www.monitor.co.ug/uganda/magazines/people-power/housefly-could-be-the-star-of-21st-century-1563796 [Accessed September 12, 2023].

Steinhaus, E. A. (1940). The microbiology of insects: with special reference to the biologic relationships between bacteria and insects. *Bacteriological Reviews* 4, 17–57.

St-Hilaire, S., et al. (2007). Fly prepupae as a feedstuff for rainbow trout, oncorhynchus mykiss. *Journal of the World Aquaculture Society* 38, 59–67. https://doi.org/10.1111/j.1749-7345.2006.00073.x.

Tan, X., et al. (2020). Amino acid digestibility in housefly and black soldier fly prepupae by growing pigs. *Animal Feed Science and Technology* 263, 114446. https://doi.org/10.1016/j.anifeedsci.2020.114446.

Tang, R., et al. (2016). Identification and testing of oviposition attractant chemical compounds for Musca domestica. *Scientific Reports* 6, 33017. https://doi.org/10.1038/srep33017.

Tavares, M. N., et al. (2022). Economic viability of insect meal as a novel ingredient in diets for broiler chickens. *Journal of Insects as Food and Feed* 8, 1015–1025. https://doi.org/10.3920/JIFF2021.0179.

Teotia, J. S. and Miller, B. F. (1973). Fly pupae as a dietary ingredient for starting chicks1. *Poultry Science* 52, 1830–1835. https://doi.org/10.3382/ps.0521830.

Teotia, J. S. and Miller, B. F. (1974). Nutritive content of house fly pupae and manure residue. *British Poultry Science* 15, 177–182. https://doi.org/10.1080/00071667408416093.

Tobin, P. C. and Pitts, C. W. (1999). Flotation method for extracting insects from poultry manure samples. *Journal of Medical Entomology* 36, 121–123. https://doi.org/10.1093/jmedent/36.1.121.

van Huis, A., et al. (2020). Insects as feed: house fly or black soldier fly? *Journal of Insects as Food and Feed* 6, 221–229. https://doi.org/10.3920/JIFF2020.x003.

Van Huis, A., et al. (2021). Nutritional qualities and enhancement of edible insects. *Annual Review of Nutrition* 41, 551–576. https://doi.org/10.1146/annurev-nutr-041520-010856.

Wakayama, E. J., et al. (1985). Occurrence and metabolism of arachidonic acid in the housefly, Musca domestica (L.). *Insect Biochemistry* 15, 367–374. https://doi.org/10.1016/0020-1790(85)90028-9.

Wang, Y., et al. (2018). Development of Musca domestica at constant temperatures and the first case report of its application for estimating the minimum postmortem interval. *Forensic Sci Int* 285, 172–180. https://doi.org/10.1016/j.forsciint.2018.02.004.

Yakti, W., et al. (2022). The effect of rearing scale and density on the growth and nutrient composition of Hermetia illucens (L.) (Diptera: Stratiomyidae) Larvae. *Sustainability* 14, 1772. https://doi.org/10.3390/su14031772.

Yang, F. and Tomberlin, J. K. (2020). Comparing selected life-history traits of black soldier fly (Diptera: Stratiomyidae) Larvae produced in industrial and bench-top-sized containers. *Journal of Insect Science* 20, 25. https://doi.org/10.1093/jisesa/ieaa113.

Yang, S., et al. (2014). Biodiesel production from swine manure via housefly larvae (Musca domestica L.). *Renewable Energy* 66, 222–227. https://doi.org/10.1016/j.renene.2013.11.076.

Zhou, F., et al. (2013). Developmental and waste reduction plasticity of three black soldier fly strains (Diptera: Stratiomyidae) raised on different livestock manures. *Journal of Medical Entomology* 50, 1224–1230. https://doi.org/10.1603/ME13021.

Zhuo, L.-C., et al. (2016). Apparent digestibility of soybean meal and Lactobacillus spp. fermented soybean meal in diets of grouper, Epinephelus coioides. *Aquaculture Research* 47, 1009–1012.

Zi-zhe, C., et al. (2017). Conversion of poultry manure to biodiesel, a practical method of producing fatty acid methyl esters via housefly (Musca domestica L.) Larval lipid. *Fuel* 210, 463–471. https://doi.org/10.1016/j.fuel.2017.08.109.

Zurek, L., et al. (2000). Diversity and contribution of the intestinal bacterial community to the development of Musca domestica (Diptera: Muscidae) Larvae. *Journal of Medical Entomology* 37, 924–928. https://doi.org/10.1603/0022-2585-37.6.924.

Chapter 12

Advances in the production of crickets as food and animal feed

Chrysantus M. Tanga, International Centre of Insect Physiology and Ecology (icipe), Kenya; Margaret O. Kababu, International Centre of Insect Physiology and Ecology (icipe) and Jaramogi Oginga Odinga University of Science and Technology (JOOUST), Kenya; and Dennis Beesigamukama, International Centre of Insect Physiology and Ecology (icipe), Kenya

1 Introduction

The global human population is expected to grow to 9.7 billion by 2050 (Grafton et al., 2015; Kemsawasd et al., 2022). However, climate change threatens production of the quantities of food required to feed this growing population, leading to increased food insecurity (Kousar et al., 2021). It is estimated that approximately 500 million people, especially in Africa and Asia, are suffering from malnutrition (Kogan et al., 2019). Insects present a suitable alternative food source that can alleviate increasing food insecurity, given advantages such as nutrient profile, availability, low production cost and reduced vulnerability to climate change (van Huis et al., 2013).

Crickets belong to the family Gryllidae (Orthoptera) which contains more than 500 species worldwide (van Huis, 2020). Crickets are among the most commonly consumed edible insects either as human food or livestock feed

http://dx.doi.org/10.19103/AS.2024.0139.14

(van Huis, 2013; Magara et al., 2021). They are nutrient dense with high levels of proteins, fats, lipids, minerals and vitamins necessary for human nutrition and health (Magara et al., 2021). They also have a short life cycle compared with conventional livestock and are relatively easy to produce in large quantities (Gahukar, 2011). Given low production costs, cricket farming can form a source of livelihood for many households (Halloran et al., 2018). However, consumption of crickets is more widespread in Africa and Asia compared with western countries (Raheem et al., 2018). The cost of a kilogram of crickets varies widely between US$1.86 and US$40.0 (with premium pricing in many developed countries).

2 Edible cricket species

Although the consumption of crickets is an ancient practice that is widespread across Africa, Asia and Latin America, continuing research on their nutritional value and increasing food insecurity has led to the recognition of crickets as a food source in Australia, the Americas and Europe in recent years (Magara et al., 2021). This has been further enhanced by legislation in, for example, the European Union (EU) that recognizes edible crickets as a novel food source capable of reducing malnutrition and food insecurity (EFSA Scientific Committee, 2015; Magara et al., 2021).

There are approximately 62 edible cricket species utilized as human food or livestock feed across the globe (Magara et al., 2021). Of these, Asian, African and American consumers can choose between 41, 26 and 5 species of crickets, respectively, with 4 species consumed in both Europe and Australia. The regional distribution of cricket consumption is 25 countries in Africa, 13 in Asia, 5 in the Americas and 4 in Europe as well as Australia and New Zealand.

3 Crickets as food and feed

Crickets are reared for use as human food or feed for domesticated animals, pets, insectivorous amphibians and reptiles (van Huis, 2020). The most consumed species include *Acheta domesticus*, *Gryllodes sigillatus*, *Gryllus assimilis*, *Gryllus bimaculatus*, *Teleogryllus testaceus (Gryllus testaceus)* and *Scapsipedus icipe* (Verner et al., 2021).

3.1 Crickets as food

A. domesticus and *G. bimaculatus* are domesticated for use as food in different countries in Asia and Africa (Halloran et al., 2018). These two species are preferred for rearing for a number of reasons. *A. domesticus* can be reared on diverse diets, require low maintenance, and exhibit low disease incidence and high nutritional value, whereas *G. bimaculatus* mature quickly, are large in size and easy to sell (Halloran et al., 2017). In Thailand, there are approximately 20 000 farmers who produce more than 7000 tonnes of crickets annually (Hanboonsong et al., 2013). The most reared species in Thailand include *G. bimaculatus, Teleogryllus occipitalis, G. testaceus* and *A. domesticus*, while *G. testaceus* is also farmed as food in Indonesia (Fuah et al., 2015; van Huis, 2020).

Other cricket species used as food include *Brachytrupes portentosus, Gymnogryllus* sp., *Velarifictorus* sp. and *Modicogryllus confirmatus* in Thailand; *Brachytrupes membranaceus colosseus, Modicogryllus* sp. and *Fryerius* sp. in Madagascar; and *B. membranaceus* in Kenya (Fuah et al., 2015; Magara et al., 2021; van Itterbeeck et al., 2019). These species are, however, collected from the wild.

Crickets can be consumed whole or dried and processed into powder to be used as an ingredient to fortify products (Halloran et al., 2016). In Thailand, for example, cricket powder is exported to European countries, while whole crickets are consumed locally. Crickets have been used in the formulation of bread, cookies and porridge flour with high consumer acceptability in developing countries (Bartkiene et al., 2022). Different industries in western countries formulate various types of cricket-based foods in the form of dried cricket and cricket flour (Kemsawasd et al., 2022).

3.2 Crickets as feed

Different species of crickets are used as animal feed or feed ingredients across the globe (Table 1). Although crickets can be consumed directly as human food, their use as animal feed is highly profitable when they are reared using organic waste side streams and weeds (Armansyah Handayani, 2020; Miech et al., 2016). Crickets can be given to animals whole or crushed into powder and used as an ingredient in the production of animal feed. The inclusion of

Table 1 Cricket species used as animal feed

Cricket species	Animal feed	References
*A. domesticus**	Fish, poultry, pet food	Kovitvadhi et al. (2019); Lee et al. (2017); Miech et al. (2016); Ververis et al. (2022)
B. portentosus	Poultry	Abdul et al. (2012)
*G. assimilis**	Fish, pet food	Alfaro et al. (2019); Weissman et al. (2012)
G. bimaculatus	Fish, poultry, pet food	Kovitvadhi et al. (2019); Taufek et al. 2018); Weissman et al. (2012)
*G. sigillatus**	Pet food, fish	Jarett et al. (2019)
G. testaceus	Pigs, poultry	Miech et al. (2017); Wang et al. (2005)
G. veletis	Not specified	Magara et al. (2021)
T. emma	Not specified	Ghosh et al. (2017)

*Approved for the production of aqua feed in the EU (Commission Regulation (EU), 2017).

crickets in livestock feed provides a cheaper alternative feed source that can significantly reduce the cost of livestock production (van Huis et al., 2013). *A. domesticus*, *G. sigillatus* and *G. assimilis* are e.g. authorized for use in the production of processed animal protein for aquaculture feeds in the EU (Bertola & Mutinelli, 2021).

4 Production systems for crickets

As noted earlier, Thailand is among the leading nations involved in the production of crickets as food, with more than 20 000 farmers involved in the enterprise. However, cricket farming remains a small-scale activity here (Halloran et al., 2016). Halloran et al. (2018) describe the different types of cricket farming systems found in Thailand, Kenya, the Democratic Republic of Congo (DRC), Cambodia and the Lao People's Democratic Republic (Lao PDR). The concrete block pen system is the most commonly used system in Thailand. Pens can be rectangular or cylindrical, constructed with lightweight concrete blocks and a concrete floor, usually covered with nylon netting to prevent the escape of crickets and entry of predators. In Cambodia, medium-scale farmers use rectangular brick and cement pens that are covered with tiles to prevent crickets from escaping. The upper part of the pen is covered with a mosquito net while the base has a water tunnel to deter predators. Small-scale farmers on the other hand use 60 L containers to breed crickets.

Rectangular and small cylindrical concrete pens, rectangular boxes with wooden frames and walls made of plywood, gypsum board, plastic sheets or furniture backing board are used for rearing crickets in Lao PDR. Plastic boxes of different sizes and 100 L plastic buckets have been tested in DRC and Kenya, respectively. Concrete pens, pens built with mud, wash basins and rearing crate systems are also used in Kenya. The rearing systems are fitted with egg cartons to provide shelter to crickets. These cartons are usually elevated to prevent accumulation of food and faecal material.

Larger-scale, automated cricket production occurs in western countries. The Aspire Food Group in North America e.g. has established a 100 000 square feet, fully automated cricket production and processing plant that is expected to produce 10 000 tonnes of crickets per year (Liceaga, 2022). Other cricket farming companies such as Cricket Crackers, Cricket Foods and The Cricket Girl among others have been established in Europe and North America with annual production capacity ranging between 200 and 12 000 metric tonnes (Kemsawasd et al., 2022). There is increasing recognition that, in addition to food and feed, crickets can produce multiple high-value market products such as protein, oils, chitin/chitosan and frass fertilizer.

5 Cricket-rearing diets

Crickets have been reared on conventional livestock feed, weeds, industrial and agricultural by-products and food waste (Kuo & Fisher, 2022; Magara et al., 2019; Sorjonen et al., 2019). Cricket rearing commonly uses cereal grain feed, soybean and poultry feed (Jucker et al., 2022). The growth of cricket farming in Thailand has led to the formulation of around five different brands of specialist cricket feed (Halloran et al., 2016). However, these feeds are very similar in composition to chicken feed (Halloran et al., 2017). The feed is supplemented with different vegetable leaves such as pumpkin, cassava and amaranth, in addition to fruit peelings such as those of pineapple, banana, mango, pawpaw and avocado (Halloran et al., 2018).

In order to reduce feed costs, alternative dietary ingredients have been tested for the production of crickets with varied results. Agro-industrial by-products present a viable feed source for the sustainable production of farmed insects (Jucker et al., 2022; Kuo & Fisher, 2022). Human waste by-products are inexpensive, available in large quantities and have a good nutrient profile, and can lower the cost of insect production and minimize environmental impact (Jucker et al., 2022). Orthopterans such as crickets are well adapted to converting these by-products into accessible sources of nutrients (Van Peer et al., 2021).

Crickets reared on diets of pure food waste resulted in a low survival yield of 6%, while diets supplemented in part with food waste yielded 77% survival. Pure weed diets and commercial feed supplemented with weeds resulted in 18% and 85% survival of crickets, respectively. Diets comprising pure agricultural by-products and diets comprising conventional feed supplemented with agricultural by-products led to cricket survival yields of 34% and 60%, respectively (Kuo & Fisher, 2022). These variations arose as a result of the diverse nutrient composition of diet types which influenced nutrient accessibility. As such, appropriate diets for optimal cricket performance should supply high-quality nutrition to enhance fitness parameters and cricket survival rates.

Optimal cricket performance has been observed for diets containing 20–30% crude protein and 4–20% crude fats (Capparos Megido et al., 2016; Lundy & Parrella, 2015; Magara et al., 2019; Vaga et al., 2019). Significant amounts of oleic and linoleic fatty acids and a high carbohydrate content are also critical for improved cricket performance (Oonincx et al., 2015; Orinda et al., 2017; Sorjonen et al., 2019; Vaga et al., 2019). High amounts of indigestible fibre in the feed is negatively correlated with cricket growth and survival, thus highlighting the need for inclusion of considerable amounts of fibre in cricket diets (Lundy & Parrella, 2015; Magara et al., 2019; Orinda et al., 2017; Vaga et al., 2020).

6 Nutrient composition of edible cricket species

Crickets have a rich profile of nutrients such as proteins, carbohydrates, lipids, vitamins and minerals. Crickets are a rich source of proteins (50–72%), fats (10–46%) and energy (452–547 kcal/100 g) (Table 2). They also contain considerable amounts of fibre, ash and carbohydrates. The nutritional content of crickets is influenced by species, developmental stage, rearing substrate/diet, sex, habitat and processing methods (Musundire et al., 2014; van Huis et al., 2013).

Edible crickets exhibit a rich profile of fatty acids (Table 3). The most abundant saturated fatty acids recorded in edible cricket species include palmitic acid (23–31%) and stearic acid (7–17%); monounsaturated fatty acids include oleic acid (11–30%) and palmitoleic acid (0.3–11%), while linoleic acid (23–41%) was the most prevalent polyunsaturated fatty acid. The inclusion of crickets in diets can thus supply polyunsaturated acids that cannot be produced by mammals (Govorushko, 2019). Polyunsaturated acids have been shown to provide protective effects against cardiovascular diseases, cancer and diabetes (Govorushko, 2019; Yorek, 2018).

Edible cricket species also contain varying amounts of amino acids (Table 4). All of the essential amino acids have been reported in *A. domesticus* and *G. assimilis* (Bednářová et al., 2013; Magara et al., 2021). Leucine was the most abundant essential amino acid recorded in edible cricket species except for *G. assimilis*, which had higher levels of lysine. Some of the essential amino acids found in crickets are absent in commonly consumed plant protein sources, indicating that inclusion of crickets in a diet can help to mitigate amino acid deficiencies that cause malnutrition in developing countries (Ghosh et al., 2017; Zielińska et al., 2015). Crickets can also be used as a source of essential amino acids in the fortification of food products to alleviate malnutrition (Köhler et al., 2019). Glutamic acid was the most abundant non-essential amino acid occurring in *A. domesticus*, *G. bimaculatus*, *G. assimilis* and *G. testaceus* (Ghosh et al., 2017; Magara et al., 2021; Wang et al., 2004; Zielińska et al., 2015).

Crickets contain diverse quantities of macro- and microminerals essential for human health (Table 5). Significant quantities of phosphorus (702–900 mg/100 g) have been reported in *A. domesticus*, *G. bimaculatus* and *G. sigillatus* while high levels of sodium (395 mg/100 g) and potassium (871 mg/100 g) occur in *S. icipe* and *G. sigillatus*, respectively (Murugu et al., 2021; Ribeiro et al., 2019; Udomsil et al., 2019). All the edible cricket species contain zinc (5–24 mg/100 g) and iron (3–12 mg/100 g). The regular consumption of crickets can decrease the large number of cases of zinc and iron deficiencies that are common among children and women of child-bearing age in developing countries (Mwangi et al., 2018).

Diverse quantities of vitamins have been reported in four species of commonly consumed cricket species (Table 6). *A. domesticus* contains high

Table 2 Proximate composition of the most commonly consumed edible cricket species

	Composition of cricket species					
Component	*A. domesticus*	*G. assimilis*	*G. bimaculatus*	*G. sigillatus*	*S. icipe*	*G. testaceus*
Crude protein (%)	71.7 ± 0.5	57.2	58.2 ± 3.7	70.0 ± 1.7	50.2 ± 2.1	63.3
Crude fat (%)	10.4 ± 0.1	34.3	46.0 ± 2.9	18.2 ± 0.7	35.7 ± 2.6	10.3
Crude fibre (%)	4.6 ± 0.2	NR	8.4 ± 2.3	3.7 ± 0.5	5.7 ± 1.4	5.2
Moisture (%)	6.3 ± 0.0	NR	3.0 ± 0.0	NR	NR	5.0
Ash (%)	5.4 ± 0.3	4.1	5.4 ± 0.1	4.7 ± 0.4	5.3 ± 0.0	5.6
Carbohydrates (%)	1.6 ± 0.1	8.6	0.1 ± 0.0	0.1 ± 0.0	NR	NR
NFE	NR	1.2	10.6	NR	NR	NR
Energy (Kcal/100 g)	NR	546.8	529.2 ± 3.2	452.0 ± 4.3	512.7 ± 1.7	NR
References	Udomsil et al. (2019)	Araujo et al. (2018); Bednářová et al. (2013)	Ghosh et al. (2017); Murugu et al. (2021); Udomsil et al. (2019)	Zielińska et al. (2015)	Murugu et al. (2021)	Shah & Wanapat (2021); Wang et al. (2004)

NR: not reported; NFE: nitrogen-free extract.

Table 3 Fatty acid composition of the most commonly consumed edible cricket species

Fatty acids (%)	*A. domesticus*	*G. assimilis*	*G. bimaculatus*	*G. sigillatus*	*S. icipe*	*T. testaceus*
Enanthic (C7:0)	NR	NR	0.21 ± 0.0	NR	0.09 ± 0.0	NR
Caprylic (C8:0)	NR	NR		NR	0.06 ± 0.0	NR
Capric (C10:0)	0.00	0.03	0.02 ± 0.0	NR	0.04 ± 0.0	NR
Undecanoic (C11:0)	NR	NR	0.35 ± 0.1	NR	0.26 ± 0.0	NR
Lauric (C12:0)	0.10 ± 0.0	0.12	0.35 ± 0.0	0.1 ± 0.0	0.31 ± 0.0	0.54 ± 0.0
Tridecanoic (C13:0)	NR	0.02	0.42 ± 0.1	NR	0.22 ± 0.1	NR
Myristic (C14:0)	0.44 ± 0.0	1.28	3.7 ± 0.6	1.65 ± 0.1	5.43 ± 0.9	0.39 ± 0.0
Myristoleic (C14:1)	0.03 ± 0.0	0.06	0.09 ± 0.0	0.09 ± 0.0	NR	NR
Pentadecanoic (C15:0)	0.11 ± 0.0	0.37	NR	0.24 ± 0.0	0.08 ± 0.0	10.18 ± 0.2
Palmitic (C16:0)	22.65 ± 0.4	25.58	31.18 ± 0.7	23.5 ± 0.7	23.89 ± 0.6	
Palmitoleic (C16:1)	0.34 ± 0.0	1.92	9.22 ± 0.6	3.78 ± 0.2	11.14 ± 1.0	3.11 ± 0.1
Margaric (C17:0)	0.12 ± 0.0	0.57	1.44 ± 0.1	0.32 ± 0.0	1.44 ± 0.1	NR
Heptadecenoic (C17:1)	0.24 ± 0.0	0.19	NR	0.29 ± 0.0	NR	NR
Stearic (C18:0)	8.54 ± 0.0	14.07	14.71 ± 1.4	7.35 ± 0.3	17.26 ± 0.6	2.63 ± 0.1
Oleic acid (C18:1)	20.18 ± 0.0	25.03	10.94 ± 0.81	29.14 ± 1.5	11.4 ± 0.8	29.58 ± 0.2
Linoleic (C18:2)	41.39 ± 0.3	26.13	23.56 ± 0.3	29.78 ± 0.8	23.56 ± 0.3	37.82 ± 0.2
Linolenic (C18:3)	NR	1.60	0.32 ± 0.0	NR	0.21 ± 0.1	10.12 ± 0.1
α-linolenic (C18:3n3)	1.11 ± 0.0	NR	NR	2.13 ± 0.0	NR	NR
Nonadecanoic (C19:0)	NR	NR	0.83 ± 0.1	NR	0.73 ± 0.1	NR

Arachidic (C20:0)	NR	0.56	0.01 ± 0.0	0.40 ± 0.0	0.00	NR
Eicosenoic (C20:1)	NR	0.24	NR	1.03 ± 0.1	NR	NR
Eicosadienoic (C20:2)	0.00	0.44	NR	NR	NR	NR
C20:3	NR	0.01	0.79 ± 0.0	NR	0.86 ± 0.1	NR
C20:4	0.01 ± 0.0	0.21	0.02 ± 0.0	NR		NR
C20:5	NR	0.38	0.07 ± 0.0	NR	0.08 ± 0.0	NR
Heneicosylic (C21:0)	0.24 ± 0.0	0.03	0.32 ± 0.1	0.13 ± 0.0	0.5 ± 0.1	NR
Behenic (C22:0)	NR	0.57	0.01 ± 0.0	0.07 ± 0.0	0.37 ± 0.0	NR
C22:1	0.52 ± 0.0	0.05	NR	NR	NR	NR
C22:2	0.11 ± 0.0	0.03	NR	NR	NR	NR
C23:0	0.02 ± 0.0	0.22	NR	NR	NR	NR
Lignoceric (C24:0)	NR	NR	0.26 ± 0.0	NR	0.28 ± 0.0	NR
others	3.84	4.80	NR	NR	NR	NR
SFA	32.22	43.72	57.60	33.74 ± 0.8	50.96	13.74
MUFA	21.72	27.49	23.24	34.33 ± 0.1	24.33	32.69
PUFA	42.64	28.80	19.16	31.91 ± 0.1	24.71	47.94
	Paul et al. (2017)	(Mlcek et al. (2018)	Murugu et al. (2024)	Zielińska et al. (2015)	Murugu et al. (2024)	Magara et al. (2021); Wang et al. (2004)

NR: not reported.

Table 4 Amino acid composition of the most commonly consumed edible cricket species

Amino acid	*A. domesticus* (g/100 g)	*G. assimilis* (g/100 g)	*G. bimaculatus* (g/100 g)	*G. sigillatus* (mg/g)	*S. icipe* (μg/100 mg)	*G. testaceus* (g/100 g)
Essential amino acids						
Histidine	2.3 ± 0.1	1.3 ± 0.4	2.5 ± 0.1	17.2 ± 0.2	10.8 ± 0.9	1.9 ± 0.0
Isoleucine	4.5 ± 0.2	2.1 ± 0.7	2.2 ± 0.0	26.6 ± 0.5	43.4 ± 10.6	3.1 ± 0.0
Leucine	9.8 ± 0.4	5.0	4.0 ± 0.1	57.8 ± 1.1	66.2 ± 12.5	5.5 ± 0.1
Lysine	5.4 ± 0.0	7.9 ± 0.6	2.4 ± 0.0	38.4 ± 0.9	18.2 ± 3.9	4.8 ± 0.1
Methionine	1.4 ± 0.1	0.6 ± 0.2	0.3 ± 0.0	15.9 ± 0.8	NR	1.3 ± 0.1
Phenylalanine	3.0 ± 0.3	0.7 ± 0.2	1.8 ± 0.0	22.0 ± 0.2	NR	2.9 ± 0.1
Threonine	3.6 ± 0.0	3.6 ± 0.6	2.0 ± 0.0	36.8 ± 0.5	NR	2.8 ± 0.1
Tryptophan	0.6 ± 0.1	1.0 ± 0.2	NR	NR	NR	NR
Valine	1.1	4.6 ± 0.6	3.2 ± 0.0	47.0 ± 1.0	29.1 ± 7.8	4.4 ± 0.0
Non-essential amino acids						
Arginine	6.1 ± 0.0	8.6	3.6 ± 0.0	46.6 ± 0.6	13.6 ± 1.1	3.7 ± 0.1
Serine	1.0	0.6 ± 0.2	2.7 ± 0.0	40.4 ± 0.5	NR	3.7 ± 0.1
Proline	1.2	1.3 ± 0.7	2.0 ± 0.0	54.2 ± 0.8	19.2 ± 6.1	4.5 ± 0.1
Alanine	8.9 ± 0.1	4.0 ± 0.6	5.6 ± 0.0	58 ± 0.8	NR	5.6± 0.1
Cysteine	0.8 ± 0.0	0.7 ± 0.1	5.1	11.1 ± 0.2	25.4 ± 8.0	1.0 ± 0.0
Tyrosine	1.0	5.4 ± 0.7	2.7 ± 0.0	31.8 ± 0.4	23.8 ± 5.8	3.9 ± 0.0
Glycine	1.0	2.4	3.3 ± 0.0	40.7 ± 0.4	NR	3.6 ± 0.1
Glutamic acid	10.5 ± 0.1	3.6	6.4 ± 0.1	106.6 ± 1.4	9.6 ± 0.3	9.1 ± 0.3
Aspartic acid	7.8 ± 0.9	3.0	3.6 ± 0.0	72.8 ± 1.0	NR	3.7 ± 0.1
References	Magara et al. (2021)	Bednářová et al. (2013)	Ghosh et al. (2017)	Zielińska et al. (2015)	Murugu et al. (2021)	Magara et al. (2021); Wang et al. (2004)

NR: not reported

Table 5 Mineral content of the most commonly consumed edible cricket species

Minerals (mg/100 g)	*A. domesticus*	*G. assimilis*	*G. bimaculatus*	*G. sigillatus*	*S. icipe*
Calcium	149.8 ± 7.2	45.3	72.7 ± 1.7	117.3 ± 12.2	66.1 ± 1.6
Potassium	389.9 ± 1.4	NR	39.5 ± 0.9	870.9 ± 18.6	66.3 ± 0.3
Magnesium	136.6 ± 4.9	27.2	29.1 ± 0.6	42.7 ± 1.8	35.6 ± 2.0
Phosphorus	899.9 ± 36.2	NR	702.0 ± 6.4	782.1 ± 86.7	NR
Sodium	101.4 ± 7.8	NR	166.5 ± 0.0	298.1 ± 18.5	395.4 ± 41.6
Iron	8.8 ± 3.9	2.8	12.3 ± 5.4	4.7 ± 0.3	10.7 ± 0.6
Zinc	19.6 ± 0.8	5.2	23.7 ±1.3	16.8 ± 0.3	19.2 ± 0.6
Manganese	4.4 ± 0.1	1.4	108.5 ± 18.3	2.9 ± 0.4	95.7 ± 17.2
Copper	4.9 ± 0.4	0.7	8.3± 1.5	4.9 ± 0.5	7.9 ± 0.7
Selenium	NR	NR	NR	0.1 ± 0.0	NR
Cobalt	NR	NR	4.4 ± 1.6	NR	5.1 ± 2.6
References	Udomsil et al. (2019)	Araujo et al. (2018)	Murugu et al., (2021); Udomsil et al. (2019)	Ribeiro et al. (2019)	Murugu et al. (2021)

NR, not reported.

Table 6 Vitamin composition of commonly consumed edible crickets

Vitamins (mg/100 g)	*A. domesticus*	*G. assimilis*	*G. bimaculatus*	*S. icipe*
Vitamin A/retinol	<67.00	2.90 ± 0.1	0.03 ± 0.0	0.04 ± 0.0
Vitamin B1/thiamin	0.04	NR	0.42 ± 0.0	0.01 ± 0.0
Vitamin B2/riboflavin	3.41	0.23 ± 0.1	0.89 ± 0.5	0.54 ± 0.1
Vitamin B3/niacin	3.84	NR	1.10 ± 0.2	NR
Vitamin B5/pantothenic acid	2.30	NR	NR	NR
Vitamin B6/pyridoxine	0.23	NR	5.28 ± 0.9	1.6 ± 0.4
Vitamin B7/biotin	0.02	NR	NR	NR
Vitamin B9/folic acid	0.15	NR	0.51 ± 0.1	0.41 ± 0.1
Vitamin B12	0.01	10.00 ± 0.0	NR	NR
Vitamin C	3.00	1.01 ± 0.6	NR	NR
Vitamin D	<17.15	NR	NR	NR
Vitamin E (α & γ tocopherol)	1.32	30.00 ± 0.0	NR	NR
References	Magara et al. (2021)	Oibiokpa et al. (2017)	Murugu et al. (2021)	Murugu et al. (2021)

NR: not reported.

quantities (≥ 3 mg/100 g) of riboflavin, niacin and vitamin C, while higher quantities of tocopherol and pantothenic acid occur in *G. assimilis* and *G. bimaculatus*, respectively (Magara et al., 2021; Murugu et al., 2021; Oibiokpa et al., 2017). Although vitamins are required for metabolism, they cannot be produced by the human body and must therefore be supplied through the diet (Alamu et al., 2013; Kekeunou et al., 2020). A deficiency of vitamins and minerals can cause adverse health outcomes that account for almost a million deaths annually (Zhou et al., 2022). The consumption of crickets can therefore supplement vitamin sources to improve human nutrition and health outcomes.

7 Safety and disease issues in cricket production

Crickets have been screened for potential hazards both in raw and processed forms and have been found to contain possible microbial hazards. Pathogenic bacteria such as *Salmonella* and *Shigella* sp., as well as spore-forming bacteria such as *Bacillus*, *Paenibacillus* and *Psychrobacillus*, have been isolated from raw crickets, as well as fungi such as *Aspergillus flavus*, *Cladosporium sphaerospermum* and *Aspergillus sydowii* (Belluco et al., 2013; Fernandez-Cassi et al., 2019; Nyangena et al., 2020). Unprocessed *A. domesticus* was found to contain high levels of aerobic mesophilic spore-forming bacteria (2.6–4.4 cfu/g), *Enterobacteriaceae*, lactic acid bacteria, *Bacillus cereus*, *Staphylococcus aureus*, yeast and mould (Ververis et al., 2022). *G. assimilis* can contain *Coprococcus*, *Lactococcus* and *Candidatus azobacteroides* (Aleknavicius et al., 2022). Mycotoxin-producing fungi such as *Aspergillus*, *Candida*, *Kodamaea*, *Lichtheimia*, *Terapisipora*, *Trichospora* and *Trichoderma* have been isolated from reared crickets (Vandeweyer et al., 2018).

Crickets are affected by insect viruses that are fatal and can annihilate the commercial rearing of some cricket species. *A. domesticus* is infected by *A. domesticus* densovirus, *A. domesticus* mini ambidensovirus, ambisense densovirus, *A. domesticus* volvovirus and *A. domesticus* Iflavirus. Viral pathogens described in *G. bimaculatus* include *Densovirus*, *Iridovirus*, *Nudivirus* and *Cripavirus*. *G. bimaculatus* shows high susceptibility to *Iridovirus*. *A. domesticus* densovirus has been isolated from *G. assimilis*, while *Iridovirus*, *A. domesticus* densovirus and *A. domesticus* volvovirus have been detected in *G. sigillatus* (Bertola & Mutinelli, 2021; Jakob et al., 2002; Szelei et al., 2011). Nudivirus is also fatal to other cricket species such as *Teleogryllus commodus*, *Teleogryllus oceanicus* and *G. campestris* (van Huis, 2020); however, these viruses are unlikely to threaten human health (Kemsawasd et al., 2022).

Limited biological hazards occur in processed crickets, though fungi can pose a risk during the production process (SLU et al., 2018). High aerobic mesophilic counts, fungi and some bacterial pathogens, such as *Bacillus* and *Staphylococcus*, have been isolated from processed crickets (Ververis et al., 2022); however, the microbial load was shown to reduce following heat processing (Aleknavicius et al., 2022; Ververis et al., 2022). Some spore-forming bacteria can, however, re-occur after processing (SLU et al., 2018).

Some crickets contain cyanogenic glycosides derived from plants and the consumption of such crickets may cause acute poisoning that can cause damage to the central nervous system and retard growth (Islamiyat et al., 2016; Magara et al., 2021; Zagrobelny et al., 2009). Allergic reactions may occur in individuals sensitive to insect chitin following the consumption of crickets (EFSA Scientific Committee, 2015). In addition, the consumption of crickets without removing the legs may result in intestinal blockage, while the tough exoskeleton found in some crickets may be indigestible by humans (Magara et al., 2021).

8 Regulation relating to edible crickets

Legislation on the use of edible insects across the world remains limited; however, some countries have legislation that approves the breeding of certain cricket species (Mariod, 2020). The production of *A. domesticus*, *G. sigillatus* and *G. assimilis* for use in fish feed is approved in the EU (Commission Regulation (EU), 2017). In 2017 the Thailand National Bureau of Agricultural Commodity and Food Standards (within the Ministry of Agriculture and Cooperatives) launched the Thai agricultural standards on Good Agricultural Practices (GAP) for Cricket Farming (TAS 8202-2017), which provides guidelines for safe cricket rearing (ACFS, 2017). *A. domesticus* produced in closed farming systems are considered safe by the EFSA panel and can be used as ingredients in several food products and snacks (EFSA Panel on Nutrition et al., 2021). In addition, some African countries, such as Botswana, Kenya, Uganda, Tanzania and Malawi, have policies that identify insects as food and feed (Niassy et al., 2022).

9 Conclusion

The occurrence of various species of edible crickets offers a diversified nutrient-dense food and feed source that is easy to produce and can supplement existing protein sources to enhance food security. The existing and expanding demand for such a valuable resource has spurred small scale cricket production in Asia and Africa and large scale automated production in western countries; culminating into circulation of high value market products. However, global cricket production remains low and concerted efforts towards intensified production has been curtailed by inadequate policies and regulations, safety concerns and insect diseases that can wipe out commercial cricket colonies. Thus, there is need for optimization of cricket production to increase productivity, limit safety concerns and minimize disease occurrence. There is also need for development of national and international policy frameworks to enhance the production and trade of crickets as food and feed globally.

10 Where to look for further information

- Ayieko, M. A., Ogola, H. J., & Ayieko, I. A. (2016). Introducing rearing crickets (gryllids) at household levels: adoption, processing and nutritional values. *Journal of Insects as Food and Feed* 2(3), 203–211.
- Hassan, S. A., Altemimi, A. B., Hashmi, A. A., Shahzadi, S., Mujahid, W., Ali, A., Bhat, Z. F., Naz, S., Nawaz, A., Abdi,, G., & Aadil, R. M. (2024). Edible crickets, as a possible way to curb protein-energy malnutrition: Nutritional status, food applications, and safety concerns. *Food Chemistry* 101533.
- Oloo, J. A., Ayieko, M., & Nyongesah, J. M. (2020). Acheta domesticus (Cricket) feed resources among smallholder farmers in Lake Victoria region of Kenya. *Food science & nutrition* 8(1), 69–78.
- Rumpold, B. A., & Schlüter, O. K. (2013). Potential and challenges of insects as an innovative source for food and feed production. *Innovative Food Science & Emerging Technologies* 17, 1–11.
- Zafar, A., Shaheen, M., Tahir, A. B., da Silva, A. P. G., Manzoor, H. Y., & Zia, S. (2023). Unraveling the nutritional, biofunctional, and sustainable food application of edible crickets: A comprehensive review. *Trends in Food Science & Technology*, 104254.

11 References

Abdul, R. I., Yusof, H. and Engku, A. E. A. 2012. Nutritional evaluation of house cricket (Brachytrupes portentosus) meal for poultry. *7th Proceedings of the Seminar in Veterinary Sciences;27 February–March 2*, 14–18.

ACFS. 2017. Good Agricultural Practices for Cricket Farm. *National Bureau of Argricultural Commodity and Food Standards, Gazette: Bangkok, Thailand* 134(Special Section 293 D).

Alamu,T. O., Amao, O. A., Oke, A. O. and Lawal, O. I. 2013. Diversity and nutritional status of edible insects in Nigeria: a review. *International Journal of Biodiversity and Conservation* 5(4), 215–222. https://doi.org/10.5897/IJBC12.121

Aleknavicius, D., Luksa, J., Strazdaite-Zieliene, Z. and Serviene, E. 2022. The bacterial microbiota of edible insects acheta domesticus and gryllus assimilis revealed by high content analysis. *Foods* 11(1073). https://doi.org/10.3390/foods11081073

Alfaro, A. O., Núñez, W., Marcia, J. and Montero Fernández, I. 2019. The cricket (Gryllus assimilis) as an alternative food versus commercial concentrate for tilapia (Oreochromis sp.) in the nursery stage. *Journal of Agricultural Science* 11, 97–104. https://doi.org/10.5539/jas.v11n6p97.

Araujo, R. R. S., Benfica, dos S. T. A. R., Ferraz, V. P. and Santos, E. M. 2018. Nutritional composition of insects gryllus assimilis and zophobas morio: potential foods harvested in Brazil. *Journal of Food Composition and Analysis*. https://doi.org/10.1016/j.jfca.2018.11.005

Armansyah, V. and Handayani, M. T. 2020. Utilization potential in Cibanteng village for integration of agriculture-animal husbandry 'cricket cultivation'. *Jurnal Pusat Inovasi Masyarakat* 2, 108–116.

Bartkiene, E., Starkute, V., Katuskevicius, K., Laukyte, N., Fomkinas, M., Vysniauskas, E., Kasciukaityte, P., Radvilavicius, E., Rokaite, S., Medonas, D., Valantinaviciute, E., Mockus, E. and Zokaityte, E. 2022. The contribution of edible cricket flour to quality parameters and sensory characteristics of wheat bread. *Food Science and Nutrition, November 2021*, 4319–4330. https://doi.org/10.1002/fsn3.3024

Bednářová, M., Borkovcová, M., Mlček, J., Rop, O. and Zeman, L. 2013. Edible insects – species suitable for entomophagy under condition of Czech Republic. *Acta Universitatis Agriculturae Et Silviculturae Mendelianae Brunensis* LXI(3), 587–594. https://doi.org/http://dx.doi.org/10.11118/acraun201361030587

Belluco, S., Losasso, C., Maggioletti, M., Alonzi, C. C., Paoletti, M. G. and Ricci, A. 2013. Edible insects in a food safety and nutritional perspective: a critical review. *Comprehensive Reviews in Food Science and Food Safety* 12(3), 296–313. https://doi.org/10.1111/1541-4337.12014

Bertola, M. and Mutinelli, F. 2021. A systematic review on viruses in mass-reared edible insect species. *Viruses* 13(2280), 1–31.

Capparos Megido, R., Alabi, T., Nieus, C., Blecker, C., Danthine, S., Bogaert, J., Haubruge, E. and Francis, F. 2016. Optimisation of a cheap and residential small-scale production of edible crickets with local by-products as an alternative protein-rich human food source in Ratanakiri Province, Cambodia. *Journal of the Science of Food and Agriculture* 96(2), 627–632.

Commission Regulation (EU). 2017. 2017/893 of 24 May 2017 amending Annexes I and IV to Regulation (EC) No 999/2001 of the European Parliament and of the Council and Annexes X, XIV and XV to Commission Regulation (EU) No 142/2011 as regards the provisions on processed animal protein. *Official Journal of the European Union, L* 138, 92–115.

EFSA Panel on Nutrition, Food, A., Turck, D., Bohn, T., Castenmiller, J., De Henauw, S., Hirsch-Ernst, K. I., Maciuk, A., Mangelsdorf, I. and McArdle, H. J. 2021. Safety of frozen and dried formulations from whole house crickets (Acheta domesticus) as a Novel food pursuant to Regulation (EU) 2015/2283. *EFSA Journal* 19(e06779). https://doi.org/10.2903/j.efsa.2021.6779

EFSA Scientific Committee., Hardy, A., Benford, D., M, N. P. J., Halldorsson, I. T., Schlatter, J., Solecki, R. A., Jeger, M., Knutsen, H. K., Simon More, S., Mortensen, A., Naegeli, H., Ockleford, C., Ricci, A., Rychen, G., Silano, V. and Turck, D. 2015. Risk profile related to production and consumption of insects as food and feed. *EFSA Journal* 13(10), 4257. https://doi.org/10.2903/j.efsa.2015.4257

Fernandez-Cassi, X., Supeanu, A., Vaga, M., Jansson, A., Boqvist, S. and Vagsholm, I. 2019. The house cricket (Acheta domesticus) as a novel food: a risk profile. *Journal of Insects as Food and Feed* 5(2), 137–157. https://doi.org/10.3920/JIFF2018.0021

Fuah, A. M., Siregar, H. C. H. and Endrawati, C. 2015. Cricket farming for animal protein as profitable business for small farmers in Indonesia. *Journal of Agricultural Science and Technology, A* 5, 296–304. https://doi.org/10.17265/2161-6256/2015.04.008

Gahukar, R. T. 2011. Entomophagy and human food security. *International Journal of Tropical Insect Science* 31, 129–144.

Ghosh, S., Lee, S., Jung, C. and Meyer-Rochow, V. B. 2017. Nutritional composition of five commercial edible insects in South Korea. *Journal of Asia-Pacific Entomology* 20(2016), 686–694. https://doi.org/10.1016/j.aspen.2017.04.003

Govorushko, S. 2019. Global status of insects as food and feed source: a review. *Trends in Food Science and Technology* 91(July 2018), 436–445. https://doi.org/10.1016/j.tifs.2019.07.032

Grafton, R. Q., Daugbjerg, C. and Qureshi, M. E. 2015. Towards food security by 2050. *Food Security* 7, 179–183.

Halloran, A., Hanboonsong, Y., Roos, N. and Bruun, S. 2017. Life cycle assessment of cricket farming in north-eastern Thailand. *Journal of Cleaner Production* 156, 83–94. https://doi.org/10.1016/j.jclepro.2017.04.017

Halloran, A., Megido, R. C., Oloo, J., Weigel, T., Nsevolo, P. and Francis, F. 2018. Comparative aspects of cricket farming in Thailand, Cambodia, Lao People's democratic republic, democratic republic of the Congo and Kenya. *Journal of Insects as Food and Feed* 4(2), 101–114. https://doi.org/10.3920/JIFF2017.0016

Halloran, A., Roos, N., Flore, R. and Hanboonsong, Y. 2016. The development of the edible cricket industry in Thailand. *Journal of Insects as Food and Feed* 2(2), 91–100. https://doi.org/10.3920/JIFF2015.0091

Hanboonsong, Y., Jamjanya, T. and Durst, P. B. 2013. *Six-legged livestock: edible insect farming, collection and market in Thailand.* Food and Agriculture Organization of the United Nations Regional Office for Asia and Pacific. www.fao.org/contact-us/licencerequest

Islamiyat, F. B., Moruf, O. O., Sulaiman, A. O. and Adeladun, S. A. 2016. A review of cyanogenic glycosides in edible plants. In M. Larramendy & S. Soloneski (Eds.), *Toxicology-New Aspects to This Scientific Conundrum* (pp. 180–186). http://dx.doi.org/10.5772/62600

Jakob, N. J., Kleespies, R. G., Tidona, C. A., Müller, K., Gelderblom, H. R. and Darai, G. 2002. Comparative analysis of the genome and host range characteristics of two insect iridoviruses: chilo iridescent virus and a cricket iridovirus isolate. *Journal of General Virology* 83(2), 463–470.

Jarett, J. K., Carlson, A., Rossoni Serao, M., Strickland, J., Serfilippi, L. and Ganz, H. H. 2019. Diets with and without edible cricket support a similar level of diversity in the gut microbiome of dogs. *PeerJ* 7. https://doi.org/10.7717/peerj.7661

Jucker, C., Belluco, S., Oddon, S. B., Ricci, A., Bonizzi, L., Lupi, D., Savoldelli, S., Biasato, I., Caimi, C., Mascaretti, A. and Gasco, L. 2022. Impact of some local organic by-products on Acheta domesticus growth and meal production Abstract. *Journal of Insects as Food and Feed* 8(6), 631–640. https://doi.org/10.3920/JIFF2021.0121

Kekeunou, S., Simeu-Noutchom, A., Mbadjoun-Nziké, M., Achu-Loh, M. B., Akono-Ntonga, P., Wandji, A. C. and Tamesse, J. L. 2020. Nutritional composition of African edible acridians. In A. A. Mariod (Ed.), *African Edible Insects As Alternative Source of Food, Oil, Protein and Bioactive Components* (pp. 169–193). Springer Nature. https://doi.org/10.1007/978-3-030-32952-5_12/COVER

Kemsawasd, V., Inthatchat, W., Suttisansanee, U. and Temviriyanukul, P. 2022. Road to the red carpet of edible crickets through integration into the human food chain with biofunctions and sustainability: a review. *International Journal of Molecular Sciences* 23(1801). https://doi.org/https:// doi.org/10.3390/ijms23031801

Kogan, F., Guo, W. and Yang, W. 2019. Drought and food security prediction from NOAA new generation of operational satellites. *Geomatics, Natural Hazards and Risk* 10(1), 651–666.

Köhler, R., Kariuki, L., Lambert, C. and Biesalski, H. K. 2019. Protein, amino acid and mineral composition of some edible insects from Thailand. *Journal of Asia-Pacific Entomology* 22(1), 372–378. https://doi.org/10.1016/j.aspen.2019.02.002

Kousar, S., Ahmed, F., Pervaiz, A. and Bojnec, Š. 2021. Food insecuriy, population growth, urbanization and water availability: the role of government stability. *Sustainability* 13(12336). https://doi.org/10.3390/su132212336

Kovitvadhi, A., Chundang, P., Luapan, J., Amarapitak, P., Sriyaphai, P., Buahom, R., Cham-iam, T., Leelehapongsathon, K., Tirawattanawanich, C. and Thongprajukaew, K. 2019. Cricket powder as an alternative protein source for broilers based on in vitro di. *Journal of Insects as Food and Feed* 5, 185–191. https://doi.org/10.3920/JIFF2018.0043

Kuo, C. and Fisher, B. L. 2022. A literature review of the use of weeds and agricultural and food industry by-products to feed farmed crickets (Insecta; Orthoptera; Gryllidae). *Frontiers in Sustainable Food Systems* 5(January), 1–14. https://doi.org/10.3389/fsufs.2021.810421

Lee, S. W., Tey, H. C., Wendy, W. and Zahari, M. W. 2017. The effect of house cricket (Acheta domesticus) meal on growth performance of red hybrid tilapia (Oreochromis sp.). *International Journal of Aquatic Science* 8, 78–82.

Liceaga, A. M. 2022. Edible insects, a valuable protein source from ancient to modern times. *Advances in Food and Nutrition Research* 101(129). https://doi.org/10.1016/bs.afnr.2022.04.002

Lundy, M. E. and Parrella, M. P. 2015. Crickets are not a free lunch: protein capture from scalable organic side-streams via high-density populations of acheta domesticus. *Plos One* 10. https://doi.org/doi: 10.1371/journal.pone. 0118785

Magara, H. J. O., Niassy, S., Ayieko, M. A., Mukundamago, M., Egonyu, J. P., Tanga, C. M., Kimathi, E., Ongere, J., Fiaboe, K. K. M., Hugel, S., Orinda, M., Roos, N. and Ekesi, S. 2021. Edible crickets (orthoptera) around the world: distribution, nutritional value, and other benefits: a review. *Frontiers in Nutrition* 7(January), 1–23. https://doi.org/10.3389/fnut.2020.537915

Magara, H. J. O., Tanga, C. M., Ayieko, M. A., Hugel, S., Mohamed, S. A., Khamis, F. M., Salifu, D., Niassy, S., Sevgan, S., Fiaboe, K. K. M., Roos, N. and Ekesi, S. 2019. Performance of newly described native edible cricket scapsipedus icipe (Orthoptera: gryllidae) on various diets of relevance for farming. *Journal of Economic Entomology* 112(2), 653–664. https://doi.org/10.1093/jee/toy397

Mariod, A. A. 2020. The legislative status of edible insects in the world. In A. A. Mariod (Ed.), *African Edible Insects as Alternative Source of Food, Oil, Protein and Bioactive Components* (pp. 141–148). Springer International Publishing. https://doi.org/10.1007/978-3-030-32952-5_9

Miech, P., Berggren, Lindberg, J. E., Chhay, T., Khieu, B. and Jansson, A. 2016. Growth and survival of reared Cambodian field crickets (Teleogryllus testaceus) fed weeds, agricultural and food industry by-products. *Journal of Insects as Food and Feed* 2(4), 285–292. https://doi.org/10.3920/JIFF2016.0028

Miech, P., Lindberg, J. E., Berggren, Å., Chhay, T. and Jansson, A. 2017. Apparent faecal digestibility and nitrogen retention in piglets fed whole and peeled Cambodian field cricket meal. *Journal of Insects as Food and Feed* 3, 279–288. https://doi.org/10.3920/jiff2017.0019

Mlcek, J., Adámková, A., Adámek, M., Borkovcová, M. and Kourimska, L. 2018. Selected nutritional values of field cricket (Gryllus assimilis) and its possible use as a human food. *Indian Journal of Traditional Knowledge* 17(July), 518–524.

Murugu, D. K., Onyango, A. N., Ndiritu, A. K., Nyangena, D. N., Osuga, I. M., Cheseto, X., Subramanian, S., Ekesi, S., Tanga, C. M., Osuga, M., Cheseto, X., Subramanian, S., Ekesi, S. and Tanga, C. M. 2024. Physicochemical properties of edible cricket oils: implications for use in pharmaceutical and food industries. *Future Foods* 9(July 2023), 100316. https://doi.org/10.1016/j.fufo.2024.100316

Murugu, D. K., Onyango, A. N., Ndiritu, A. K., Osuga, I. M., Xavier, C., Nakimbugwe, D. and Tanga, C. M. 2021. From farm to fork: crickets as alternative source of protein, minerals, and vitamins. *Frontiers in Nutrition* 8(August). https://doi.org/10.3389/fnut.2021.704002

Musundire, R., Chidewe, C., Samende, B. K. and Manditsera, F. A. 2014. Nutrient and anti-nutrient composition of Henicus whellani (Orthoptera: stenopelmatidae), an edible ground cricket, in south-eastern Zimbabwe. *International Journal of Tropical Insect Science* 34, 223–31.

Mwangi, M. N., Oonincx, D. G. A. B., Stouten, T., Veenenbos, M., Melse-Boonstra, A., Dicke, M. and Van Loon, J. J. A. 2018. Insects as sources of iron and zinc in human nutrition. *Nutrition Research Reviews* 31(2), 248–255. https://doi.org/10.1017/S0954422418000094

Niassy, S., Omuse, E., Roos, N., Halloran, A., Eilenberg, J., Egonyu, J. P., Tanga, C. M., Meutchieye, F., Mwangi, R., Subramanian, S. V., Musundire, R., Nkunika, P. O. Y., Anankware, J. P., Kinyuru. J, N., Yusuf, A. and Ekesi, S. 2022. Safety, regulatory and environmental issues related to breeding and international trade of edible insects in Africa. *Scientific and Technical Review* 41(1). https://doi.org/10.20506/rst.41.1.3309

Nyangena, D. N., Mutungi, C., Imathiu, S., Kinyuru, J., Affognon, H., Ekesi, S., Nakimbugwe, D. and Fiaboe, K. K. M. 2020. Effects of Traditional Processing Techniques on the Nutritional and Microbiological Quality of Four East Africa. *Foods* 9, 574.

Oibiokpa, F. I., Akanya, H. O., Jigam, A. A. and Saidu, A. 2017. Nutrient and antinutrient compositions of some edible insect species in Northern Nigeria. *Fountain Journal of Natural and Applied Sciences* 6(1), 9–24.

Oonincx, D. G. A. B., van Broekhoven, S., van Huis, A. and van Loon, J. J. A. 2015. Feed conversion, survival and development, and composition of four insect species on diets composed of food by-products. *Plos One* 10(12), e0144601. https://doi.org/10.1371/journal.pone.0144601

Orinda, M. A., Mosi, R. O., Ayieko, M. A. and Amimo, F. A. 2017. Growth performance of Common house cricket (Acheta domesticus) and field cricket (Gryllus bimaculatus) fed on agro-byproducts. *Journal of Entomology and Zoology Studies* 5(6), 1664–1668.

Paul, A., Frederich, M., Megido, C. R., Malik, P., Uyttenbroeck, R., Francis, F., Blecker, C., Haubruge, E., Lognay, G. and Danthine, S. 2017. Insect fatty acids: a comparison of lipids from three Orthopterans and Tenebrio molitor L. larvae. *Asia Pacific Entomology* 20, 337–340. https://doi.org/10.1016/j.aspen.2017.02.001

Raheem, D., Carrascosa, C., Oluwole, O. B., Nieuwland, M., Saraiva, A., Millán, R. and Raposo, A. 2018. Traditional consumption of and rearing edible insects in Africa, Asia and Europe. *Critical Reviews in Food Science and Nutrition* 59(14), 2169–2188. https://doi.org/10.1080/10408398.2018.1440191

Ribeiro, J. C., Lima, C. R., Maia, M. R. G., Almeida, A. A., Fonseca, A. J. M., Cabrita, A. R. J. and Cunha, L. M. 2019. Impact of defatting freeze-dried edible crickets (Acheta domesticus and Gryllodes sigillatus) on the nutritive value, overall liking and sensory profile of cereal bars. *LWT - Food Science and Technology* 113(May), 108335. https://doi.org/10.1016/j.lwt.2019.108335

Shah, A. A. and Wanapat, M. 2021. Gryllus testaceus walker (crickets) farming management, chemical composition, nutritive profile, and their effect on animal digestibility. *Entomological Research* 51, 639–649. https://doi.org/10.1111/1748-5967.12557

SLU, Swedish University of Agricultural Sciences; Department of Biomedical Sciences and Veterinary, Fernandez-Cassi, X., Supeanu, A., Jansson, A., Boqvist, S. and Vagsholm, I. 2018. Novel foods: a risk profile for the house cricket (Acheta domesticus). *EFSA Journal* 16, e16082.

Sorjonen, J. M., Valtonen, A., Hirvisalo, E., Karhapää, M., Lehtovaara, V. J., Lindgren, J., Marnila, P., Mooney, P., Mäki, M., Siljander-Rasi, H., Tapio, M., Tuiskula-Haavisto, M. and Roininen, H. 2019. The plant-based by-product diets for the mass-rearing of Acheta domesticus and Gryllus bimaculatus. *Plos One* 14(6). https://doi.org/10.1371/journal.pone.0218830

Szelei, J., Woodring, J., Goettel, M. S., Duke, G., Jousset, F. X., Liu, K. Y., Zadori, Z., Li, Y., Styer, E., Boucias, D. G. and Kleespies, R. G. 2011. Susceptibility of North-American and European crickets to Acheta domesticus densovirus (AdDNV) and associated epizootics. *Journal of Invertebrate Pathology* 106(3), 394–399.

Taufek, N. M., Simarani, K., Muin, H., Aspani, F., Raji, A. A., Alias, Z. and Razak, S. A. 2018. Inclusion of cricket (Gryllus bimaculatus) meal in African catfish (Clarias gariepinus) feed influences disease resistance. *Journal of Fisheries*. https://dx.doi.org/10.17017/jfish. v6i2.2018.264

Udomsil, N., Imsoonthornruksa, S., Gosalawit, C. and Ketudat-Cairns, M. 2019. Nutritional values and functional properties of house cricket (acheta domesticus) and field cricket (gryllus bimaculatus). *Food Science and Technology Research* 25(4), 597–605. https://doi.org/10.3136/fstr.25.597

Vaga, M., Berggren, Å. and Jansson, A. 2020. Growth, survival and development of house crickets (Acheta domesticus) fed flowering plants. *Journal of Insects as Food and Feed* 7, 151–161. https://doi.org/10.3920/JIFF2020.0048

Vaga, M., Berggren, Å., Pauly, T. and Jansson, A. 2019. Effect of red clover-only diets on house crickets (Acheta domesticus) growth and survival. *Journal of Insects as Food and Feed* 6, 179–189. https://doi.org/doi: 10.3920/JIFF20 19.0038

van Huis, A. 2013. Potential of insects as food and feed in assuring food security. In *Annual Review of Entomology* (Vol. 58, Issue 1, pp. 563–583). https://doi.org/10.1146/annurev-ento-120811-153704

van Huis, A. 2020. Edible crickets, but which species? *Journal of Insects as Food and Feed* 6(2), 91–94. https://doi.org/10.3920/JIFF2020.x001

van Huis, A., van Itterbeeck, J., Klunder, H., Mertens, E., Halloran, A., Muir, G. and Vantomme, P. 2013. *Edible Insects: Future Prospects for Food and Feed Security*. Food and Agriculture Organization of the United Nations, Rome.

van Itterbeeck, J., Rakotomalala Andrianavalona, I. N., Rajemison, F. I., Rakotondrasoa, J. F., Ralantoarinaivo, V. R., Hugel, S. and Fisher, B. L. 2019. Diversity and use of edible grasshoppers, locusts, crickets, and katydids (Orthoptera) in Madagascar. *Foods* 8(12), 1–19. https://doi.org/10.3390/foods8120666

Van Peer, M., Frooninckx, L., Coudron, C., Berrens, S., Carlos, Á., Deruytter, D., Verheyen, G. and Van Miert, S. 2021. Valorisation potential of using organic side streams as feed for tenebrio molitor, Acheta domesticus and Locusta migratoria. *Insects* 12(9), 796.

Vandeweyer, D., Wynants, E., Crauwels, S., Verreth, C., Viaene, N., Claes, J., Lievens, B. and Van Campenhout, L. 2018. Microbial dynamics during industrial rearing, processing and storage of the tropical house cricket (Gryllodes sigillatus) for human consumption. *Applied and Environmental Microbiology*. https://doi.org/10.1128/AEM.00255-18

Verner, D., Roos, N., Halloran, A., Surabian, G., Tebaldi, E., Ashwill, M., Vellani, S. and Konishi, Y. 2021. Insect and hydroponic farming in africa:the new circular food economy. *Agriculture and Food Series*. https://doi.org/10.1596/978-1-4648-1766-3

Ververis, E., Boue, G., Poulsen, M., Niforou, A., Theresa, S., Tesson, V., Federighi, M. and Naska, A. 2022. A systematic review of the nutrient composition, microbiological and toxicological profile of Acheta domesticus (house cricket). *Food Composition and Analysis* 114(June). https://doi.org/10.1016/j.jfca.2022.104859

Wang, D., Bai, Y., Li, J. and Zhang, C. 2004. Nutritional value of the field cricket (Gryllus testaceus walker). *Entomologia Sinica* 11(December), 275–283.

Wang, D., Zhai, S. W., Zhang, C. X., Bai, Y. Y., An, S. H. and Xu, Y. N. 2005. Evaluation on nutritional value of field crickets as a Poultry Feedstuff. *Asian-Australasian Journal of Animal Sciences* 18(5), 667–670. https://doi.org/10.5713/ajas.2005.667

Weissman, D. B., Gray, D. A., Pham, H. T. H. I. and Tijssen, P. 2012. Billions and billions sold: pet-feeder crickets (Orthoptera: gryllidae), commercial cricket farms, an epizootic densovirus, and government regulations make for a potential disaster. *Zootaxa* 3504, 67–88.

Yorek, M. A. 2018. The potential role of fatty acids in treating diabetic neuropathy. *Current Diabetes Reports* 18(10), 1–10.

Zagrobelny, M., Dreon, A. L., Gomiero, T., Marcazzan, G. L., Glaring, M. A., MøLler, B. L. and Paoletti, M. G. 2009. Toxic moths: source of a truly safe delicacy. *Journal of Ethnobiology* 29(1), 64–76.

Zhou, Y., Wang, D., Zhou, S., Duan, H., Guo, J. and Yan, W. 2022. Nutritional composition, health benefits, and application value of edible insects: a review. *Foods* 11(24), 3961. https://doi.org/10.3390/foods11243961

Zielińska, E., Baraniak, B., Karaś, M., Rybczyńska, K. and Jakubczyk, A. 2015. Selected species of edible insects as a source of nutrient composition. *Food Research International* 77, 460–466. https://doi.org/10.1016/j.foodres.2015.09.008

Part 3

General issues in use of insect protein

Chapter 13

The chemical and microbial safety of insect-based feed and food

Christelle Planche and Erwan Engel, INRAE - UR370 Animal Products Quality Unit, France; and Christina Nielsen-LeRoux, Université Paris-Saclay – INRAE-UMR1319 Micalis-AgroParisTech, France

1 Introduction

All human and animal food and feed must meet minimum safety standards, both to avoid the direct risks of chemical and microbial hazards to health as well as indirect effects in reducing the overall quality of food. The development of 'novel' food/feed sources (such as insects) has led to intensive recent research to identify, prevent and manage potential hazards relating to chemicals, pathogens and allergens.

A number of insect species have been part of the human diet for many centuries in different regions of the world. These were typically wild caught (rather than farmed) which meant that the safety, flavour and nutritional quality of these species largely depended on factors such as what plant/foodstuff the insect had consumed as a substrate as well as the stage of the life cycle (larvae, pupae, adult). With potential hazards varying from one species to another, over time different communitees learned which local species were safe to eat and what food preparation techniques were required to make them safe and palatable (Olivadese and Dindo, 2023).

http://dx.doi.org/10.19103/AS.2024.0139.16

The recent interest in mass production of insects as an alternative, more sustainable protein source for human food and animal feed has focussed on a small number of species suitable for commercial production. Currently, in the European Union, eight species of insects are authorized for use in animal feed black soldier fly (*Hermetia illucens)*, house fly (*Musca domestica)*, yellow mealworm (*Tenebrio molitor*), buffalo worm (*Alphitobius diaperinus*), silkworm (*Bombyx mori*), house cricket (*Acheta domesticus*) and two other cricket species (*Gryllodes sigillatus, Gryllus assimilis*) while only four species are authorized as novel foods for human consumption (*T. molitor*, migratory locust (*Locusta migratoria)* in addtion to *A. domesticus* and *A. diaperinus*).

This chapter focuses on chemical and pathogen hazards from a few insect species, in particular those mass-produced in Europe such as the yellow mealworm (*T. molitor*) and house cricket (*A. domesticus*). We will also consider the black soldier fly larvae (BSFL) (*H. illucens),* which is only allowed for use as animal feed in Europe, and which raises particular issues regarding chemical and pathogen hazards as it can be reared on a large variety of waste streams, making this species particularly relevant when looking at insects from circular economy and nutritional food chain perspectives (Hosseindoust et al., 2024). Our focus will be on hazards associated with rearing, processing and storage of the insect and the derived products. The chapter will deal with chemicals and pathogens known to be of risk for human consumption, which are subject to regulation and for which acceptable threshold levels are known. The adverse effects of these hazards on insect rearing and health will also be mentioned. Finally, the chapter discusses techniques related to the detection and monitoring of chemicals and pathogens.

2 Hazards across the insect food/feed production chain

Throughout the insect chain, there are a wide range of chemical, pathogenic and allergenic hazards that can result in acute or long-term effects on consumer health. Figure 1 provides an overview of sources, transfer and potential adverse effects of various chemical and microbial hazards throughout the insect food/feed production chain. Table 1 summarizes a range of hazards linked to different potential substrates used in commercial insect rearing (the role of substrates is discussed later in this section).

Insect-derived products, like other food and feed, can be contaminated by a wide range of chemicals. They include

- mycotoxins, which are metabolites produced by moulds;
- heavy metals and persistent organic pollutants (POPs), which are environmental chemical hazards;
- residues of phytosanitary treatments and veterinary drugs, which may persist in substrates;

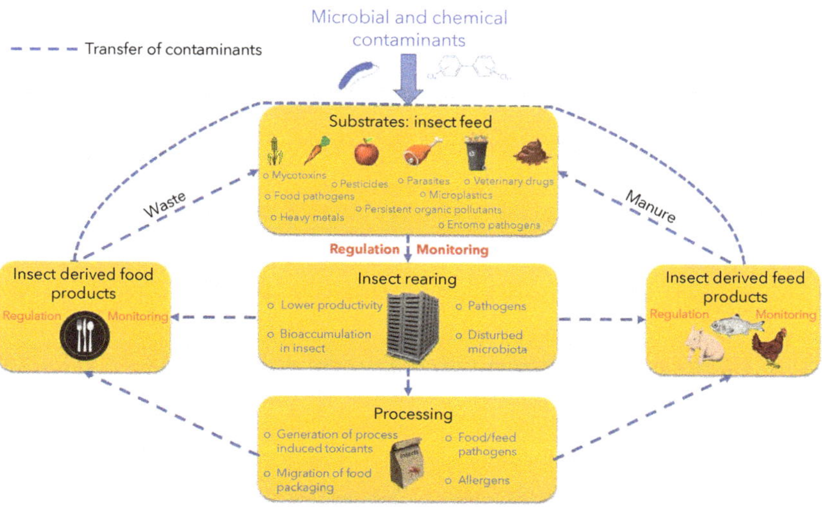

Figure 1 Overview of the transfer and potential adverse effects of various chemical and microbial hazards throughout the insect food/feed production chain.

- process-induced toxicants which are generated during the processing of foods; and
- migration of particles from contact materials like packaging.

It is generally agreed that diet, especially protein-based foods, plays a major role in exposing the consumer to various toxic chemical hazards present in food products (Engel et al., 2015). Chemical hazards can be found at trace levels in food and feed either individually or in combination. Chronic exposure to combined chemicals even at low doses has been associated with increased risk of health problems such as cancer, disruption of the endocrine system, developmental problems, immune and neurodevelopmental disorders (van der Burg et al., 2022). Depending on their physico-chemical properties, these chemical hazards can be transmitted or even bioaccumulated throughout the food chain and may react differently to transformation processes such as cooking (Planche et al., 2017). To ensure consumer safety, the European Commission sets maximum levels corresponding to the highest levels of chemical contaminant residue that are legally tolerated in food or feed.

Looking at biological hazards, food-associated pathogens like bacteria, viruses and parasites associated with intestinal/digestive tract diseases are of concern for human health. In younger, elder and immune compromised individuals, even low amounts of a biological hazard can result in systemic infection. In Europe, the production of insect-based products for feed and food follows the same rules for microbial safety and hygiene as any

Table 1 Key biological and chemical hazards in insect rearing based on substrate used

Substrate	Most important biological hazards	Most important chemical hazards
Plant products	• Bacteria: *Clostridia, Bacillus* • Fungus: *Aspergillus*	• Pesticides • Mycotoxins: *Aflatoxins...* • Heavy metals: *Lead, Cadmium, Arsenic, Mercury...*
Animal products	• Virus • Bacteria: *Salmonella, Listeria, Staphylococcus, Clostridia, Klebsiella, Cambylobacter* • Parasites • Antibiotics	• Persistent organic pollutants: *Polychlorobiphenyls, Dioxins...*
Household, canteen or supermarket waste	• Fungus • Bacteria: *Bacillus, Clostridia, Staphylococcus*	• Microplastics • Packaging migrants • Process-induced toxicants: *Polycyclic aromatic hydrocarbons, Acrylamide...*
Manure	• Bacteria • Virus • Antibiotics	• Veterinary drugs

other food/feed production system (e.g. in the use of good manufacturing practices/good hygiene practices (GMP/GHP) and hazard analysis and critical control point (HACCP) systems to ensure microbiological safety (regulations relating to issues such as biosecurity, hygiene and GMP include EC 178/2002, EC 183/2005 and EC 106/2009). Many microbial hazards can be controlled by processing techniques such as heating, exposure to UV light or gamma irradiation, or, in some cases, by freezing and drying. These may act as critical control points (CCPs) in eliminating a hazard if properly implemented. However, with novel food/feed sources and production systems, like insects, it is important both to identify and assess the effectiveness of potential CCPs as well as to identify and understand both well-known and less-studied pathogens that potentially occur during insect rearing and processing (Kooh et al., 2020). Good practices to avoid pathogens in insect-rearing facilities are described in the guide proposed by International Platform of Insects for Food and Feed (IPIFF) (IPIFF, 2022), which can help identify CCPs in each rearing system.

Although this chapter focuses on chemical and microbiological hazards, the potential allergenicity of insect-derived food products is also an important issue to consider. Food allergic reactions are defined as 'adverse reactions to an otherwise harmless food or food component that involves an abnormal response of the body's immune system to specific protein(s) in foods' (FAO and WHO, 2001). Various insect allergens have been identified including tropomyosin and arginine kinase, which are both pan-allergens known for their cross-reactivity with homologous proteins in crustaceans and house dust mite, indicating that insect allergens in insect-food are on about the same risk level as seafood (De Gier and Verhoeckx, 2018). However, the impact of processing is crucial for risk assessment of allergenic potential, since processing may affect the solubility and the immunoreactivity of insect allergens, depending on species and type of proteins as well as the process used (De Marchi et al., 2021).

The substrate on which insects are reared provides the main source of exposure to chemical and microbial hazards, depending on the substrate type (van der Fels-Klerx et al., 2018) (Table 1). In the European Union, only substrates of plant origin may currently be used to feed insects, with a few exceptions such as processed egg products or milk-derived products. The use of waste streams from supermarkets or school canteens or manure is prohibited as a precaution given the greater likelihood of high initial microbial and chemical contamination (see e.g. EC 767/2009 and EC 142/2011). However, such waste streams are allowed for the rearing of black soldier fly larvae in other regions of the world, such as Thailand and Vietnam. The absence of antinutritional or toxic compounds and pathogenic organisms, which may inhibit larval growth, are also important issues in maintaining production.

3 Chemical hazards in insect production

Maximum residue limits (MRLs) for contaminants of concern in animal feed are regulated by the Directive 2002/32/EC of the European Parliament and of the Council of 7 May 2002 (European Commission, 2002). Table 2 summarizes relevant EU legislation/regulation governing maximum levels of contaminants such as heavy metals, mycotoxins, dioxins and polychlorobiphenyls (PCBs).

Insects produced in Europe are likely to be exposed, *via* plant-based feed substrates permitted by EU regulation, to contaminants such as pesticide residues, mycotoxins and heavy metals (Poma et al., 2017). Outside Europe, regulations vary on permitted substrates for feeding insects. In South America and South Asia e.g. despite the fact that insects are regarded as traditional foods, there is little or no regulation on the production of edible insects (Baiano, 2020). A wide variety of different substrates can therefore be used, creating different types and levels of contamination risk. Insects raised on animal products, for example, could be more exposed to POPs such as PCBs, while insects raised on manure could be more exposed to residues of veterinary substances. The presence of plastic residues such as microplastics (plastic particles <5 mm) in waste streams used as substrates (household or supermarket waste, for example) should also not be overlooked. These residues could act as carriers of various harmful adsorbed chemical hazards including heavy metals or POPs (Godoy et al., 2019; Golwala et al., 2021; Zhang et al., 2022). Several recent studies have e.g. identified the potential health risks of microplastics entering the food chain (Fournier et al., 2023; Osman et al., 2023; Wang et al., 2021).

In addition to substrates, rearing, processing and storage conditions can also have an impact on the level of contamination of insect products. Regarding insect rearing, an example is the use of veterinary drugs which may be sprayed on or mixed with the substrate to prevent or fight bacterial infections (Kolakowski et al., 2021; van der Fels-Klerx et al., 2018). Concerning processing,

Table 2 European legislation/regulation relating to insects as novel foods

Dried *T. molitor* larvae	Commission Implementing Regulation (EU)2021/882 of 1 June 2021
Frozen, dried and powdered forms of *T.o molitor* larvae	Commission Implementing Regulation (EU)2022/169 of 8 February 2022
Frozen, dried and powder forms of *L. migratoria*	Commission Implementing Regulation (EU)2021/1975 of 12 November 2021
Frozen, dried and powdered forms of *A. domesticus*	Commission Implementing Regulation (EU)2022/188 of 10 February 2022
Partially defatted powder of *A. domesticus*	Commission Implementing Regulation (EU)2023/5 of 3 January 2023
Frozen, paste, dried and powdered forms of *A. diaperinus* larvae	Commission Implementing Regulation (EU) 2023/58 of 5 January 2023

although there is currently limited data available, many hazardous chemicals are relatively stable and may be difficult to remove or reduce by processing steps such as cooking (Meyer et al., 2021). However, some processing methods involving heat treatment can lead to additional process-induced toxicants such as polycyclic aromatic hydrocarbons (Belluco et al., 2018). The use of a grinding step can also lead to contamination by trace metallic elements (Cappelli et al., 2020). Other processing steps such as industrial seasoning could influence the levels of some pollutants (organophosphorus flame retardants and plasticizers, for example) in insect-based products (Poma et al., 2021, 2019). The migration of organophosphorus flame retardants and plasticizers from insect packaging can also explain the presence of these chemicals in edible insects (Poma et al., 2018, 2017). Risks from the storage of edible insects must also be considered since, in contrast to other foods of animal origin, edible insects may be contaminated by mycotoxins if they are stored in humid conditions promoting the growth of mycotoxin-producing fungi (Belluco et al., 2018). More data are currently needed on these different risks to define good practices aimed at risk reduction.

Finally, it is important to note that insects can synthesize natural toxins such as quinones and alkanes produced by darkling beetles (Schlüter et al., 2017; van der Fels-Klerx et al., 2018). Since research on the chemical safety of edible insects is relatively recent compared with other traditional food sectors, the presence of unknown or unsuspected intrinsic chemical hazards needs to be investigated in future research.

Data in the literature show great variability in levels of contaminants found in edible insects. A study carried out by Poma et al. (2017) e.g. assessed the presence of chemical hazards in four species of edible insects (greater wax moth – *Galleria mellonella*, migratory locust – *L. migratoria*, mealworm beetle – *T. molitor*, and buffalo worm – *A. diaperinus*) and four insect-based food products commercialized in Belgium. They found that some organochlorine pesticides were detected in almost all the analyzed samples, probably accumulated by insects from the substrate used for feeding. Levels were below regulatory MRLs and lower than reported levels for fish, seafood, meat and eggs. The authors observed significant variations in levels of PCBs, ranging from 26.5 to 2065 pg/g wet weight. This could be explained by the fact that the analyzed samples came from different suppliers, suggesting that insects were probably fed with different substrates, exposed to different rearing conditions and processed differently. Even greater variability was observed in a further study by Poma et al. (2018) who compared samples of edible insects from five European and three Asian countries. Even though the overall levels of chemical hazards were generally low and comparable with other commonly consumed animal products, a higher presence of organochlorine compounds, halogenated flame retardants, legacy plasticizers and emerging phosphorous flame retardants was found in Asian compared to European samples.

Significant differences in contamination can also be observed between insect species. For example, Poma et al. (2019) found a positive correlation between some insect orders (Orthoptera, Coleoptera, Lepidoptera, Hemiptera, Odonata and Hymenoptera) and the presence of some chemical hazards such as PCBs. Another study focused on edible insects available to Canadian consumers revealed that, among the samples analyzed, none of the silkworm pupae–based products and all cricket-based samples contained detectable levels of cadmium and mercury (Kolakowski et al., 2021). These results can be explained at least in part by the different substrates, physiology and life cycle of the insect species. The fat content of the insects is also a key factor: insects with higher fat content and a longer lifespan will tend to accumulate more lipophilic chemical hazards such as POPs.

The level of chemical contamination can also vary throughout an insect's life cycle. For example, unlike the larval stage, pupal/nymph stages of the yellow mealworm and black soldier fly do not involve eating, the gut contents are therefore reduced and the levels of chemical hazards are potentially less. It also appears that during the metamorphosis of yellow mealworm and black soldier fly into the pupal life stage, some chemical hazards such as heavy metals accumulated in the exoskeleton could be expelled with the exuviae (Lievens et al., 2021).

The potential bioaccumulation of chemical hazards in insects during rearing is a serious concern, especially if the insects are subsequently used as feed or food (Surendra et al., 2020). To assess the bioaccumulation of a contaminant, bioaccumulation factors (BAFs) can be calculated. BAF = concentration of contaminant in insect / concentration of contaminant in the substrate, with BAF greater than 1 indicating bioaccumulation from the substrate into the insect. Large differences can be observed in terms of BAF between different types of chemical hazards and different insect species, depending on physiology, composition and life cycle. The highest bioaccumulation described in the literature concerns heavy metals with BAF values reaching 34 for lead in yellow mealworms (Truzzi et al., 2019). However, it is worth noticing that there are major discrepancies between studies. For example, BAFs from 2.5 (Schmidt et al., 2019) to 20.4 (van der Fels-Klerx et al., 2020) were reported for cadmium in BSFL. Moreover, fasting the insects before harvesting seems to be an effective lever for reducing their heavy metal content. Cardoso et al. (2023) observed that BSFL can eliminate half of the accumulated mercury in 1 day (after 5 days of exposure) and yellow mealworms in 4–5 days (after 21 days of exposure).

Unlike heavy metals, other types of chemical hazards such as residues of phytosanitary treatments (Meijer et al., 2021; Purschke et al., 2017) or mycotoxins (Bosch et al., 2017; Meijer et al., 2022) do not appear to be bioaccumulated by insects during rearing. In the case of lipophilic compounds such as PCBs, Ratel et al. (2023) have observed that drying can lead to an increase in the concentration of PCBs in the lipid fraction of dried larvae by a factor of almost 3. However, it may be possible to use the lipid fraction (where PCBs are accumulated) in non-food/feed applications.

A related issue is the potential biomagnification of chemical hazards i.e. the progressive accumulation of residues at different stages of the food supply chain. Biancarosa et al. (2019) have shown that, when Atlantic salmon (*Salmo salar*) were fed diets containing more insect meal (BSFL) and less fish meal, this led to a marked decrease in the total arsenic content of fillets because arsenic in BSFL was in a form that is not as readily bio-accumulated in the fillets. Another study focused on laying hens receiving diets including BSFL found that transfer of heavy metals was low, resulting in levels in meat and eggs below maximum levels permitted in Europe, except for the kidneys (Heuel et al., 2023). However, studies on biomagnification are limited and more studies are needed.

4 Pathogenic hazards in insect production

Microbiological safety of insects as food and feed is a key parameter in order to produce high quality and safe insects and derived products. At the end of rearing, the larvae, nymph/pupae or adult stages are likely to contain various microorganisms in the insect digestive tract derived from consuming substrate (FAO, 2021), including pathogenic bacteria, yeasts and moulds, viruses and parasites (Kooh et al., 2019).

Pathogenic risks are typically assessed using microbiological risk assessment which involves (i) hazard identification, (ii) hazard characterization, (iii) exposure assessment, and (iv) risk characterization (FAO-WHO, 1999). Risk assessment then determines acceptable limits (European Commission, 2005). There is a range of potential pathogenic risks in insect production related to bacteria found in the natural environment and in the food chain, whether from sources such as soil or present in the digestive tract of livestock or humans, from which it can find its way into the substrates. These have been discussed in a number of reviews of the literature on bacteria associated with edible insects collected from the wild or reared for feed/food, but the presence and the quantity varies a lot among the studies (Garofalo et al., (2019), Vandeweyer et al. (2021), Ververis et al. (2022) and Toalá et al. (2022)). Examples of food-borne bacterial pathogens to be considered include:

- Enterobacteria, *Salmonella enteritica*, *Campylobacter jejuni* or *Escherichia coli*, which are commonly found in the intestinal microbiota of livestock such as chickens and are known to persist in soil.

- *Klebsiella*, an opportunistic pathogen present on skin and in intestines and faeces of mammals, as well as in soil, which can cause severe illnesses in immunocompromised patients, the elderly and infants (Osimani et al., 2018).
- *Pseudomonas aeruginosa* which can pose a serious health risk of pulmonary infection.
- *Listeria monocytognes*, traditionally known to be associated with farmed products such as milk, and is particularly dangerous e.g. for pregnant women (Koopmans et al., 2023).
- *Staphylococcus aureus,* part of the skin microbiota of most people and livestock such as cows, and therefore likely to be present in the food chain (Abril et al., 2020).
- Sporulating bacteria like *Clostridium* sp. and *Bacillus*, in particular *Bacillus cereus* (Dietrich et al., 2021), which are of concern as spores are resistant to treatments such as heating, freezing or drying (Walia et al., 2018).

A recent overview of different microbial hazards in the food chain can be found in Pakdel et al. (2023). To determine their contamination level, samples are typically ground into a powder to make a homogeneous suspension, which can then be diluted and suspended in a specified bacterial/fungal growth medium (Fasolato et al., 2018). The number of resulting bacteria or fungal units is referred to colony forming units (CFU) often measured for convenience in (log 10) logarithmic units/gram (e.g. 10^5 is 100 000 CFU).

An example of potential levels of contamination is a study by Ververis et al. (2022) of fresh and processed samples of house cricket (*A. domesticus*). The authors found viable counts of aerobic mesophilic bacteria in fresh or frozen samples from 10^7 CFU/gram (7 log10 colony forming units/gram) to a maximum 10^{10} CFU/g, while in dried samples they found from $10^{0.8}$ to a maximum $10^{8.8}$ CFU/g. The number of spore-forming bacteria in fresh samples was highest at $10^{4.3}$ CFU/g. Among the spore-forming bacteria, the majority was determined to be of the *B. cereus* group, with some *Clostridium perfringens* spoilage bacteria and among non-sporulating bacteria, *S. aureus* and *Pseudomonas* spp. were found in the highest number. However, some key foodborne pathogens (*Salmonella* spp., *E. coli, L. monocytognes* and *Campylobacter)* were not found.

Fasolato et al. (2018) investigated levels of fungi (moulds), *Clostridium* spp., *S. aureus* and *Salmonella* in processed edible insects such as mealworms, crickets and silkworms. The authors found generally low levels of moulds (25% of total samples $<10^4$ CFU/g), *Staphylococcus* spp. (12.5% of total samples $<10^4$ CFU/g) and lactic acid bacteria (6% of total samples varying from $10^{2.3}$ to $10^{5.6}$ CFU/g). Similar results were found in other studies such as Garofalo et al. (2017) and Osimani et al. (2017). Fasolato et al. (2018) found higher levels of aerobic spore-forming bacteria ($\approx 10^8$ CFU/g). *Bacillus cereus* counts were between 10^4

and $10^{6.6}$ CFU/g. Knowing that *B. cereus* is responsible for intoxication at high doses (>10^5 CFU/g) this number is within the range of what could be of risk (Walia et al., 2018).

A good example of how to manage potential risks from pathogens in insect production is provided by Kooh et al. (2020) who outline a risk-based HACCP plan for the production of mealworm (*T. molitor*) powders. Factors to consider include the initial concentration of a microorganism (bacteria or fungi) in the substrate and rearing environment conditions (humidity, temperature, insect density, etc.) which may favour colonization and create a high microbial load that may then be more challenging to eliminate during subsequent steps in insect production and processing (Gold et al., 2018; Cohen, 2018; Eilenberg et al., 2015; Maciel-Vergara et al., 2021). Vandeweyer et al. (2021) e.g. found that high initial loads of *S. aureus* and spore-forming bacteria like *C. perfringens* and *B. cereus* in substrates can be difficult to eliminate in subsequent processing of powdered insect products. In addition, research has indicated the ability of pathogens such as *S. enterica* and *Serratia marcescens* to persist in substrates and insects (Jensen et al., 2020; Wynants et al., 2019; Dupriez et al., 2022). The low pH (presence of lactic acid bacteria) of a substrate has been found to be a potential limiting factor in pathogen growth (Gorrens et al., 2021).

A distinctive characteristic of BSFL (*H. illucens*) is the capacity of its immune system to produce an extraordinarily high varity of antimicrobial peptides (AMPs) capable of supressing pathogens such as *S. aureus, S. enterica, E. coli* 0157:H7, *B. cereus* and *Clostridium* (Awasthi et al., 2020; Erickson et al., 2004), although this suppressive effect did not extend to the parasite *Ascaris suum* and Enterococcus species (Lalander et al., 2013). Further research is needed to identify and isolate potential antimicrobial compounds from insect species (Gorrens et al., 2021; Chevrette et al., 2019).

5 Potential adverse effects of chemical contaminants and pathogens on insect production

As noted earlier, contaminants are not just a potential safety risk for humans or livestock consuming insect food/feed products, but can significantly impact the insects themselves and insect production. The exposure of insects to chemicals can have a negative impact e.g. on survival, development and weight. Exposure to some insecticides (e.g. present in plant-based substrates) may impact the survival of insects during rearing (Meijer et al., 2021). Other chemical hazards can also increase insect mortality. Research on common housefly larvae (*M. domestica*) have shown that some heavy metals are immunotoxins and induce immunosuppression, resulting in a reduction of the number of haemocytes in insects and a consequent decrease in survival (Lievens et al., 2021).

Bulak et al. (2018) e.g. reported that BSFL mortality was four times higher (up to 16.7%) when larvae were exposed to cadmium at 50 mg/kg of dry weight. Cadmium has also been found to delay insect development. In the case of BSFL, measuring the time between <24 hour-old larvae and the day the first prepupae were observed resulted in 2.0 days with exposure to cadmium (0.25 mg/kg in diet) compared to 12.3 days for the control (van der Fels-Klerx et al., 2016). In contrast, cadmium exposure had no impact on the development time of mealworms, suggesting that the effect of exposure varies, depending on the insect species.

It has been found that, when BSFL were exposed to some antibiotics (sulfonamides), the body weight of larvae was significantly lower than that of controls and this impact was greater when the concentration of sulfonamide increased (from 0.1 to 10 mg/kg in the substrate) (Gao et al., 2019). However, Ratel et al. (2023) observed that a PCB contamination of the substrate, up to almost 15 times the MRL authorized in feed materials of plant origin, had no impact on *T. molitor* weight after 20 days of exposure even if PCBs are transferred from the substrate to the larvae during rearing.

An exposure to chemical hazards can also have an impact on insect biochemical composition. Exposure of BSFL to polypropylene microplastics can change fatty acid composition with a significant increase of some short-chain fatty acids (Romano and Fischer, 2021). De Filippis et al. (2023) also demonstrated that exposure of BSFL to polyethylene or polystyrene significantly changed the composition of insect microbiota in terms of species and strain levels.

Mass rearing of insects potentially increases levels of stress in individuals, potential susceptibility to disease and risk of rapid transmission of disease within insect populations. The main pathogens affecting insects are bacteria, fungi and viruses. Some pathogens are specific entomopathogens, like the bacteria *Bacillus thuringiensis*, the fungus *Bauveria brassiana* or Baculovirus (Eilenberg et al., 2015). House crickets are particularly prone to the Denso virus. Bacteria like *S. marcescens*, and some species of the *Pseudomonas* genera or Microsporidia, can also be a problem for some insects.

In the case of BSFL (*H. illucens*), no pathogens is currently known to impact rearing (Joosten et al., 2020). However, BSFLs are vulnerable to the so-called bacterial 'soft rot' (caused by *Paenibacillus thiaminolyticus*) which targets the gut, arresting growth and significantly increasing mortality (She et al., 2023). An insect-specific totivirus associated with BSFL has been found, though it does not appear to be pathogenic (Pienaar et al., 2022).

Mealworm (*T. molitor*) is vulnerable to *Bacillus thuringiensis* sv *morrisoni* biovar *tenebrionis*, which produces toxins killing coleopteran larvae. A recent laboratory study revealed that the smaller the larvae, the higher the risk of reduced growth and death. *B. thuringiensis* can also persist in low amounts in live larvae (Savio et al. 2024a). Addition of the probiotic bacteria *Pediococcus pentosaceus* has been found to reduce the impact of *B. thuringiensis* and fungal *Metarhizium* infection (Savio et al., 2024b). Among parasites, gregarines can reduce the longevity of mealworms (Rodriguez et al., 2007). Denso viruses have been identified in several mealworm farming outbreaks in the United States in 2020 (Armién et al., 2023). By penetrating the host cuticle, particular strains of the fungal genera *Beauveria* and *Metarhizium* are also virulent and lethal to *T. molitor* (Praprotnik et al., 2021).

An emerging area of research is the study of microbiomes: the communities of micro-organisms (bacteria, fungi, viruses, etc.) interacting with a living organism (particularly in the gut) in ways that can be commensal, mutualist or pathogenic (Engel and Moran, 2013). Investigating insect gut microbiomes promises to be an important area of research in better understanding insect health, growth and susceptibility to disease as well as potential risks from insect food/feed products (Cifuentes et al., 2022; Eke et al., 2023). A broader knowledge of the composition of insect microbiota can be achieved via analyzing 16S ribosomal DNA of the V3/V4 region of bacteria using barcoding techniques, Illumina sequencing or, more recently, nanopore technology. These approaches can track the relative abundance of species, depending on the larval stage, substrate or other treatments like addition of probiotics (Savio et al., 2024a).

An example of microbiome metagenomic analysis shows that the core bacterial genome of BSFL (*H. illucens*) contains species from the genera of *Enterococcus*, *Morganella*, *Providencia*, *Scrofimicrobium* and *Klebsiella,* (Eke et al., 2023). Using 16S rRNA gene sequencing on the Illumina MiSeq platform, Li et al. (2016) analyzed differences in microbial community structures in *T. molitor* reared in open and controlled environments. Compared with samples from the closed ecosystem, those from the open environment had higher abundance of *Bacillus, Lactococcus, Weissella, Escherichia* and *Clostridiaceae*, while Enterococcus and some unclassified Enterobacteriaceae were identified as the predominant bacteria in samples from the closed ecosystem. Interestingly, Acidobacteria, Planctomycetacia and Sphingobacteria were found in mealworm feed but almost disappeared in insect samples in the two groups.

6 Improving contaminant monitoring and risk assessment

Data on the chemical safety of edible insects is relatively recent and often incomplete. One of the challenges for the next few years will therefore be to better assess which priority chemical hazards need to be monitored in the insect sector. This will require more insect-rearing experiments in which substrates are intentionally contaminated with particular contaminants at different concentrations (Lievens et al., 2023; Ratel et al., 2023). This will make it easier to track uptake, transformation and excretion during rearing. In addition, simulation of the fate of chemical hazards in the digestive tract of insects or humans can be undertaken using *in vitro* digesters (Planche et al., 2022). This will then provide data for more robust risk assessments to inform acceptable limits (Tressou et al., 2017).

There will also be a need to improve surveillance and monitoring by both food safety authorities and the insect-rearing industry itself. Current analytical methods are highly sensitive but are expensive with low throughput. Options for improvement include analyzing samples in pools/batches rather than individually to improve both speed and cost (Ratel et al., 2024). New technologies such as electrochemical sensors, which evaluate differences in electrical signals related to particular chemicals, offer cheaper, more rapid methods with high accuracy and repeatability (Manikandan et al., 2024). One example is the use of volatile organic compound (VOC) sensors (e-noses), which seem relevant for monitoring insect exposure to chemical hazards (Fernandes et al., 2023; Jalal et al., 2018; Khatib and Haick, 2022). Other innovations include non-targeted analysis to detect and identify potentially unknown chemical hazards using techniques such as nuclear magnetic resonance and high-resolution mass spectrometry (McCord et al., 2022; Poma et al., 2022; Manz et al., 2023). Such approaches can also be coupled with effect-directed analysis in which a sample is first tested with bioassays for toxicity (Engel et al., 2022). If the sample shows toxicity, step-wise fractionation of the chemical mixture leads to purification of the chemical responsible for the bioactivity of interest. When sufficiently pure, the unknown chemical can then be identified using advanced analytics (Li and Guo, 2023; van der Burg et al., 2022).

Information related to the maximum tolerated CFU/g of particular pathogens or food-spoiling bacteria and protocols can be found at Codex Alimentarius Commission (CXC 1-1969) and IPIFF (2022). If the aim is to get a broader knowledge on the composition of the insect microbiota, studies should be done with 16S ribosomal DNA of the V3/V4 region of bacteria, using barcoding and Illumina sequencing or, more recently, using nanopore technology. This approach can be used to track the relative abundance of species, depending on larval stage, feed or processing conditions. If the objective is to evaluate the presence and to establish a threshold of a particular species, this can be done on DNA or cDNA with a targeted PCR primer. The strength of molecular techniques such as whole-genome sequencing or 16S rDNA is that many species can be monitored from each sample. It is important to note that the presence of the DNA does not mean the microbe is alive (Dupriez et al., 2022). Analysing cDNA from extracted mRNA will also help understand the metabolic activity of microbes (Mugo-Kamiri et al., 2024).

To improve knowledge on a particular pathogen dynamics and to determine threshold values in a substrate or insect, the best way is to set up a challenge test with a given amount of the pathogen (Jensen et al., 2020; Wynants et al., 2019, Dupriez et al., 2022). It is recommended to develop specific procedures adapted to chitin-rich insect-based products to achieve good extractions and optimal outgrowth in order to estimate the contamination/hazard level (Kooh et al., 2019, Brulé et al., 2024).

7 Conclusion and future trends

This chapter has reviewed the various chemical and microbial hazards that can be found in the insect industry, their potential negative effects as well as techniques for monitoring, avoiding or mitigating them. Future changes may include authorization of new substrates in Europe (e.g. animal products, waste, manure, etc.) which will then increase the range of hazards. It is therefore important to anticipate these changes. There is a need to better understand the fate of chemicals and pathogens throughout the insect rearing and processing chain, to clarify toxicity thresholds, to focus on molecules that have been little studied so far in the insect sector (process-induced toxicants for example), and also to propose new detection methods that are easier and/or less costly to implement on a routine basis.

It is also important to note that insect production could have a very beneficial effect on hazard management, particularly in a circular economy context. As mentioned, BSFL can reduce some pathogens during substrate conversion (Awasthi et al., 2020; Erickson et al., 2004) and preliminary studies have shown that insects may degrade some chemicals such as mycotoxins (Niermans et al., 2019; Van Broekhoven et al., 2017) or pesticides (de Almeida et al., 2017) or even that they could be a source of antimicrobials that might be used as new antibiotics (Patyra and Kwiatek, 2023). These are promising areas for future research.

8 References

Abril, A. G., et al., 2020. *Staphylococcus aureus* exotoxins and their detection in the dairy industry and mastitis. *Toxins* 12, 537. https://doi.org/10.3390/toxins12090537

Aguilar-Toalá, J. E., et al., 2022. Beyond human nutrition of edible insects: health benefits and safety aspects. *Insects* 13, 1007. https://doi.org/10.3390/insects13111007

Armién, A. G., et al., 2023. Outbreak of densovirus with high mortality in a commercial mealworm (*Tenebrio molitor*) farm: A molecular, bright-field, and electron microscopic characterization. *Vet. Pathol.* 60, 689–703. https://doi.org/10.1177/03009858231180488

Awasthi, M. K., et al., 2020. Manure pretreatments with black soldier fly *Hermetia illucens* L. (Diptera: Stratiomyidae): a study to reduce pathogen content. *Sci. Total Environ.* 737, 139842. https://doi.org/10.1016/j.scitotenv.2020.139842

Baiano, A., 2020. Edible insects: an overview on nutritional characteristics, safety, farming, production technologies, regulatory framework, and socio-economic and ethical implications. *Trends Food Sci. Technol.* 100, 35–50. https://doi.org/10.1016/j.tifs.2020.03.040

Belluco, S., et al., 2018. Edible Insects in a Food Safety Perspective, In: Halloran, A., et al. (Eds.), *Edible Insects in Sustainable Food Systems*. Springer International Publishing, Cham, pp. 109–126. https://doi.org/10.1007/978-3-319-74011-9_7

Biancarosa, I., et al., 2019. Replacing fish meal with insect meal in the diet of Atlantic salmon (*Salmo salar*) does not impact the amount of contaminants in the feed and it lowers accumulation of arsenic in the fillet. *Food Addit. Contam. Part A* 36, 1191–1205. https://doi.org/10.1080/19440049.2019.1619938

Bosch, G., et al., 2017. Aflatoxin B1 tolerance and accumulation in black soldier fly Larvae (*Hermetia illucens*) and yellow Mealworms (*Tenebrio molitor*). Toxins 9, 185. https://doi.org/10.3390/toxins9060185

Bulak, P., et al., 2018. *Hermetia illucens* as a new and promising species for use in entomoremediation. *Sci. Total Environ.* 633, 912–919. https://doi.org/10.1016/j.scitotenv.2018.03.252

Brulé, L., et al., 2024. Evaluation of the microbial quality of *Hermetia illucens* Larvae for animal feed and human consumption: study of different type of rearing substrates. *Foods* 20, 1587. https://doi.org/10.3390/foods13101587

Cappelli, A., et al., 2020. Insects as food: a review on risks assessments of Tenebrionidae and Gryllidae in relation to a first machines and plants development. *Food Control* 108, 106877. https://doi.org/10.1016/j.foodcont.2019.106877

Cardoso, D. N., et al., 2023. Improving product safety for edible insects: toxicokinetics of Hg in *Tenebrio molitor* and *Hermetia illucens*. *ACS Food Sci. Technol.* 3, 790–798. https://doi.org/10.1021/acsfoodscitech.3c00051

Chevrette, M. G., et al., 2019. The antimicrobial potential of Streptomyces from insect microbiomes. *Nat. Commun.* 10, 516. https://doi.org/10.1038/s41467-019-08438-0

Cifuentes, Y., et al., 2022. Isolation of *Hermetia illucens* larvae core gut microbiota by two different cultivation strategies. *Antonie Van Leeuwenhoek* 115, 821–837. https://doi.org/10.1007/s10482-022-01735-7

Cohen, A. C., 2018. Ecology of insect rearing systems: a mini-review of insect rearing papers from 1906-2017. *Adv. Entomol* 6, 86–115. https://doi.org/10.4236/ae.2018.62008

de Almeida, L. G., et al., 2017. The gut microbiota of insecticide-resistant insects houses insecticide-degrading bacteria: a potential source for biotechnological exploitation. *PLoS One* 12, e0174754. https://doi.org/10.1371/journal.pone.0174754

De Filippis, F., et al., 2023. Plastics shape the black soldier fly larvae gut microbiome and select for biodegrading functions. *Microbiome* 11, 205. https://doi.org/10.1186/s40168-023-01649-0

De Gier, S. and Verhoeckx, K., 2018. Insect (food) allergy and allergens. *Mol. Immunol.* 100, 82–106. https://doi.org/10.1016/j.molimm.2018.03.015

De Marchi, L., et al., 2021. Allergens from edible insects: cross-reactivity and effects of processing. *Curr. Allergy Asthma Rep.* 21, 35. https://doi.org/10.1007/s11882-021-01012-z

Dietrich, R., et al., 2021. The food poisoning toxins of *Bacillus cereus*. *Toxins* 13, 98. https://doi.org/10.3390/toxins13020098

Dupriez, F., et al., 2022. Impact and persistence of *Serratia marcescens* in *Tenebrio molitor* Larvae and feed under optimal and stressed mass rearing conditions. *Insects* 13, 458. https://doi.org/10.3390/insects13050458
Eilenberg, J., et al., 2015. Diseases in insects produced for food and feed. *J. Insects Food Feed* 1, 87–102. https://doi.org/10.3920/JIFF2014.0022
Eke, M., et al., 2023. Deciphering the functional diversity of the gut microbiota of the black soldier fly (*Hermetia illucens*): recent advances and future challenges. *Anim. Microbiome* 5, 40. https://doi.org/10.1186/s42523-023-00261-9
Engel, E., et al., 2015. Novel approaches to improving the chemical safety of the meat chain towards toxicants. *Meat Sci.* 109, 75–85. https://doi.org/10.1016/j.meatsci.2015.05.016
Engel, E., et al., 2022. Safe food for infants: an EU-China project to enhance the control of safety risks raised by microbial and chemical hazards all along the infant food chains. *Glob. Pediatr.* 2, 100009. https://doi.org/10.1016/j.gpeds.2022.100009
Engel, P. and Moran, N. A., 2013. The gut microbiota of insects – diversity in structure and function. *FEMS Microbiol. Rev.* 37, 699–735. https://doi.org/10.1111/1574-6976.12025
Erickson, M. C., et al., 2004. Reduction of *Escherichia coli* O157:H7 and *Salmonella enterica* serovar enteritidis in chicken manure by larvae of the black soldier fly. *J. Food Prot.* 67, 685–690. https://doi.org/10.4315/0362-028X-67.4.685
European Commission, 2002. Directive 2002/32/EC of the European Parliament and of the Council of 7 May 2002 on undesirable substances in animal feed.
European Commission, 2005. Commission Regulation (EC) No 2073/2005 of 15 November 2005 on microbiological criteria for foodstuffs.
FAO-WHO, 1999. Principles and guidelines for the conduct of microbiological risk assessment.
FAO and WHO, 2001. Evaluation of allergenicity of genetically modified foods.
FAO, 2021. *Looking at edible insects from a food safety perspective. Challenges and opportunities for the sector*. Rome. https://doi.org/10.4060/cb4094en
Fasolato, L., et al., 2018. Edible processed insects from e-commerce: food safety with a focus on the *Bacillus cereus* group. *Food Microbiol.* 76, 296–303. https://doi.org/10.1016/j.fm.2018.06.008
Fernandes, V., et al., 2023. A GABA receptor modulator and semiochemical compounds evidenced using volatolomics as candidate markers of chronic exposure to fipronil in *Apis mellifera*. Metabolites 13, 185. https://doi.org/10.3390/metabo13020185
Fournier, E., et al., 2023. Microplastics: what happens in the human digestive tract? First evidences in adults using in vitro gut models. *J. Hazard. Mater.* 442, 130010. https://doi.org/10.1016/j.jhazmat.2022.130010
Gao, Q., et al., 2019. Effect of sulfonamide pollution on the growth of manure management candidate *Hermetia illucens*. *PLoS One* 14, e0216086. https://doi.org/10.1371/journal.pone.0216086
Garofalo, C., et al., 2017. The microbiota of marketed processed edible insects as revealed by high-throughput sequencing. *Food Microbiol.* 62, 15–22. https://doi.org/10.1016/j.fm.2016.09.012
Garofalo, C., et al., 2019. Current knowledge on the microbiota of edible insects intended for human consumption: a state-of-the-art review. *Food Res. Int.* 125, 108527. https://doi.org/10.1016/j.foodres.2019.108527

Godoy, V., et al., 2019. The potential of microplastics as carriers of metals. *Environ. Pollut.* 255, 113363. https://doi.org/10.1016/j.envpol.2019.113363

Gold, M., et al., 2018. Decomposition of biowaste macronutrients, microbes, and chemicals in black soldier fly larval treatment: a review. *Waste Manag.* 82, 302–318. https://doi.org/10.1016/j.wasman.2018.10.022

Golwala, H., et al., 2021. Solid waste: an overlooked source of microplastics to the environment. *Sci. Total Environ.* 769, 144581. https://doi.org/10.1016/j.scitotenv.2020.144581

Gorrens, E., et al., 2021. *Staphylococcus aureus* in substrates for black soldier fly Larvae (*Hermetia illucens*) and its dynamics during rearing. *Microbiol. Spectr.* 9, e02183-21. https://doi.org/10.1128/spectrum.02183-21

Heuel, M., et al., 2023. Low transfer of cadmium, lead and aflatoxin B1 to eggs and meat of laying hens receiving diets with black soldier fly larvae reared on contaminated substrates. *Anim. Feed Sci. Technol.* 304, 115733. https://doi.org/10.1016/j.anifeedsci.2023.115733

Hosseindoust, A., et al., 2024. A metanalysis to evaluate the effects of substrate sources on the nutritional performance of black soldier fly larvae: implications for sustainable poultry feed. *Poult. Sci.* 103, 103299. https://doi.org/10.1016/j.psj.2023.103299

IPIFF, 2022. IPIFF Guide on Good Hygiene Practices for European Union (EU) producers of insects as food and feed. https://ipiff.org/good-hygiene-practices/

Jalal, A. H., et al., 2018. Prospects and challenges of volatile organic compound sensors in human healthcare. *ACS Sens.* 3, 1246–1263. https://doi.org/10.1021/acssensors.8b00400

Jensen, A. N., et al., 2020. *Salmonella Typhimurium* Level in Mealworms (*Tenebrio molitor*) after exposure to contaminated substrate. *Front. Microbiol.* 11, 1613. https://doi.org/10.3389/fmicb.2020.01613

Joosten, L., et al., 2020. Review of insect pathogen risks for the black soldier fly (*Hermetia illucens*) and guidelines for reliable production. *Entomol. Exp. Appl.* 168, 432–447. https://doi.org/10.1111/eea.12916

Khatib, M. and Haick, H., 2022. Sensors for volatile organic compounds. *ACS Nano* 16, 7080–7115. https://doi.org/10.1021/acsnano.1c10827

Kolakowski, B. M., et al., 2021. Analysis of microbiological and chemical hazards in edible insects available to canadian consumers. *J. Food Prot.* 84, 1575–1581. https://doi.org/10.4315/JFP-21-099

Kooh, P., et al., 2019. Entomophagy and public health: a review of microbiological hazards. *Health (NY)* 11, 1272–1290. https://doi.org/10.4236/health.2019.1110098

Kooh, P., et al., 2020. Control of biological hazards in insect processing: application of HACCP method for yellow Mealworm (*Tenebrio molitor*) powders. *Foods* 9, 1528. https://doi.org/10.3390/foods9111528

Koopmans, M. M., et al., 2023. Human listeriosis. *Clin. Microbiol. Rev* 36, e00060-19. https://doi.org/10.1128/cmr.00060-19

Lalander, C., et al., 2013. Faecal sludge management with the larvae of the black soldier fly (*Hermetia illucens*) – from a hygiene aspect. *Sci. Total Environ.* 458–460, 312–318. https://doi.org/10.1016/j.scitotenv.2013.04.033

Li, L., et al., 2016. Can closed artificial ecosystem have an impact on insect microbial community? A case study of yellow mealworm (*Tenebrio molitor* L.). *Ecol. Eng.* 86, 183–189. https://doi.org/10.1016/j.ecoleng.2015.09.015

Li, M. and Guo, L.-H., 2023. Chemo/biosensors towards effect-directed analysis: an overview of current status and future development. *TrAC Trends Anal. Chem*. 158, 116824. https://doi.org/10.1016/j.trac.2022.116824

Lievens, S., et al., 2021. Chemical safety of black soldier fly larvae (*Hermetia illucens*), knowledge gaps and recommendations for future research: a critical review. *J. Insects Food Feed* 7, 383–396. https://doi.org/10.3920/JIFF2020.0081

Lievens, S., et al., 2023. Ingestion and excretion dynamics of microplastics by black soldier fly larvae and correlation with mouth opening size. *Sci. Rep*. 13, 4341. https://doi.org/10.1038/s41598-023-31176-9

Maciel-Vergara, G., et al., 2021. Diseases in edible insect rearing systems. *J. Insects Food Feed* 7, 621–638. https://doi.org/10.3920/JIFF2021.0024

Manikandan, R., et al., 2024. Recent advances in miniaturized electrochemical analyzers for hazardous heavy metal sensing in environmental samples. *Coord. Chem. Rev*. 499, 215487. https://doi.org/10.1016/j.ccr.2023.215487

Manz, K. E., et al., 2023. Non-targeted analysis (NTA) and suspect screening analysis (SSA): a review of examining the chemical exposome. *J. Expo. Sci. Environ. Epidemiol*. 33, 524–536. https://doi.org/10.1038/s41370-023-00574-6

McCord, J. P., et al., 2022. Quantitative non-targeted analysis: bridging the gap between contaminant discovery and risk characterization. *Environ. Int*. 158, 107011. https://doi.org/10.1016/j.envint.2021.107011

Meijer, N., et al., 2021. Effects of insecticides on mortality, growth and bioaccumulation in black soldier fly (*Hermetia illucens*) larvae. *PLoS One* 16, e0249362. https://doi.org/10.1371/journal.pone.0249362

Meijer, N., et al., 2022. Aflatoxin B1 metabolism of reared *Alphitobius diaperinus* in different life-stages. *Insects* 13, 357. https://doi.org/10.3390/insects13040357

Meyer, A. M., et al., 2021. Chemical food safety hazards of insects reared for food and feed. J. Insects Food Feed 7, 823–831. https://doi.org/10.3920/JIFF2020.0085

Mugo-Kamiri, L., et al., 2024. The effect of diet composition on the diversity of active gut bacteria and on the growth of *Spodoptera exigua* (Lepidoptera: Noctuidae). *J. Insect Sci*. 24, 13. https://doi.org/10.1093/jisesa/ieae031

Niermans, K., et al., 2019. Feeding study for the mycotoxin zearalenone in yellow mealworm (*Tenebrio molitor*) larvae–investigation of biological impact and metabolic conversion. *Mycotoxin Res*. 35, 231–242. https://doi.org/10.1007/s12550-019-00346-y

Olivadese, M. and Dindo, M. L., 2023. Edible insects: a historical and cultural perspective on entomophagy with a focus on western societies. *Insects* 14(8), 690. https://doi.org/10.3390/insects14080690

Osimani, A., et al., 2017. Insight into the proximate composition and microbial diversity of edible insects marketed in the European Union. *Eur. Food Res. Technol*. 243, 1157–1171. https://doi.org/10.1007/s00217-016-2828-4

Osimani, A., et al., 2018. The bacterial biota of laboratory-reared edible mealworms (*Tenebrio molitor* L.): from feed to frass. *Int. J. Food Microbio*l. 272, 49–60. https://doi.org/10.1016/j.ijfoodmicro.2018.03.001

Osman, A. I., et al., 2023. Microplastic sources, formation, toxicity and remediation: a review. *Environ. Chem. Lett*. 21, 2129–2169. https://doi.org/10.1007/s10311-023-01593-3

Pakdel, M., et al., 2023. A review of food contaminants and their pathways within food processing facilities using open food processing equipment. *J Food Prot*. 86, 100184. https://doi.org/10.1016/j.jfp.2023.100184

Patyra, E. and Kwiatek, K., 2023. Insect meals and insect antimicrobial peptides as an alternative for antibiotics and growth promoters in livestock production. *Pathogens* 12, 854. https://doi.org/10.3390/pathogens12060854

Pienaar, R. D., et al., 2022. First evidence of past and present interactions between viruses and the black soldier fly, *Hermetia illucens*. *Viruses* 14, 1274. https://doi.org/10.3390/v14061274

Planche, C., et al., 2017. Effects of pan cooking on micropollutants in meat. *Food Chem.* 232, 395–404. https://doi.org/10.1016/j.foodchem.2017.03.049

Planche, C., et al., 2022. *In vitro* assessment of polychlorinated biphenyl bioaccessibility in meat: influence of fat content, cooking level and consumer age on consumer uptake. *Food Chem.* 374, 131623. https://doi.org/10.1016/j.foodchem.2021.131623

Poma, G., et al., 2017. Evaluation of hazardous chemicals in edible insects and insect-based food intended for human consumption. *Food Chem. Toxicol.* 100, 70–79. https://doi.org/10.1016/j.fct.2016.12.006

Poma, G., et al., 2018. Occurrence of organophosphorus flame retardants and plasticizers (PFRs) in Belgian foodstuffs and estimation of the dietary exposure of the adult population. *Environ. Sci. Technol.* 52, 2331–2338. https://doi.org/10.1021/acs.est.7b06395

Poma, G., et al., 2019. Occurrence of selected organic contaminants in edible insects and assessment of their chemical safety. *Environ. Health Perspect.* 127, 127009. https://doi.org/10.1289/EHP5782

Poma, G., et al., 2021. Occurrence, patterns, and sources of hazardous organic chemicals in edible insects and insect-based food from the Japanese market. *Food Chem. Toxicol.* 154, 112311. https://doi.org/10.1016/j.fct.2021.112311

Poma, G., et al., 2022. Edible insects in the metabolomics era. First steps towards the implementation of entometabolomics in food systems. *Trends Food Sci. Technol.* 119, 371–377. https://doi.org/10.1016/j.tifs.2021.12.018

Praprotnik, E., et al., 2021. Testing virulence of different species of insect associated fungi against *Yellow Mealworm* (Coleoptera: Tenebrionidae) and their potential growth stimulation to maize. *Plants* 10, 2498. https://doi.org/10.3390/plants10112498

Purschke, B., et al., 2017. Impact of substrate contamination with mycotoxins, heavy metals and pesticides on the growth performance and composition of black soldier fly larvae (*Hermetia illucens*) for use in the feed and food value chain. *Food Addit. Contam. Part A* 34, 1410–1420. https://doi.org/10.1080/19440049.2017.1299946

Ratel, J., et al., 2023. Fate of polychlorobiphenyls in *Tenebrio molitor* larvae: consequences for further use as food and feed. *J. Insects Food Feed* 9, 781–788. https://doi.org/10.3920/JIFF2022.0104

Ratel, J., et al., 2024. Implementation of sample pooling to strengthen the surveillance of food chemical safety: case study of nDL-PCBs in pork meat. *Food Chem.* 433, 137111. https://doi.org/10.1016/j.foodchem.2023.137111

Rodriguez, Y., et al., 2007. Individual and population effects of eugregarine, Gregarina niphandrodes (Eugregarinida: Gregarinidae), on *Tenebrio molitor* (Coleoptera: Tenebrionidae). *Environ. Entomol.* 36, 689–693. https://doi.org/10.1603/0046-225X(2007)36[689:IAPEOE]2.0.CO;2

Romano, N. and Fischer, H., 2021. Microplastics affected black soldier fly (*Hermetia illucens*) pupation and short chain fatty acids. *J. Appl. Entomol.* 145, 731–736. https://doi.org/10.1111/jen.12887

Savio, C., et al., 2024a. Minor impact of probiotic bacteria and egg white on *Tenebrio molitor* growth, microbial composition and pathogen infection. *Front. Insect Sci.* 4. https://doi.org/10.3389/finsc.2024.1334526

Savio, C., et al., 2024b. *Bacillus thuringiensis* sv morrisoni biovar tenebrionis impact and persistence in *Tenebrio molitor* larvae. *J. Insects Food Feed* 1–15. https://doi.org/10.1163/23524588-00001028

Schlüter, O., et al., 2017. Safety aspects of the production of foods and food ingredients from insects. *Mol. Nutr. Food Res.* 61, 1600520. https://doi.org/10.1002/mnfr.201600520

Schmidt, A., et al., 2019. Determination of vitamin B12 in four edible insect species by immunoaffinity and ultra-high performance liquid chromatography. *Food Chem.* 281, 124–129. https://doi.org/10.1016/j.foodchem.2018.12.039

She, W., et al., 2023. Isolated and identified pathogenic bacteria from black soldier fly larvae with "soft rot" reared in mass production facilities and its incidence characteristics. *Waste Manag.* 163, 85–95. https://doi.org/10.1016/j.wasman.2023.03.023

Surendra, K. C., et al., 2020. Rethinking organic wastes bioconversion: evaluating the potential of the black soldier fly (*Hermetia illucens* (L.)) (Diptera: Stratiomyidae) (BSF). *Waste Manag.* 117, 58–80. https://doi.org/10.1016/j.wasman.2020.07.050

Tressou, J., et al., 2017. Exposure assessment for dioxin-like PCBs intake from organic and conventional meat integrating cooking and digestion effects. *Food Chem. Toxicol.* 110, 251–261. https://doi.org/10.1016/j.fct.2017.10.032

Truzzi, C., et al., 2019. Influence of feeding substrates on the presence of toxic metals (Cd, Pb, Ni, As, Hg) in Larvae of *Tenebrio molitor*: Risk assessment for human consumption. *Int. J. Environ. Res. Public. Health* 16, 4815. https://doi.org/10.3390/ijerph16234815

Van Broekhoven, S., et al., 2017. Degradation and excretion of the *Fusarium* toxin deoxynivalenol by an edible insect, the Yellow mealworm (*Tenebrio molitor* L.). *World Mycotoxin J.* 10, 163–169. https://doi.org/10.3920/WMJ2016.2102

van der Burg, B., et al., 2022. Improving infant food safety by avoiding hazards of chemical mixture effects using novel integrated methods based on bioassays and analytical chemistry. *Glob. Pediatr.* 2, 100012. https://doi.org/10.1016/j.gpeds.2022.100012

van der Fels-Klerx, H. J., et al., 2016. Uptake of cadmium, lead and arsenic by *Tenebrio molitor* and *Hermetia illucens* from Contaminated Substrates. *PLoS One* 11, e0166186. https://doi.org/10.1371/journal.pone.0166186

van der Fels-Klerx, H. J., et al., 2018. Food safety issues related to uses of insects for feeds and foods. *Compr. Rev. Food Sci. Food Saf.* 17, 1172–1183. https://doi.org/10.1111/1541-4337.12385

van der Fels-Klerx, H. J., et al., 2020. Chemical food safety of using former foodstuffs for rearing black soldier fly larvae (*Hermetia illucens*) for feed and food use. *J. Insects Food Feed* 6, 475–488. https://doi.org/10.3920/JIFF2020.0024

Vandeweyer, D., et al., 2021. Biological contaminants in insects as food and feed. *J. Insects Food Feed* 7, 807–822. https://doi.org/10.3920/JIFF2020.0060

Ververis, E., et al., 2022. A systematic review of the nutrient composition, microbiological and toxicological profile of *Acheta domesticus* (house cricket). *J. Food Compos. Anal.* 114, 104859. https://doi.org/10.1016/j.jfca.2022.104859

Walia, K., et al., 2018. Qualitative risk assessment of cricket powder to be used to treat undernutrition in infants and children in Cambodia. *Food Control* 92, 169–182. https://doi.org/10.1016/j.foodcont.2018.04.047

Wang, C., et al., 2021. Environmental source, fate, and toxicity of microplastics. *J. Hazard. Mater.* 407, 124357. https://doi.org/10.1016/j.jhazmat.2020.124357

Wynants, E., et al., 2019. Risks related to the presence of *Salmonella* sp. during rearing of mealworms (*Tenebrio molitor*) for food or feed: survival in the substrate and transmission to the larvae. *Food Control* 100, 227–234. https://doi.org/10.1016/j.foodcont.2019.01.026

Zhang, Q., et al., 2022. Recent advances in toxicological research and potential health impact of microplastics and nanoplastics in Vivo. *Environ. Sci. Pollut. Res.* 29, 40415–40448. https://doi.org/10.1007/s11356-022-19745-3

Chapter 14

Understanding and changing attitudes toward eating protein derived from insects

Ashkan Pakseresht, Brunel University London, Brunel Business School, UK; and Lynn Frewer, Newcastle University, School of Natural and Environmental Sciences, UK

1 Introduction

In recent times, there has been a global increase in research interest toward edible insects (Lange and Nakamura, 2021). Edible insects have garnered attention as a valuable source of future nutrients on a global level of consumption (Megido et al., 2018; Sun-Waterhouse et al., 2016; Van Huis et al., 2013) because of their rich content of high-quality proteins, vitamins, and other minerals (Orkusz, 2021; Rumpold and Schlüter, 2013). In addition, the utilization of edible insects may deliver economic and environmental advantages (Dobermann et al., 2017; van Huis and Oonincx, 2017).

Insect farming offers environmental advantages, such as reduced greenhouse gas and ammonia emissions, compared to most livestock systems (Lange and Nakamura, 2021; Van Huis et al., 2013). Moreover, insect rearing needs fewer land and water resources compared to other livestock production systems (Halloran et al., 2017; Nadeau et al., 2015). *Figure* 1 compares the energy, water, and land use, as well as global warming impacts, of commercial

http://dx.doi.org/10.19103/AS.2024.0139.18

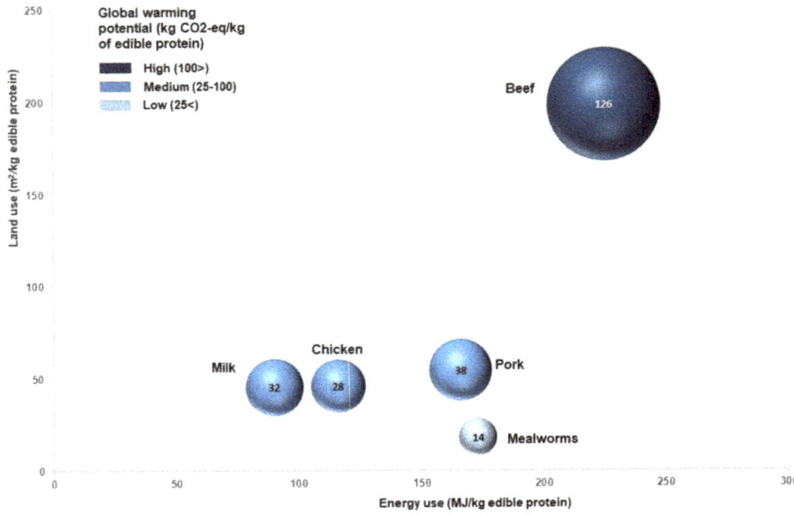

Figure 1. Environmental impacts of insect farming (commercial scale) compared with other livestock raising. The size of the bubbles indicates water use (ranges between 100 and 9700 L/kg live weight). Source: Data extracted from: Nadeau et al. (2015) and Govorushko (2019).

scale insect farming in comparison with conventional livestock production (Govorushko, 2019; Nadeau et al., 2015). Additionally, due to their cold-blooded nature, insects exhibit high efficiency in converting feed into protein in comparison with other livestock systems (Oonincx and de Boer, 2012; Van Huis, 2013; van Huis and Oonincx, 2017). Furthermore, insects have the capability to be reared on organic waste streams, thereby contributing to the mitigation of environmental contamination (Kröger et al., 2022).

The significance of insects as food sources has led to a notable growth in technological advancements in utilizing insects as novel food and feed sources in recent years (Alhujaili et al., 2023; Govorushko, 2019). However, the future success of insects as a novel food source in Western societies depends largely on the consumer acceptance (Alhujaili et al., 2023). Consumer acceptance of insect-based food is a multifaceted phenomenon that encompasses various contextual and attitudinal factors (Alhujaili et al., 2023; Hartmann and Siegrist, 2017; Kröger et al., 2022; Mancini et al., 2019; Onwezen et al., 2021).

2 Changing attitudes toward insect-based food

In some regions of the world, insects are regarded as a source of food (Kröger et al., 2022; Van Huis et al., 2013). However, in Western societies where protein is primarily derived from livestock, insects are commonly associated with being a nuisance or pest (Van Huis et al., 2013). Beyond this association, insects are also being recognized as a viable and sustainable option as animal feed. Their high nutritional quality and effective conversion rates make them an attractive alternative to conventional feed sources. Research indicates that insects as animal feed are more acceptable to consumers than as human food (Pakseresht et al., 2023).

Despite general EU consumers' reluctance to directly consume insects as food (Sogari, 2015; Van Huis, 2016; Van Huis et al., 2013), increased consumer awareness about the advantages of rearing insects for food production (Piha et al., 2018), coupled with industry efforts to enhance sensory properties of edible insects (Looy et al., 2014; Van Huis et al., 2013), has lessened the negative attitudes among some EU consumers (Kim et al., 2019; Lange and Nakamura, 2021). For instance, consumers in Belgium increasingly support the idea of insects as a sustainable food source (Van Thielen et al., 2019). The Netherlands' insect industry has also successfully marketed insect-derived foods to consumers, with freeze-dried insect powder sold as a meat substitute (Raheem et al., 2019). In the context of emerging food technologies, the characteristics of early adopters of insect-based food reveal a consumer segment primarily composed of young and highly educated males (Mancini et al., 2019; Onwezen et al., 2021). However, the characteristics of early adopters of insect-based food, particularly the demographic of young males and highly educated individuals, may vary depending on aspects such as cultural context, regional preferences, and other socioeconomic factors. Therefore, further empirical research is warranted to examine these differences comprehensively, enabling a more nuanced understanding of this consumer segment and its potential inferences for sustainable food systems.

Advocates of insect-based food development have identified health and environmental advantages of these foods compared to other sources of protein (Kröger et al., 2022). Research has emphasized the role of increasing knowledge about edible insects on consumers' willingness to purchase insect-based foods (Piha et al., 2018). However, various social, cognitive, and contextual factors affect consumers' acceptance of edible insects (House, 2016). These factors can be broadly classified as psychological, product-related, and sociocultural determinants (Siegrist, 2008). Psychological factors comprise knowledge and familiarity, neophobia, feelings of disgust, and risk-benefit perception. Product-related issues such as price and (perceived) quality, the availability of products, and sensory appeal may impact consumer acceptance of insect-based products. Cultural differences, social norms, and sociodemographic factors may also be influential (for details refer to e.g. Alhujaili et al. (2023); Kröger et al. (2022); Mancini et al. (2019); Onwezen et al. (2021); Siegrist, (2008)).

3 Psychological factors affecting attitudes toward insect-based food

Attitudes toward eating insects can be a strong determinant of consumer acceptance of insect-based food (Dupont and Fiebelkorn, 2020; Elorinne et al., 2019; Fischer and Steenbekkers, 2018; La Barbera et al., 2018; Lensvelt and Steenbekkers, 2014; Lombardi et al., 2019; Megido et al., 2016; Menozzi et al., 2017; Onwezen et al., 2019; Vartiainen et al., 2020; Wilkinson et al., 2018; Zielińska et al., 2020). Attitude is an individual's predisposed state of mind toward a stimulus object, which might be reflected in different cognitive or affective responses (Kröger et al., 2022).

The cognitive element of an attitude refers to the opinion, perception, or knowledge of the person toward the stimulus object which shapes the attitude. Provision of information regarding entomophagy and associated rearing technologies is contributing to the growing familiarity with them and the reduction in negative attitudes toward them (Barsics et al., 2017; Circus and Robison, 2019; de-Magistris et al., 2015; Laureati et al., 2016). However, research has indicated that consumer responses to different types of information clues may vary. Research which has included different forms of information intervention (Ardoin and Prinyawiwatkul, 2020; Barsics et al., 2017; Laureati et al., 2016; Lombardi et al., 2019; Mancini et al., 2019; Naranjo-Guevara et al., 2021; Schouteten et al., 2016; Verneau et al., 2016) has reported a positive impact on the acceptance of entomophagy on consumers' acceptance, however Lensvelt and Steenbekkers (2014), Arena et al. (2020), and Pascucci and de-Magistris (2013) did not confirm this effect. It seems that consumers are more positive when they receive information regarding the safety and nutrient content (Elorinne et al., 2019; Petersen et al., 2020; Ruby et al., 2015; Wilkinson et al., 2018) of insects-based food compared to the cases where the focus of information is on the long-term health or environmental advantages (Kröger et al., 2022). Cavallo and Materia (2018) observed that participants who received information about the content of the insect products (high protein level) reduced willingness to purchase silkworm protein-based drinks, although this may reflect the specific research parameters presented to participants.

Affective factors such as disgust, food neophobia, and curiosity can influence attitudes and the evaluation of insects as food (Alhujaili et al., 2023). Disgust is an emotional reaction characterized by intense feelings of aversion toward certain stimuli, including those that are food-related (Hartmann and Siegrist, 2018) and it has been associated with an unwillingness to try insects-based food (Castro and Chambers, 2019a; Circus and Robison, 2019; Fischer and Steenbekkers, 2018; Kornher et al., 2019; La Barbera et al., 2018; La Barbera et al., 2020; Lammers et al., 2019; Liu et al., 2020; Menozzi et al., 2017; Orkusz et al., 2020; Orsi et al., 2019; Powell et al., 2019; Ruby and Rozin,

2019; Sogari et al., 2017; Szendrő et al., 2020; Tuccillo et al., 2020; Videbæk and Grunert, 2020; Woolf et al., 2019). The experience of disgust toward insect foods differs among individuals based on their varying levels of 'mindfulness' (i.e. consciousness about what is going on within oneself and the environment) (Chan, 2019).

Feelings of disgust have been reported to be associated with unfamiliar foods (Rozin and Fallon, 1980). Lack of knowledge regarding entomophagy, and the lack of previous exposure to insect-based foods (Hartmann et al., 2015b) contribute to unfamiliarity and hence the sense of disgust (Alhujaili et al., 2023). Familiarity refers to whether individuals have heard about the idea of consuming insect-based foods and differs from knowledge (e.g. about benefits) of entomophagy, despite their frequent interchangeable use in various studies (Woolf et al., 2019). Familiarity with food from insects may reduce the negative disgust feelings toward insect-based food (for reviews refer to Hartmann and Siegrist, 2017; Kröger et al., 2021; Mancini et al., 2019; Onwezen et al., 2021; Wendin and Nyberg, 2021). A number of intervention studies have indicated that previous experience of tasting edible insect products improves consumers' familiarity and willingness to try new food products developed using insect-based ingredients (Barton et al., 2020; Lensvelt and Steenbekkers, 2014; Mancini et al., 2019). For instance, Van Thielen et al. (2019) conducted a survey in Belgium a few years after edible insects were introduced to the Belgian market and found that 57% of the participants were unwilling to eat food containing insects mainly due to disgust, and that reluctance in countries where consumers had not previously been exposed to such proteins was much higher than those where exposure had occurred (Gómez-Luciano et al., 2019).

Experiences with other new and novel foods, as well as greater preference for dietary variation, have been found to improve the acceptance of insect-based food (Gere et al., 2017; House, 2016; La Barbera et al., 2020; Modlinska et al., 2020; Nyberg et al., 2020; Verneau et al., 2020). Insects as an innovative food source could be acceptable among certain consumers, driving them to explore insect-based food options (Harms and Pirolet, 2018; House, 2016; Videbæk and Grunert, 2020). An example of this was reported in Videbæk and Grunert (2020), where it was found that curiosity and interest in trying 'novel food' led to an increase in the willingness of Dutch consumers to try insect-based food. The results indicate that, for those consumers, willingness to try was influenced by their curiosity and interest rather than any feelings of disgust (Videbæk and Grunert, 2020). Offering consumers the opportunity to taste insect-based products and providing them with information could reduce emotional reactions they may have toward consuming such products (for reviews, refer to Hartmann and Siegrist, 2017; Kröger et al., 2021; Lundén et al., 2020; Mancini et al., 2019; Onwezen et al., 2021; Wendin and Nyberg, 2021).

A large body of research underscores the existence of consumer concerns for novel foods and emerging technologies in general (Matin et al., 2012; Ronteltap et al., 2007). Food sector innovations are often not well-received by consumers, partly due to food neophobia, which is the aversion to novel or unfamiliar foods (Kröger et al., 2021). A high level of food neophobia negatively affects people's willingness to try novel food (Henriques et al., 2009; Olabi et al., 2009). There is substantial evidence within the literature that there is a negative impact of food neophobia on consumer acceptance of insect-based foods (Chang et al., 2019; Dupont and Fiebelkorn, 2020; Gere et al., 2017; Lammers et al., 2019; Laureati et al., 2016; Liu et al., 2020; Lombardi et al., 2019; Mancini et al., 2019; Olum et al., 2021; Orsi et al., 2019; Palmieri et al., 2019; Ruby and Rozin, 2019; Schäufele et al., 2019; Schlup and Brunner, 2018; Sogari et al., 2019a, 2019b; Tuccillo et al., 2020). The consistency of findings on the effect of neophobia across different countries suggests a universal pattern of consumer response (Brunner and Nuttavuthisit, 2020; de Koning et al., 2020; Hartmann et al., 2015b; Piha et al., 2018).

Modlinska et al. (2020) found that individuals with lower neophobia and a stronger preference for trying new foods were more willing to try products labeled as being produced using insects compared to consumers who do not have these traits. One form of neophobia that has received relatively less attention is food technology neophobia, which refers to the aversion toward food products that are produced using new technologies (Lammers et al., 2019). Schlup and Brunner (2018) reported that food technology neophobia negatively impacts the acceptance of edible insects among Swiss consumers. Similar findings were found in Belgium (Lammers et al., 2019; Verbeke, 2015) and Italy (Palmieri et al., 2019). Food technology neophobia reduced acceptance of edible insects among individuals who had not previously consumed insects (Schlup and Brunner, 2018).

Risk beliefs associated with novel food technologies have also been reported to affect individual's food choice behaviour (Magnan et al., 2021). Perceived risk is closely associated with health and nutritional considerations. Allergenicity, intoxication, and diseases transmitted by microbes are common perceived risks linked with the consumption of edible insects (Castro and Chambers, 2019b). The literature presents varying interpretations of the effects of perceived risks associated with insect-based foods and consumption. The greater perceived risk may reduce consumers' acceptance of insects as a novel food source (Baker et al., 2016; Legendre and Baker, 2022; Orsi et al., 2019; Palmieri et al., 2019; Ruby et al., 2015), however Ruby and Rozin (2019) and Russell and Knott (2021) found no relation between perceived risk and willingness to consumer insect-based food. Risk perception can also vary based on the product type and the risk behavior of consumers. For example, Orsi et al. (2019) found that perceived risk had a limited impact on acceptance of processed insect-based products. In addition, Ruby et al. (2015) found that individuals who demonstrated a greater inclination for participating in risky behaviors (potentially representing a higher tolerance for risks) displayed more openness toward consuming edible insects. Individuals' perception of self-infectability and their perceived vulnerability to infections, may also influence their risk perception regarding insect-based foods.

4 Product-related determinants of consumer acceptance of insect-based food

There is evidence that consumers are willing to pay for insect-based food if they perceive these to have positive attributes for health, sustainability, and high quality (Florença et al., 2022; Kröger et al., 2022; Mina et al., 2023a; Onwezen et al., 2021). Nevertheless, price is likely to represent a determinant of acceptability, especially for individuals with lower incomes and limited knowledge about insect-based food production processes (Florença et al., 2022; Kröger et al., 2022; Mina et al., 2023). While the price of products is important, consumers are also sensitive toward product characteristics, particularly sensory attributes (Biró et al., 2020; Castro and Chambers, 2019a; Cheseto et al., 2020; Elhassan et al., 2019; Mishyna et al., 2020; Schouteten et al., 2016; Tan et al., 2016, 2017; Tzompa-Sosa et al., 2021; Wendin et al., 2019).

Sensory elements, including taste, texture, and visual appeal, affect consumer acceptance of insect-based products (Mishyna et al., 2020; Tan et al., 2016). The appearance and the degree of visibility of insects are often viewed as a more significant deterrent to food choice than the taste (Sogari et al., 2018), leading to an approach of including insect proteins discreetly in familiar products such as energy bars, energy shakes, and burgers (Megido et al., 2016; Van Thielen et al., 2019; Wilkinson et al., 2018). For instance, Tan et al. (2017) assessed the perceived ingredient appropriateness and sensory perceptions associated with familiar food (burger patties) containing different levels of unusual novel food ingredients (e.g. lamb brain, frog meat, and mealworms) before and after tasting. It was found that food's taste perception was influenced by species-related associations and tended to be unfavorable when the food was unfamiliar. In addition, the amount of insect ingredients used in a product may affect consumer acceptance of these products (Detilleux et al., 2020; Dupont and Fiebelkorn, 2020; Fischer and Steenbekkers, 2018). Consumers showed a higher liking for products where no insects or insect parts were visible (Biró et al., 2020; Castro Delgado et al., 2020; Delicato et al., 2020). These results indicate that the sensory perceptions after tasting insect-based foods are dependent on product attributes. Incorporating unfamiliar ingredients into a familiar product may induce some positive perceptions and feelings. However, foods produced using insects remain less appealing than the original products with which consumers are already acquainted (Tan et al., 2017).

The perception of sensory attributes may also be influenced by the type of insect species used in producing the novel foods (Mishyna et al., 2020). Different insect species may potentially elicit sensory expectations with different impacts on consumer attitudes toward the products produced using them (Kröger et al., 2022). For example, Olum et al. (2021) reported that participants in northern Uganda displayed greater acceptance toward long-horned grasshoppers, flying African termites, and wingless red termites compared to other insect species. In contrast, participants in central Uganda were predominantly receptive only to long-horned grasshoppers, possibly due to their familiarity with this particular species (Olum et al., 2021). However, most research has investigated the overall consumer acceptance of insects as food without specifying the species involved in production (Le Goff and Delarue, 2017; Myers and Pettigrew, 2018; Vanhonacker et al., 2013; Verneau et al., 2020). This lack of specificity may potentially lead to misinterpretation of consumers' responses (Hartmann and Siegrist, 2017).

The level of processing in insect-based food may directly correlate with their level of consumer appeal (Kröger et al., 2022). It has been found that processed products that incorporate grounded insects, or insects mixed with familiar foods, will reduce cognitive disgust toward these products (Elhassan et al., 2019; Mancini et al., 2019; Mina et al., 2023; Tan et al., 2016, 2017).

5 The impact of sociocultural impacts on consumer acceptance of insect-based food

Social norms and food cultures may have a significant influence on food preferences (Hartmann et al., 2015a; Mina et al., 2023; Motoki et al., 2020; Tan et al., 2015). While entomophagy continues to play a major role in parts of Asia, Latin America, and Africa (Van Huis, 2013), Western societies are still reluctant to adopt insects as human foods (Ardoin and Prinyawiwatkul, 2021). A systematic review conducted by Mina et al. (2023) confirmed that the incompatibility with local food culture serves as a significant barrier to consumption of insect-based foods.

There is inconsistent evidence regarding the impact of sociodemographic factors as predictors of consumer acceptance. For example, according to the reviews conducted by Alhujaili et al. (2023) and Hartmann and Siegrist (2017), there is some limited evidence that males exhibit a higher level of acceptance of various forms of insect-based food compared to females, although this finding is not consistent. There is also inconsistency regarding the influence of other sociodemographic factors on consumer acceptance of insect-based foods, including age, education, and household income (Alhujaili et al., 2023; Florença et al., 2022; Hartmann and Siegrist, 2017). Socio-demographic factors which impact consumer acceptance of insect-based foods may be dependent on the context in which the foods are presented and the type of foods into which insects are incorporated.

It has been established that differences in individual characteristics and personality traits contribute to differences in consumers' food decision-making (Alhujaili et al., 2023). For instance, individuals with a tendency to share their unusual eating experiences with others are more receptive to insect products (Ruby et al., 2015). Onwezen et al. (2019) reported a positive impact of ambivalence (i.e. condition of experiencing conflicting attitudes simultaneously toward an object) on the willingness to try an insect-based burger. This implies that consumers who experience mixed or conflicting emotions regarding insect-based food are significantly more inclined to consume it.

Russell and Knott (2021) explored the relationship between consumer acceptance of insect food and the 'Big Five' personality traits, namely openness (i.e. willingness to embrace new ideas), conscientiousness (i.e. self-disciplined), extraversion (i.e. sociability), agreeableness (person's inclination to be cooperative), and neuroticism (i.e. more prone to emotional volatility). The findings indicated that extraversion and openness have a positive impact on the acceptance of insect food, while agreeableness showed no significant effect (Russell and Knott, 2021). However, conscientiousness was found to act as a barrier to acceptance, and the impact of neuroticism on acceptance remained uncertain (for a review refer to Kröger et al. (2022)).

6 Conclusion and future directions of research

The European Union's (EU) Farm to Fork Strategy, launched in 2020, encourages sustainable food systems and acknowledges the significance of insects used within the food system as food and ingredients in achieving these goals (Schebesta and Candel, 2020). EU member states have initiated the development of legislation and regulations aimed at governing insect-based food production and marketing. The EU has funded research initiatives to investigate the safety, quality, and nutritional aspects of insect-based food. In line with this, there has been a significant growth in patent filings and research focused on insects as food technologies (Alhujaili et al., 2023). Kim et al. (2019) examined the life cycle assessment of edible-insect-related technologies through a time-series analysis (1998–2018) of patent applications. The results suggest that the research in this technological area is increasing. Nevertheless, the sustainability of insects as a novel food source is contingent on factors such as the species used, the type of substrate used to rear insects, and the (potentially high levels of) energy used in producing insect-based food (Dagevos, 2020).

Within the EU, there has been a gradual shift in consumer attitudes toward insects as a novel source of food in recent years, and insect-based food products have already been introduced to the market in some EU countries (Kim et al., 2019). However, given the existing culture associated with perceptions, attitudes, and food choices, the adoption rate of insect-based food in the EU is expected to be relatively slow (Sogari, 2015; Van Huis, 2016). The potential incorporation of entomophagy into Western diets may depend on the availability of edible insects and producers' ability to address negative consumers attitudes and emotions toward insect-based foods (van Huis, 2020). The social, practical, and contextual factors that influence insect-based food consumption may need to be addressed when new products are being developed (Alhujaili et al., 2023; van Huis, 2020). Consumer education and advocacy may raise awareness about the potential advantages of edible insects in addressing environmental and sustainability problems and food security challenges (Kim et al., 2019). Improving accessibility and familiarity with these insect-based foods may reduce feelings of disgust and undesirable associations experienced by consumers. A further possible approach to promote familiarity, and hence alleviate aversion toward insect-based foods, is incorporating them into familiar foods (Kröger et al., 2022; Naranjo-Guevara et al., 2021; Tan et al., 2016).

The availability of high-quality insect-based food also requires the development of large-scale rearing systems to achieve economic competitiveness with traditional livestock meat production (Lange and Nakamura, 2021). The most promising insect species proposed for industrial-scale production are common housefly larvae, silkworms, yellow mealworms, and black soldier flies (Van Huis et al., 2013). However, consumers have shown considerable concern regarding food safety and potential health risks associated with consuming insect-based food, so the implementation of suitable food safety measures is needed to ensure effective hygiene practices in production systems that safeguard consumer health (Cadinu et al., 2020; Lange and Nakamura, 2021). Further research is needed to identify consumer preferences for different types of insect-based foods and their preferences for consuming different insect species. A systematic comparison of these factors across different cultural and demographic contexts is required, as research to date has tended to focus on specific food choice environments. Moreover, legal frameworks and regulations associated with the use of insects as novel food sources may need further development and harmonization across regions and countries to promote the industrial-scale development of insect farming and processing (Lange and Nakamura, 2021). Understanding the factors that determine consumer perceptions of health (nutrition and food safety), sensory preferences, and optimal insect-base product design is necessary for broader consumer acceptance of insect-based foods to occur in line with current policy objectives related to food security.

7 References

Alhujaili, A., Nocella, G. and Macready, A. (2023). Insects as food: consumers' acceptance and marketing. *Foods* 12(4). https://doi.org/10.3390/foods12040886.

Ardoin, R. and Prinyawiwatkul, W. (2020). Product appropriateness, willingness to try and perceived risks of foods containing insect protein powder: a survey of U.S. consumers. *International Journal of Food Science and Technology* 55(9), 3215–3226. https://doi.org/10.1111/IJFS.14612.

Ardoin, R. and Prinyawiwatkul, W. (2021). Consumer perceptions of insect consumption: a review of western research since 2015. *International Journal of Food Science and Technology* 56(10), 4942–4958. https://doi.org/10.1111/ijfs.15167.

Arena, E., Mazzaglia, A., Selvaggi, R., Pecorino, B., Fallico, B., Serranò, M. and Pappalardo, G. (2020). Exploring consumer's propensity to consume insect-based foods. Empirical evidence from a study in Southern Italy. *Applied System Innovation* 3(3), 38. https://doi.org/10.3390/ASI3030038.

Baker, M. A., Shin, J. T. and Kim, Y. W. (2016). An exploration and investigation of edible insect consumption: the impacts of image and description on risk perceptions and purchase intent. *Psychology and Marketing* 33(2), 94–112. https://doi.org/10.1002/MAR.20847.

Barsics, F., Caparros Megido, R., Brostaux, Y., Barsics, C., Blecker, C., Haubruge, E. and Francis, F. (2017). Could new information influence attitudes to foods supplemented with edible insects? *British Food Journal* 119(9), 2027–2039. https://doi.org/10.1108/BFJ-11-2016-0541.

Barton, A., Richardson, C. D. and McSweeney, M. B. (2020). Consumer attitudes toward entomophagy before and after evaluating cricket (Acheta domesticus)-based protein powders. *Journal of Food Science* 85(3), 781–788. https://doi.org/10.1111/1750-3841.15043.

Biró, B., Sipos, M. A., Kovács, A., Badak-Kerti, K., Pásztor-Huszár, K. and Gere, A. (2020). Cricket-enriched oat biscuit: technological analysis and sensory evaluation. *Foods (Basel, Switzerland)* 9(11). https://doi.org/10.3390/FOODS9111561.

Brunner, T. A. and Nuttavuthisit, K. (2020). A consumer-oriented segmentation study on edible insects in Switzerland and Thailand. *British Food Journal* 122(2), 482–488. https://doi.org/10.1108/BFJ-08-2018-0526/FULL/PDF.

Cadinu, L. A., Barra, P., Torre, F., Delogu, F. and Madau, F. A. (2020). Insect rearing: potential, challenges, and circularity. *Sustainability* 12(11), 4567. https://doi.org/10.3390/SU12114567.

Castro, M. and Chambers, E. (2019a). Consumer avoidance of insect containing foods: primary emotions, perceptions and sensory characteristics driving consumers considerations. *Foods (Basel, Switzerland)* 8(8). https://doi.org/10.3390/FOODS8080351.

Castro, M. and Chambers, E. (2019b). Willingness to eat an insect based product and impact on brand equity: a global perspective. *Journal of Sensory Studies* 34(2). https://doi.org/10.1111/JOSS.12486.

Castro Delgado, M., Chambers, E., Carbonell-Barrachina, A., Noguera Artiaga, L., Vidal Quintanar, R. and Burgos Hernandez, A. (2020). Consumer acceptability in the USA, Mexico, and Spain of chocolate chip cookies made with partial insect powder replacement. *Journal of Food Science* 85(6), 1621–1628. https://doi.org/10.1111/1750-3841.15175.

Cavallo, C. and Materia, V. C. (2018). Insects or not insects? Dilemmas or attraction for young generations: a case in Italy. *International Journal on Food System Dynamics* 9(3), 226–239. https://doi.org/10.18461/IJFSD.V9I3.932.

Chan, E. Y. (2019). Mindfulness and willingness to try insects as food: the role of disgust. *Food Quality and Preference* 71, 375–383. https://doi.org/10.1016/J.FOODQUAL.2018.08.014.

Chang, H. P., Ma, C. C. and Chen, H. S. (2019). Climate change and consumer's attitude toward insect food. *International Journal of Environmental Research and Public Health* 16(9), 1606. https://doi.org/10.3390/IJERPH16091606.

Cheseto, X., Baleba, S. B. S., Tanga, C. M., Kelemu, S. and Torto, B. (2020). Chemistry and sensory characterization of a bakery product prepared with oils from African edible insects. *Foods (Basel, Switzerland)* 9(6). https://doi.org/10.3390/FOODS9060800.

Circus, V. E. and Robison, R. (2019). Exploring perceptions of sustainable proteins and meat attachment. *British Food Journal* 121(2), 533–545. https://doi.org/10.1108/BFJ-01-2018-0025.

Dagevos, H. (2020). A literature review of consumer research on edible insects: recent evidence and new vistas from 2019 studies. *Journal of Insects as Food and Feed* 7(3), 249–259. https://doi.org/10.3920/JIFF2020.0052.

de Koning, W., Dean, D., Vriesekoop, F., Aguiar, L. K., Anderson, M., Mongondry, P., Oppong-Gyamfi, M., Urbano, B., Luciano, C. A. G., Jiang, B., Hao, W., Eastwick, E., Jiang, Z. V. and Boereboom, A. (2020). Drivers and inhibitors in the acceptance of meat alternatives: the case of plant and insect-based proteins. *Foods* 9(9), 1292. https://doi.org/10.3390/FOODS9091292.

Delicato, C., Schouteten, J. J., Dewettinck, K., Gellynck, X. and Tzompa-Sosa, D. A. (2020). Consumers' perception of bakery products with insect fat as partial butter replacement. *Food Quality and Preference* 79, 103755. https://doi.org/10.1016/J.FOODQUAL.2019.103755.

de-Magistris, T., Pascucci, S. and Mitsopoulos, D. (2015). Paying to see a bug on my food: how regulations and information can hamper radical innovations in the European Union. *British Food Journal* 117(6), 1777–1792. https://doi.org/10.1108/BFJ-06-2014-0222.

Detilleux, L., Wittock, G., Dogot, T., Francis, F. and Caparros Megido, R. (2020). Edible insects, what about the perceptions of Belgian youngsters? *British Food Journal* 123(6), 1985–2002. https://doi.org/10.1108/BFJ-08-2020-0754/FULL/PDF.

Dobermann, D., Swift, J. A. and Field, L. M. (2017). Opportunities and hurdles of edible insects for food and feed. *Nutrition Bulletin* 42(4), 293–308.

Dupont, J. and Fiebelkorn, F. (2020). Attitudes and acceptance of young people toward the consumption of insects and cultured meat in Germany. *Food Quality and Preference* 85, 103983. https://doi.org/10.1016/J.FOODQUAL.2020.103983.

Elhassan, M., Wendin, K., Olsson, V. and Langton, M. (2019). Quality aspects of insects as food-nutritional, sensory, and related concepts. *Foods (Basel, Switzerland)* 8(3). https://doi.org/10.3390/FOODS8030095.

Elorinne, A. L., Niva, M., Vartiainen, O. and Väisänen, P. (2019). Insect consumption attitudes among vegans, non-vegan vegetarians, and omnivores. *Nutrients* 11(2), 67–68. https://doi.org/10.3390/NU11020292.

Fischer, A. R. H. and Steenbekkers, L. P. A. B. (2018). All insects are equal, but some insects are more equal than others. *British Food Journal (Croydon, England)* 120(4), 852–863. https://doi.org/10.1108/BFJ-05-2017-0267.

Florença, S. G., Guiné, R. P. F., Gonçalves, F. J. A., Barroca, M. J., Ferreira, M., Costa, C. A., Correia, P. M. R., Cardoso, A. P., Campos, S., Anjos, O. and Cunha, L. M. (2022). The motivations for consumption of edible insects: a systematic review. *Foods (Basel, Switzerland)* 11(22). https://doi.org/10.3390/foods11223643.

Gere, A., Székely, G., Kovács, S., Kókai, Z. and Sipos, L. (2017). Readiness to adopt insects in Hungary: a case study. *Food Quality and Preference* 59, 81–86. https://doi.org/10.1016/J.FOODQUAL.2017.02.005.

Gómez-Luciano, C. A., Vriesekoop, F. and Urbano, B. (2019). Towards food security of alternative dietary proteins: a comparison between Spain and the Dominican Republic. *Www.Amfiteatrueconomic.Ro* 21(51), 393–407. https://doi.org/10.24818/EA/2019/51/393.

Govorushko, S. (2019). Global status of insects as food and feed source: a review. *Trends in Food Science and Technology* 91, 436–445. https://doi.org/10.1016/J.TIFS.2019.07.032.

Halloran, A., Hanboonsong, Y., Roos, N. and Bruun, S. (2017). Life cycle assessment of cricket farming in north-eastern Thailand. *Journal of Cleaner Production* 156, 83–94. https://doi.org/10.1016/J.JCLEPRO.2017.04.017.

Harms, E. and Pirolet, N. (2018). Consumer acceptance of insect-based burgers. *Appetite* 130, 306. https://doi.org/10.1016/J.APPET.2018.05.195.

Hartmann, C., Shi, J., Giusto, A. and Siegrist, M. (2015). The psychology of eating insects: a cross-cultural comparison between Germany and China. *Food Quality and Preference* 44, 148–156. https://doi.org/10.1016/j.foodqual.2015.04.013.

Hartmann, C. and Siegrist, M. (2017). Consumer perception and behaviour regarding sustainable protein consumption: a systematic review. *Trends in Food Science and Technology* 61, 11–25. https://doi.org/10.1016/J.TIFS.2016.12.006.

Hartmann, C. and Siegrist, M. (2018). Development and validation of the Food Disgust Scale. *Food Quality and Preference* 63, 38–50.

Henriques, A. S., King, S. C. and Meiselman, H. L. (2009). Consumer segmentation based on food neophobia and its application to product development. *Food Quality and Preference* 20(2), 83–91. https://doi.org/10.1016/J.FOODQUAL.2008.01.003.

House, J. (2016). Consumer acceptance of insect-based foods in the Netherlands: academic and commercial implications. *Appetite* 107, 47–58. https://doi.org/10.1016/j.appet.2016.07.023.

Kim, T. K., Yong, H. I., Kim, Y. B., Kim, H. W. and Choi, Y. S. (2019). Edible insects as a protein source: a review of public perception, processing technology, and research trends. *Food Science of Animal Resources* 39(4), 521–540. https://doi.org/10.5851/KOSFA.2019.E53.

Kornher, L., Schellhorn, M. and Vetter, S. (2019). Disgusting or innovative-consumer willingness to pay for insect based burger patties in Germany. *Sustainability* 11(7), 1878. https://doi.org/10.3390/SU11071878.

Kröger, T., Dupont, J., Büsing, L. and Fiebelkorn, F. (2021). Acceptance of insect-based food products in western societies: a systematic review. *Frontiers in Nutrition* 8, 759885. https://doi.org/10.3389/FNUT.2021.759885.

La Barbera, F., Verneau, F., Amato, M. and Grunert, K. (2018). Understanding Westerners' disgust for the eating of insects: the role of food neophobia and implicit associations. *Food Quality and Preference* 64, 120–125. https://doi.org/10.1016/J.FOODQUAL.2017.10.002.

La Barbera, F., Verneau, F., Videbæk, P. N., Amato, M. and Grunert, K. G. (2020). A self-report measure of attitudes toward the eating of insects: construction and validation of the Entomophagy Attitude Questionnaire. *Food Quality and Preference* 79, 103757. https://doi.org/10.1016/J.FOODQUAL.2019.103757.

Lammers, P., Ullmann, L. M. and Fiebelkorn, F. (2019). Acceptance of insects as food in Germany: is it about sensation seeking, sustainability consciousness, or food disgust? *Food Quality and Preference* 77, 78–88. https://doi.org/10.1016/J.FOODQUAL.2019.05.010.

Lange, K. W. and Nakamura, Y. (2021). Edible insects as future food: chances and challenges. *Journal of Future Foods* 1(1), 38–46. https://doi.org/10.1016/J.JFUTFC.2021.10.001.

Laureati, M., Proserpio, C., Jucker, C. and Savoldelli, S. (2016). New sustainable protein sources: consumers' willingness to adopt insects as feed and food. *Italian Journal of Food Science* 28(4), 652–668.

Legendre, T. S. and Baker, M. A. (2022). Legitimizing edibLe insects for Human consumption: the impacts of trust, risk-benefit, and purcHase activism. *Journal of Hospitality and Tourism Research* 46(3), 467–489. https://doi.org/10.1177/1096348020914375.

Le Goff, G. and Delarue, J. (2017). Non-verbal evaluation of acceptance of insect-based products using a simple and holistic analysis of facial expressions. *Food Quality and Preference* 56, 285–293. https://doi.org/10.1016/J.FOODQUAL.2016.01.008.

Lensvelt, E. J. S. and Steenbekkers, L. P. A. (2014). Exploring consumer acceptance of entomophagy: a survey and experiment in Australia and the Netherlands. *Ecology of Food and Nutrition* 53(5), 543–561. https://doi.org/10.1080/03670244.2013.879865.

Liu, A. J., Li, J. and Gómez, M. I. (2019). Factors influencing consumption of edible insects for Chinese consumers. *Insects* 11(1). https://doi.org/10.3390/INSECTS11010010.

Lombardi, A., Vecchio, R., Borrello, M., Caracciolo, F. and Cembalo, L. (2019). Willingness to pay for insect-based food: the role of information and carrier. *Food Quality and Preference* 72, 177–187. https://doi.org/10.1016/J.FOODQUAL.2018.10.001.

Looy, H., Dunkel, F. V. and Wood, J. R. (2014). How then shall we eat? Insect-eating attitudes and sustainable foodways. *Agriculture and Human Values* 31(1), 131–141. https://doi.org/10.1007/s10460-013-9450-x.

Lundén, S., Hopia, A., Forsman, L. and Sandell, M. (2020). Sensory and conceptual aspects of ingredients of sustainable sources–Finnish consumers' opinion. *Foods* 9(11), 1669. https://doi.org/10.3390/FOODS9111669.

Magnan, R. E., Gibson, L. P. and Bryan, A. D. (2021). Cognitive and affective risk beliefs and their association with protective health behavior in response to the novel health threat of COVID-19. *Journal of Behavioral Medicine* 44(3), 285–295. https://doi.org/10.1007/S10865-021-00202-4.

Mancini, S., Moruzzo, R., Riccioli, F. and Paci, G. (2019). European consumers' readiness to adopt insects as food: a review. *Food Research International* 122, 661–678. https://doi.org/10.1016/j.foodres.2019.01.041.

Mancini, S., Sogari, G., Menozzi, D., Nuvoloni, R., Torracca, B., Moruzzo, R. and Paci, G. (2019). Factors predicting the intention of eating an insect-based product. *Foods* 8(7), 270. https://doi.org/10.3390/foods8070270.

Matin, A. H., Goddard, E., Vandermoere, F., Blanchemanche, S., Bieberstein, A., Marette, S. and Roosen, J. (2012). Do environmental attitudes and food technology neophobia affect perceptions of the benefits of nanotechnology? *International Journal of Consumer Studies* 36(2), 149–157. https://doi.org/10.1111/J.1470-6431.2011.01090.X.

Megido, R. C., Gierts, C., Blecker, C., Brostaux, Y., Haubruge, É., Alabi, T., Francis, F., Caparros Megido, R., Gierts, C., Blecker, C., Brostaux, Y., Haubruge, É., Alabi, T. and Francis, F. (2016). Consumer acceptance of insect-based alternative meat products in Western countries. *Food Quality and Preference* 52, 237–243. https://doi.org/10.1016/j.foodqual.2016.05.004.

Megido, R. C., Haubruge, É. and Francis, F. (2018). Insects, the next European foodie craze? *Edible Insects in Sustainable Food Systems*, 353–361. https://doi.org/10.1007/978-3-319-74011-9_21/COVER.

Menozzi, D., Sogari, G., Veneziani, M., Simoni, E. and Mora, C. (2017). Eating novel foods: an application of the Theory of Planned Behaviour to predict the consumption of an insect-based product. *Food Quality and Preference* 59, 27–34. https://doi.org/10.1016/J.FOODQUAL.2017.02.001.

Mina, G., Peira, G. and Bonadonna, A. (2023). The potential future of insects in the European food system: a systematic review based on the consumer point of view. *Foods (Basel, Switzerland)* 12(3). https://doi.org/10.3390/FOODS12030646.

Mishyna, M., Chen, J. and Benjamin, O. (2020). Sensory attributes of edible insects and insect-based foods – future outlooks for enhancing consumer appeal. *Trends in Food Science and Technology* 95, 141–148. https://doi.org/10.1016/J.TIFS.2019.11.016.

Modlinska, K., Adamczyk, D., Goncikowska, K., Maison, D. and Pisula, W. (2020). The effect of labelling and visual properties on the acceptance of foods containing insects. *Nutrients* 12(9), 2498. https://doi.org/10.3390/NU12092498.

Motoki, K., Ishikawa, S., Spence, C. and Velasco, C. (2020). Contextual acceptance of insect-based foods. *Food Quality and Preference* 85, 103982. https://doi.org/10.1016/J.FOODQUAL.2020.103982.

Myers, G. and Pettigrew, S. (2018). A qualitative exploration of the factors underlying seniors' receptiveness to entomophagy. *Food Research International* 103, 163–169. https://doi.org/10.1016/J.FOODRES.2017.10.032.

Nadeau, L., Nadeau, I., Franklin, F. and Dunkel, F. (2015). The potential for entomophagy to address undernutrition. *Ecology of Food and Nutrition* 54(3), 200–208. https://doi.org/10.1080/03670244.2014.930032.

Naranjo-Guevara, N., Fanter, M., Conconi, A. M. and Floto-Stammen, S. (2021). Consumer acceptance among Dutch and German students of insects in feed and food. *Food Science and Nutrition* 9(1), 414–428. https://doi.org/10.1002/fsn3.2006.

Nyberg, M., Olsson, V. and Wendin, K. (2020). Reasons for eating insects? Responses and reflections among Swedish consumers. *International Journal of Gastronomy and Food Science* 22, 100268. https://doi.org/10.1016/J.IJGFS.2020.100268.

Olabi, A., Najm, N. E. O., Baghdadi, O. K. and Morton, J. M. (2009). Food neophobia levels of Lebanese and American college students. *Food Quality and Preference* 20(5), 353–362. https://doi.org/10.1016/J.FOODQUAL.2009.01.005.

Olum, S., Wesana, J., Mawadri, J., Nakiranda, J. K. and Odongo, W. (2021). Insects as food: illuminating the food neophobia and socio-cultural dynamics of insect consumption in Uganda. *International Journal of Tropical Insect Science* 41(3), 1–10. https://doi.org/10.1007/s42690-020-00309-2.

Onwezen, M. C., Bouwman, E. P., Reinders, M. J. and Dagevos, H. (2021). A systematic review on consumer acceptance of alternative proteins: pulses, algae, insects, plant-based meat alternatives, and cultured meat. *Appetite* 159, 105058. https://doi.org/10.1016/J.APPET.2020.105058.

Onwezen, M. C., Van den Puttelaar, J., Verain, M. C. D. and Veldkamp, T. (2019). Consumer acceptance of insects as food and feed: the relevance of affective factors. *Food Quality and Preference* 77, 51–63. https://doi.org/10.1016/j.foodqual.2019.04.011.

Oonincx, D. G. A. B. and de Boer, I. J. M. (2012). Environmental impact of the production of mealworms as a protein source for humans: a life cycle assessment. *PLoS ONE* 7(12), e51145. https://doi.org/10.1371/JOURNAL.PONE.0051145.

Orkusz, A. (2021). Edible Insects versus Meat–nutritional Comparison: knowledge of Their Composition Is the Key to Good Health. *Nutrients* 13(4), 1207. https://doi.org/10.3390/NU13041207.

Orkusz, A., Wolańska, W., Harasym, J., Piwowar, A. and Kapelko, M. (2020). Consumers' attitudes facing entomophagy: polish case perspectives. *International Journal of Environmental Research and Public Health* 17(7), 2427. https://doi.org/10.3390/IJERPH17072427.

Orsi, L., Voege, L. L. and Stranieri, S. (2019). Eating edible insects as sustainable food? Exploring the determinants of consumer acceptance in Germany. *Food Research International (Ottawa, Ont.)* 125, 108573. https://doi.org/10.1016/J.FOODRES.2019.108573.

Pakseresht, A., Vidakovic, A. and Frewer, L. J. (2023). Factors affecting consumers' evaluation of food derived from animals fed insect meal: a systematic review. *Trends in Food Science and Technology* 138, 310–322.

Palmieri, N., Perito, M. A., Macrì, M. C. and Lupi, C. (2019). Exploring consumers' willingness to eat insects in Italy. *British Food Journal* 121(11), 2937–2950. https://doi.org/10.1108/BFJ-03-2019-0170.

Pascucci, S. and De-Magistris, T. (2013). Information bias condemning radical food innovators? The case of insect-based products in the Netherlands. *International Food and Agribusiness Management Review* 16(3), 1–16. https://doi.org/10.22004/AG.ECON.156421.

Petersen, M., Olson, O. and Rao, S. (2020). University student perspectives of entomophagy: positive attitudes lead to observability and education opportunities. *Journal of Insect Science* 20(5), 1–6. https://doi.org/10.1093/JISESA/IEAA120.

Piha, S., Pohjanheimo, T., Lähteenmäki-Uutela, A., Křečková, Z. and Otterbring, T. (2018). The effects of consumer knowledge on the willingness to buy insect food: an exploratory cross-regional study in Northern and Central Europe. *Food Quality and Preference* 70, 1–10. https://doi.org/10.1016/J.FOODQUAL.2016.12.006.

Powell, P. A., Jones, C. R. and Consedine, N. S. (2019). It's not queasy being green: the role of disgust in willingness-to-pay for more sustainable product alternatives. *Food Quality and Preference* 78, 103737. https://doi.org/10.1016/J.FOODQUAL.2019.103737.

Raheem, D., Carrascosa, C., Oluwole, O. B., Nieuwland, M., Saraiva, A., Millán, R. and Raposo, A. (2019). Traditional consumption of and rearing edible insects in Africa,

Asia and Europe. *Critical Reviews in Food Science and Nutrition* 59(14), 2169–2188. https://doi.org/10.1080/10408398.2018.1440191.
Ronteltap, A., van Trijp, J. C. M., Renes, R. J. and Frewer, L. J. (2007). Consumer acceptance of technology-based food innovations: lessons for the future of nutrigenomics. *Appetite* 49(1), 1–17. https://doi.org/10.1016/j.appet.2007.02.002.
Rozin, P. and Fallon, A. (1980). The psychological categorization of foods and non-foods: a preliminary taxonomy of food rejections. *Appetite* 1(3), 193–201. https://doi.org/10.1016/S0195-6663(80)80027-4.
Ruby, M. B. and Rozin, P. (2019). Disgust, sushi consumption, and other predictors of acceptance of insects as food by Americans and Indians. *Food Quality and Preference* 74, 155–162. https://doi.org/10.1016/J.FOODQUAL.2019.01.013.
Ruby, M. B., Rozin, P. and Chan, C. (2015). Determinants of willingness to eat insects in the USA and India. *Journal of Insects as Food and Feed* 1(3), 215–225. https://doi.org/10.3920/JIFF2015.0029.
Rumpold, B. A. and Schlüter, O. K. (2013). Nutritional composition and safety aspects of edible insects. *Molecular Nutrition and Food Research* 57(5), 802–823.
Russell, P. S. and Knott, G. (2021). Encouraging sustainable insect-based diets: the role of disgust, social influence, and moral concern in insect consumption. *Food Quality and Preference* 92, 104187. https://doi.org/10.1016/J.FOODQUAL.2021.104187.
Schäufele, I., Barrera Albores, E. and Hamm, U. (2019). The role of species for the acceptance of edible insects: evidence from a consumer survey. *British Food Journal* 121(9), 2190–2204. https://doi.org/10.1108/BFJ-01-2019-0017.
Schebesta, H. and Candel, J. J. L. (2020). Game-changing potential of the EU's Farm to Fork Strategy. *Nature Food* 1(10), 586–588. https://doi.org/10.1038/s43016-020-00166-9.
Schlup, Y. and Brunner, T. (2018). Prospects for insects as food in Switzerland: a Tobit regression. *Food Quality and Preference* 64, 37–46. https://doi.org/10.1016/J.FOODQUAL.2017.10.010.
Schouteten, J. J., De Steur, H., De Pelsmaeker, S., Lagast, S., Juvinal, J. G., De Bourdeaudhuij, I., Verbeke, W. and Gellynck, X. (2016). Emotional and sensory profiling of insect-, plant- and meat-based burgers under blind, expected and informed conditions. *Food Quality and Preference* 52, 27–31. https://doi.org/10.1016/j.foodqual.2016.03.011.
Siegrist, M. (2008). Factors influencing public acceptance of innovative food technologies and products. *Trends in Food Science and Technology* 19(11), 603–608. https://doi.org/10.1016/J.TIFS.2008.01.017.
Sogari, G. (2015). Entomophagy and Italian consumers: an exploratory analysis. *Progress in Nutrition* 17(4), 311–316.
Sogari, G., Bogueva, D. and Marinova, D. (2019). Australian consumers' response to insects as food. *Agriculture* 9(5), 108. https://doi.org/10.3390/AGRICULTURE9050108.
Sogari, G., Menozzi, D. and Mora, C. (2017). Exploring young foodies' knowledge and attitude regarding entomophagy: a qualitative study in Italy. *International Journal of Gastronomy and Food Science* 7, 16–19. https://doi.org/10.1016/J.IJGFS.2016.12.002.
Sogari, G., Menozzi, D. and Mora, C. (2018). Sensory-liking expectations and perceptions of processed and unprocessed insect products. *International Journal on Food System Dynamics* 9 1012.
Sogari, G., Menozzi, D. and Mora, C. (2019). The food neophobia scale and young adults' intention to eat insect products. *International Journal of Consumer Studies* 43(1), 68–76. https://doi.org/10.1111/IJCS.12485.

Sun-Waterhouse, D., Waterhouse, G. I. N., You, L., Zhang, J., Liu, Y., Ma, L., Gao, J. and Dong, Y. (2016). Transforming insect biomass into consumer wellness foods: a review. *Food Research International* 89(1), 129–151. https://doi.org/10.1016/j.foodres.2016.10.001.

Szendrő, K., Tóth, K. and Nagy, M. Z. (2020). Opinions on insect consumption in Hungary. *Foods (Basel, Switzerland)* 9(12). https://doi.org/10.3390/FOODS9121829.

Tan, H. S. G., Fischer, A. R. H., Tinchan, P., Stieger, M., Steenbekkers, L. P. A. and van Trijp, H. C. M. (2015). Insects as food: exploring cultural exposure and individual experience as determinants of acceptance. *Food Quality and Preference* 42, 78–89. https://doi.org/10.1016/j.foodqual.2015.01.013.

Tan, H. S. G., Fischer, A. R. H., van Trijp, H. C. M. and Stieger, M. (2016). Tasty but nasty? Exploring the role of sensory-liking and food appropriateness in the willingness to eat unusual novel foods like insects. *Food Quality and Preference* 48, 293–302. https://doi.org/10.1016/J.FOODQUAL.2015.11.001.

Tan, H. S. G., Tibboel, C. J. and Stieger, M. (2017). Why do unusual novel foods like insects lack sensory appeal? Investigating the underlying sensory perceptions. *Food Quality and Preference* 60, 48–58. https://doi.org/10.1016/J.FOODQUAL.2017.03.012.

Tan, H. S. G., van den Berg, E. and Stieger, M. (2016). The influence of product preparation, familiarity and individual traits on the consumer acceptance of insects as food. *Food Quality and Preference* 52, 222–231. https://doi.org/10.1016/j.foodqual.2016.05.003.

Tan, H. S. G., Verbaan, Y. T. and Stieger, M. (2017). How will better products improve the sensory-liking and willingness to buy insect-based foods? *Food Research International* 92, 95–105. https://doi.org/10.1016/j.foodres.2016.12.021.

Tuccillo, F., Marino, M. G. and Torri, L. (2020). Italian consumers' attitudes towards entomophagy: influence of human factors and properties of insects and insect-based food. *Food Research International* 137, 109619. https://doi.org/10.1016/j.foodres.2020.109619.

Tzompa-Sosa, D. A., Dewettinck, K., Gellynck, X. and Schouteten, J. J. (2021). Replacing vegetable oil by insect oil in food products: effect of deodorization on the sensory evaluation. *Food Research International (Ottawa, Ont.)* 141, 110140. https://doi.org/10.1016/J.FOODRES.2021.110140.

Vanhonacker, F., Van Loo, E. J., Gellynck, X. and Verbeke, W. (2013). Flemish consumer attitudes towards more sustainable food choices. *Appetite* 62, 7–16. https://doi.org/10.1016/J.APPET.2012.11.003.

Van Huis, A. (2013). Potential of insects as food and feed in assuring food security. *Annual Review of Entomology* 58, 563–583. https://doi.org/10.1146/annurev-ento-120811-153704.

Van Huis, A. (2016). Edible insects are the future? *The Proceedings of the Nutrition Society* 75(3), 294–305. https://doi.org/10.1017/S0029665116000069.

Van Huis, A. (2020). Insects as food and feed, a new emerging agricultural sector: a review. *Journal of Insects as Food and Feed* 6(1), 27–44. https://doi.org/10.3920/JIFF2019.0017.

Van Huis, A. and Oonincx, D. G. A. B. (2017). The environmental sustainability of insects as food and feed: a review. *Agronomy for Sustainable Development* 37(5), 37. https://doi.org/10.1007/S13593-017-0452-8.

Van Huis, A., Van Itterbeeck, J., Klunder, H., Mertens, E., Halloran, A., Muir, G. and Vantomme, P. (2013). *Edible Insects: Future Prospects for Food and Feed Security* (Issue 171). Food and agriculture organization of the United Nations.

Van Thielen, L., Vermuyten, S., Storms, B., Rumpold, B. and Van Campenhout, L. (2019). Consumer acceptance of foods containing edible insects in Belgium two years after their introduction to the market. *Journal of Insects as Food and Feed* 5(1), 35–44. https://doi.org/10.3920/JIFF2017.0075.

Vartiainen, O., Elorinne, A. L., Niva, M. and Väisänen, P. (2020). Finnish consumers' intentions to consume insect-based foods. *Journal of Insects as Food and Feed* 6(3), 261–272. https://doi.org/10.3920/JIFF2019.0042.

Verbeke, W. (2015). Profiling consumers who are ready to adopt insects as a meat substitute in a Western society. *Food Quality and Preference* 39, 147–155. https://doi.org/10.1016/J.FOODQUAL.2014.07.008.

Verneau, F., La Barbera, F., Amato, M., Riverso, R. and Grunert, K. G. (2020). Assessing the role of food related lifestyle in predicting intention towards edible insects. *Insects* 11(10), 1–12. https://doi.org/10.3390/INSECTS11100660.

Verneau, F., La Barbera, F., Kolle, S., Amato, M., Del Giudice, T. and Grunert, K. (2016). The effect of communication and implicit associations on consuming insects: an experiment in Denmark and Italy. *Appetite* 106, 30–36. https://doi.org/10.1016/j.appet.2016.02.006.

Videbæk, P. N. and Grunert, K. G. (2020). Disgusting or delicious? Examining attitudinal ambivalence towards entomophagy among Danish consumers. *Food Quality and Preference* 83. https://doi.org/10.1016/J.FOODQUAL.2020.103913.

Wendin, K. M. and Nyberg, M. E. (2021). Factors influencing consumer perception and acceptability of insect-based foods. *Current Opinion in Food Science* 40, 67–71. https://doi.org/10.1016/J.COFS.2021.01.007.

Wendin, K. M., Olsson, V. and Langton, M. (2019). Mealworms as food ingredient-sensory investigation of a model system. *Foods (Basel, Switzerland)* 8(8). https://doi.org/10.3390/FOODS8080319.

Wilkinson, K., Muhlhausler, B., Motley, C., Crump, A., Bray, H. and Ankeny, R. (2018). Australian consumers' awareness and acceptance of insects as food. *Insects* 9(2), 44. https://doi.org/10.3390/insects9020044.

Woolf, E., Zhu, Y., Emory, K., Zhao, J. and Liu, C. (2019). Willingness to consume insect-containing foods: a survey in the United States. *LWT* 102, 100–105. https://doi.org/10.1016/J.LWT.2018.12.010.

Zielińska, E., Zieliński, D., Karaś, M. and Jakubczyk, A. (2020). Exploration of consumer acceptance of insects as food in Poland. *Journal of Insects as Food and Feed* 6(4), 383–392. https://doi.org/10.3920/JIFF2019.0055.

Index

http://dx.doi.org/5

www.ingramcontent.com/pod-product-compliance
Ingram Content Group UK Ltd.
Pitfield, Milton Keynes, MK11 3LW, UK
UKHW021826100926
8556IPUK00003B/172

9 781801 46584